Intelligent Systems Reference Library

Volume 206

The aim of this series is to publish a Reference Library, including novel advances and developments in all aspects of Intelligent Systems in an easily accessible and well structured form. The series includes reference works, handbooks, compendia, textbooks, well-structured monographs, dictionaries, and encyclopedias. It contains well integrated knowledge and current information in the field of Intelligent Systems. The series covers the theory, applications, and design methods of Intelligent Systems. Virtually all disciplines such as engineering, computer science, avionics, business, e-commerce, environment, healthcare, physics and life science are included. The list of topics spans all the areas of modern intelligent systems such as: Ambient intelligence, Computational intelligence, Social intelligence, Computational neuroscience, Artificial life, Virtual society, Cognitive systems, DNA and immunity-based systems, e-Learning and teaching, Human-centred computing and Machine ethics, Intelligent control, Intelligent data analysis, Knowledge-based paradigms, Knowledge management, Intelligent agents, Intelligent decision making, Intelligent network security, Interactive entertainment, Learning paradigms, Recommender systems, Robotics and Mechatronics including human-machine teaming, Self-organizing and adaptive systems, Soft computing including Neural systems, Fuzzy systems, Evolutionary computing and the Fusion of these paradigms, Perception and Vision, Web intelligence and Multimedia.

Indexed by SCOPUS, DBLP, zbMATH, SCImago.

All books published in the series are submitted for consideration in Web of Science.

More information about this series at http://www.springer.com/series/8578

Mayuri Mehta · Philippe Fournier-Viger ·
Maulika Patel · Jerry Chun-Wei Lin
Editors

Tracking and Preventing Diseases with Artificial Intelligence

Springer

Editors
Mayuri Mehta
Department of Computer Engineering
Sarvajanik College of Engineering
and Technology
Surat, Gujarat, India

Maulika Patel
Department of Computer Engineering
G. H. Patel College of Engineering
and Technology
Charutar Vidya Mandal University
Vallabh Vidyanagar, Gujarat, India

Philippe Fournier-Viger
School of Humanities and Social Sciences
Harbin Institute of Technology (Shenzhen)
Shenzhen, Guangdong, China

Jerry Chun-Wei Lin
Department of Computer Science,
Electrical Engineering and
Mathematical Sciences
Western Norway University of
Applied Sciences
Bergen, Norway

ISSN 1868-4394 ISSN 1868-4408 (electronic)
Intelligent Systems Reference Library
ISBN 978-3-030-76731-0 ISBN 978-3-030-76732-7 (eBook)
https://doi.org/10.1007/978-3-030-76732-7

This Springer imprint is published by the registered company Springer Nature Switzerland AG
The registered company address is: Gewerbestrasse 11, 6330 Cham, Switzerland

Preface

All around the world, the spread of infectious diseases is a major concern as it directly impacts the health of people. While some infectious diseases may have a minor impact on society, some can have major impacts such as the recent SARS-CoV-2 coronavirus pandemic, also known as COVID-19.

To cope with the spread of infectious diseases, some traditional approaches are used such as to study the effect of medicines and develop new ones, design appropriate vaccines, and enforce various measures such as washing hands, wearing face masks, and doing temperature checks. However, despite the usage of such measures and medical advancements, there remain several incurable diseases for which prevention is the only cure. Besides, time is often critical when coping with new diseases that are highly contagious such as COVID-19 as no vaccine or very effective medicine is initially available.

To cope with these challenges, artificial intelligence (AI) has been rapidly adopted to assist physicians in diagnosis, disease tracking, prevention, and control. Due to increasing the availability of electronic healthcare data and rapid progress of analytics techniques, a lot of research is being carried out in this area by applying machine learning and data mining techniques to assist the medical professionals for making a preliminary evaluation.

This book is a collection of 11 chapters that provides a timely overview of recent advances in this area, that is, to use artificial intelligence techniques for tracking and preventing diseases. The target audience of this book is researchers, practitioners, and students. A brief description of each chapter is given below.

In Chap. 1, four approaches to identify stress by recognizing the emotional state of a person have been proposed. Pradeep et al. have analyzed the performance of the proposed approaches using Surrey Audio-Visual Expressed Emotion (SAVEE) and ENTERFACE databases. The results illustrate the considerable reduction in computational time and show that vector quantization-based features perform better than mel-frequency cepstral coefficients feature.

In Chap. 2, Fayemiwo et al. compared various approaches for the detection of COVID-19 from X-ray images. The problem is viewed as a classification problem with two classes (normal vs COVID-19) or three classes (normal, pneumonia, and COVID-19). A fine-tuned VGG-19 convolutional neural network with deep transfer learning shows that high accuracy can be obtained (from 89% to 99% depending on the scenario.

In Chap. 3, Falguni et al. aim to develop an intelligent diagnostic system for glaucoma—an eye-related disease, from the data obtained through clinicians by various examination devices or equipment used in ophthalmology. The classification is done by using a hybrid approach using artificial neural network, Naïve Bayes algorithms, decision tree algorithms, and 18 medical examination parameters for a patient. FGLAUC-99 is developed with J48, Naïve Bayes, and MLP classifiers with accuracy of 99.18%. The accuracy is not compared with other classifiers as the dataset is exclusively developed.

In Chap. 4, Pathak et al. have introduced two approaches, one based on a simple neural network and another based on a deep convolutional neural network, for diagnosis of tuberculosis disease. To evaluate the performance of the proposed approaches, they conducted experiments using tuberculosis chest X-ray dataset available on Kaggle and received classification accuracy of 99.24%.

In Chap. 5, Sarumi and Leung proposed an adaptive Naive Bayes-based machine learning algorithm for efficient prediction of genes in the genome of eukaryotic organisms. The adaptive Naive Bayes algorithm provided a sensitivity, specificity, and accuracy of 81.52%, 94.01%, and 96.02%, respectively, on discovering the protein-coding genes from the human genome chromosome GRCh37.

In Chap. 6, Deshpande et al. presented a survey work on different areas where microscopic imaging of blood cells is used for disease detection. A small note on blood composition is first discussed, which is followed by a generalized methodology for microscopic blood image analysis for certain application of medical imaging. Several models using microscopic blood cell image analysis are also summarized for disease detection.

In Chap. 7, Mahajan and Rana presented a comprehensive review of the recent clinical named entity classification using rule-based, deep learning-based, and hybrid approaches. The efficacy of clinical named entity recognition (NER) techniques for information extraction is also discussed and several experiments are then evaluated to show the state-of-the-art models with high accuracy by combining deep learning (DL) models with a sequential model.

In Chap. 8, the topic of disease diagnosis from CT scan images is discussed. Sajja et al. present a generic and hybrid intelligent architecture for disease diagnosis. The architecture can classify images into various disease categories using a convolutional neural network and is applied for detecting the COVID-19 disease. The design of the model is presented in detail with an experimental evaluation and a discussion of applications for other disease diagnoses using radiology images, as well as possibilities for future work.

In Chap. 9, skin lesion classification problem is addressed. Rock et al. developed an online system to assist doctors to quickly diagnose skin disease through skin lesion observation. Results demonstrate 78% testing accuracy and 84% training and validation accuracy.

In Chap. 10, Oza et al. have discussed various mammogram classification techniques that are categorized based on function, probability, rule, and similarity. They have presented comparative analysis of these techniques including strengths, drawbacks, and challenges. A few mechanisms to deal with these challenges have been described. In addition, some publicly available mammogram datasets are discussed in this chapter.

In Chap. 11, Sachdev et al. have presented the state-of-the-art similarity-based and feature-based chemogenomic techniques for the prediction of interaction between drug compounds and proteins. They have illustrated comparison of these techniques including their merits and demerits.

Surat, India — Mayuri Mehta
Shenzhen, China — Philippe Fournier-Viger
Vallabh Vidyanagar, India — Maulika Patel
Bergen, Norway — Jerry Chun-Wei Lin
March 2021

Contents

Contributors

Rajanikanth Aluvalu Department of CSE, Vardhaman College of Engineering, Hyderabad, India

C. Alvino Rock Department of Computer Science and Engineering, Karunya Institute of Technology and Sciences, Coimbatore, Tamil Nadu, India

Samson A. Arekete Department of Computer Science, Redeemer's University, Ede, Osun, Nigeria

C. Arvinthan Department of Computer Science and Engineering, Karunya Institute of Technology and Sciences, Coimbatore, Tamil Nadu, India

E. Bijolin Edwin Department of Computer Science and Engineering, Karunya Institute of Technology and Sciences, Coimbatore, Tamil Nadu, India

A. D. Darji Department of Electronics Engineering, Sardar Vallabhbhai Patel National Institute of Technology, Surat, Gujrat, India;
S.V.N.I.T., Ichhanath, Surat, Gujarat, India

Nilkanth Mukund Deshpande Department of Electronics and Telecommunication, Symbiosis Institute of Technology, Lavale, Pune, India;
Dr. Vithalrao Vikhe Patil College of Engineering, Ahmednagar, India;
Symbiosis International (Deemed University), Pune, India

Michael A. Fayemiwo Department of Computer Science, Redeemer's University, Ede, Osun, Nigeria

Shilpa Shailesh Gite Department of Computer Science, Symbiosis Institute of Technology, Symbiosis Centre for Applied AI (SCAAI), Lavale, Pune, India;
Symbiosis International (Deemed University), Pune, India

Manoj K. Gupta Department of Computer Science and Engineering, Shri Mata Vaishno Devi University, Katra, India

Richard Jayaraj Department of Computer Science and Engineering, Karunya Institute of Technology and Sciences, Coimbatore, Tamil Nadu, India

R. J. S. Jeba Kumar Department of Electronics and Communication Engineering, Karunya Institute of Technology and Sciences, Coimbatore, Tamil Nadu, India

B. Kevin Joseph Paul Department of Computer Science and Engineering, Karunya Institute of Technology and Sciences, Coimbatore, Tamil Nadu, India

Swathi S. Kundaram S.C.E.T., Athwalines, Surat, Gujarat, India

Carson K. Leung University of Manitoba, Winnipeg, MB, Canada

Pranita Mahajan SIESGST, Navi Mumbai, India;
SVNIT, Surat, India

Mba O. Odim Department of Computer Science, Redeemer's University, Ede, Osun, Nigeria

Adewale O. Ogunde Department of Computer Science, Redeemer's University, Ede, Osun, Nigeria

Bosede O. Oguntunde Department of Computer Science, Redeemer's University, Ede, Osun, Nigeria

Theresa O. Ojewumi Department of Computer Science, Redeemer's University, Ede, Osun, Nigeria

Oluwabunmi O. Olaniyan Department of Computer Science, Redeemer's University, Ede, Osun, Nigeria

Toluwase A. Olowookere Department of Computer Science, Redeemer's University, Ede, Osun, Nigeria

Idowu S. Oyetade Department of Computer Science, Redeemer's University, Ede, Osun, Nigeria

Parita Oza Nirma University, Ahmedabad, India

Hemant Patel Sumandeep Vidyapeeth, Vadodara, Gujarat, India

Ketki C. Pathak S.C.E.T., Athwalines, Surat, Gujarat, India

Dipti Rana SIESGST, Navi Mumbai, India;
SVNIT, Surat, India

Falguni Ranadive Rishabh Software, Vadodara, Gujarat, India

Kanica Sachdev Department of Computer Science and Engineering, Shri Mata Vaishno Devi University, Katra, India

Priti Srinivas Sajja Sardar Patel University, Vallabh Vidyanagar, Anand, India

Oluwafemi A. Sarumi University of Manitoba, Winnipeg, MB, Canada;
The Federal University of Technology—Akure (FUTA), Akure, Nigeria

Jignesh N. Sarvaiya S.V.N.I.T., Ichhanath, Surat, Gujarat, India

Yash Shah Nirma University, Ahmedabad, India

Akil Z. Surti Enlighten Infosystems, Vadodara, Gujarat, India

Pradeep Tiwari Department of Electronics Engineering, Sardar Vallabhbhai Patel National Institute of Technology, Surat, Gujrat, India;
Department of Electronics and Telecommunication Engineering, Mukesh Patel School of Technology Management and Engineering, NMIMS University, Mumbai, India

Marsha Vegda Nirma University, Ahmedabad, India

Abbreviations

2D	Two-Dimensional
3D	Three-Dimensional
ACG	Angle Closure Glaucoma
ACS	American Chemical Society
ADBRF	Adaboost Algorithm with RF
AER	Automatic Emotion (Stress) Recognition
AI	Artificial Intelligence
ALL	Acute Lymphocytic Leukemia
AMD	Age Related Macular Degeneration
AML	Acute Myelogenous Leukemia
ANFIS	Adaptive Neuro-Fuzzy Inference System
ANN	Artificial Neural Network
API	Application Programming Interface
ARIMA	Autoregressive Integrated Moving Average
ARM	Association Rule Mining
ASR	Automatic Speech Recognition
AUC	Area Under the Curve; Area Under the ROC Curve; Automatic operating Characteristic Curve
BLSTM	Bidirectional LSTM
BMJ	British Medical Journal
BMP	Bitmap
C-EPAC	Coupled Edge Profile Active Contours
CAD	Computer-Aided Design; Computer-Aided Diagnosis
CART	Classification And Regression Tree
CASD	Computer-Aided Skin Disease
CBIS-DDSM	Curated Breast Imaging Subset of DDSM
CET	Central European Time
CHAID	Chi-square Automatic Interaction Detection
CLAMP	Clinical Language Annotation, Modeling, and Processing
CLL	Chronic Lymphocytic Leukemia
CML	Chronic Myelogenous Leukemia

CNN	Computational Neural Network; Convolutional Neural Network
CNVM	Choroidal Neo-Vascular Membrane
Conv.	Convolutional
COVID-19	Coronavirus Disease 2019
Covid-19/COVID-19	Corona Virus Disease found in 2019, linked to family of viruses as Severe Acute Respiratory Syndrome (SARS)
CRF	Conditional Random Field
CRVO	Central Retinal Vein Occlusion
CSR	Central Serous Retinopathy
CSSA	Chaotic Salp Swarm Algorithm
CT scanned images	Computerized Tomography scanned images
CT	Computed Tomography; Computerized Tomography
CUP	Cambridge University Press
CXR	Chest X-ray
CXRs	Chest Radiographs
DCNN	Deep Convolutional Neural Network
DDBJ	DNA Data Bank of Japan
DDSM	Digital Database for Screening Mammography
DEM	Diffused Expectation Maximization
DICOM	Digital Imaging and Communications in Medicine
DL	Deep Learning
DNA	Deoxyribonucleic acid
DNN	Deep Neural Network
DOST	Discrete Orthonormal Stock-well Transform
DR	Diabetic Retinopathy
DT	Decision Tree
DTL	Deep Transfer Learning
DTL-CNN	Deep Transfer Learning Convolutional Neural Network
EEG	Electroencephalogram
EMBL-EBI	European Molecular Biology Laboratory—European Bioinformatics Institute
EMBO	European Molecular Biology Organization
EMR	Electronic Medical Record
EST	Expressed Sequence Tag
FAQs	Frequently Asked Questions
FC	Fully Connected Layer
FCBC	Fast Correlation-Based Filter
FCM	Firebase Cloud Messaging
FGLAUC-99	Falguni Glaucoma 99
Fig	Figure
FN	False Negative; False Negative Value
FOA	Fly Optimization Algorithm
FP	False Positive; False Positive Value
FSG	Future Science Group

FT	Fourier Transform
GGO	Ground Glass Opacity
GIP	Gaussian Interaction Profile
GLCM	Gray-Level Co-Occurrence Matrix
GMM	Gaussian Mixture Model
GO	Gene Ontology
GPU	Graphics Processing Unit
GRCh37	Genome Reference Consortium Human Build 37
GRCh38	Genome Reference Consortium Human Build 38
GRCh37.p13	GRCh37 patch 13
GRCh38.p10	GRCh38 patch 10
HB	Hemoglobin
HD	Hausdorff Dimension
HGP	Human Genome Project
HIV	Human Immuno-deficiency Virus
HMI	Human Machine Interfacing
HMM	Hidden Markov Model
HOG	Histogram of Oriented Gradients
HRCT	High-Resolution Computed tomography
HT	Hough Transform
HTK	Hidden Markov Model Tool kKit
Ibk	Instance Based Learner
ICT	Information and Communication Technologies
ID3	Iterative Dichotomiser 3
IDE	Inverse Differentiate Method
IDM	Inverse Difference Moment
IFT	Inverse Fourier Transform
ILSVRC	ImageNet Large Scale Visual Recognition Challenge
IOP	Institute of Physics; Intraocular Pressure
iOS	iPhone Operating System
IoT	Internet of Things
ISODATA	Self-organizing Data Analysis Technique
JPEG	Joint Photographic Experts Group; Joint Picture Expert Group
KFCG	Kekre's Fast Codebook Generation
KNN	K Nearest Neighbor
LBG	Linde Buzo Gray
LBP	Local Binary Pattern
LDA	Linear Discriminant Analysis
LDP	Local Directional Pattern
Lib	Library
LOG	Laplacian of Gaussian
LPCC	Linear Prediction Cepstral Coefficients
LR	Logistic Regression
LSTM	Long and Short term Memory

MCH	Mean Corpuscular Hemoglobin
MCS	Multi-Classification
MCV	Mean Corpuscular Volume
MDPI	Multidisciplinary Digital Publishing Institute
ME	Maximum Entropy
MEA	Midpoint Ellipse Algorithm
Mel	Melody
MEMM	Maximum Entropy Hidden Markov Model
MERS	Middle East Respiratory Syndrome
MFCC	Mel-Frequency Cepstral Coefficients
MIAS	Mammographic Image Analysis Society
miRBase	microRNA (miRNA) Database
miRNA	micro-Ribonucleic Acid
ML	Machine Learning
MLP	Multi-Layer Perceptron; Multi-Level Perceptron
MM	Mathematical morphology
Mod	Moderate; it is a fuzzy value of various symptoms such as fever and joint pain
MRI	Magnetic Resonance Imaging
MUC-6	Message Understanding Conferences—6
NB	Naive Bayes
NBML	Naive Bayes-based Machine Learning
NCBI	US National Centre for Biotechnology Information
NCS	Non-Coding Sequence
NEJM	New England Journal of Medicine
NER	Named Entity Recognition
NGS	Next-Generation Sequencing
NIH	National Institutes of Health
NLP	Natural Language Processing
NN	Neural Network
NPDR	Non-Proliferative Diabetic Retinopathy
OAG	Open Angle Glaucoma
OD	Optical Disc
OFR	Open Reading Frame
OOB	Out Of Bag
OS	Operating System
PA	Prophet Algorithm
PCA	Principal Component Analysis
PCR	Polymerase Chain Reaction
PCS	protein-coding sequence
PHOG	Pyramid Histogram of Oriented Gradients
PLOS	Public Library of Science
PMC	PubMed Central is a free digital archive database of full-text scientific literature in biomedical and life sciences
PNAS	Proceedings of the National Academy of Sciences

POAG	Primary Open Angle Glaucoma
PPI	Protein-Protein Interaction
PPV	Positive Predictive Value
PSSM	Position Specific Scoring Matrix
R&D	Research and Development
RAM	Random Access Memory
RBC	Red Blood Cells
RDW	RBC Distribution Width
ReLU	Rectified Linear Unit
ResExLBP	Residual Exemplar Local Binary Pattern
ResNet	Residual Neural Network
RF	Random Forest
RFE	Recursive Feature Elimination
RGB	Red-Green-Blue
RNA	Ribonucleic Acid
ROBC	Region of blood cell
ROC	Receiver Operating Characteristic curve
ROI	Region of Interest
RT-PCR	Real-time Reverse-Transcriptase-Polymerase Chain Reaction
RUP	Rockefeller University Press
SAGE	Sarah and George Publishing
SARS	Severe Acute Respiratory Syndrome
SARS-CoV-2	Severe Acute Respiratory Syndrome Coronavirus 2
SAVEE	Surrey Audio-Visual Expressed Emotion
SCA	Sine Cosine Algorithm
SCCA	Sparse Canonical Correspondence Analysis
SDM	Stimulating discriminant measures
Sec	Seconds
SESSA	Salp Swarm Algorithm
SFTA	Segmentation based Fractal Texture Analysis
SGD	Stochastic Gradient Descent
SMACC	Sequential Maximum Angle Convex Cone
SMO	Sequential Minimal Optimization
SMTT	Self dual Multi-scale Morphological Toggle Block
SNP	Single-Nucleotide Polymorphism
SRGAE	Seed region growing area extraction
SS—SVM	Semi Supervised—SVM
SVM	Support Vector Machine
TB	Tuberculosis
TLA	Transfer learning approach
TN	True Negative; Truly Negative Value
TP	True Positive; Truly Positive Value
UCSC	University of California–Santa Cruz
UI	User Interface

USA	United States of America
V. Low	Very Low; it is a fuzzy value of various symptoms such as fever and joint pain
VAR	Vector Autoregression
VGG-16	Visual Geometry Group Convolutional Neural Network model with 16 Layers depth
VGG-19	Visual Geometry Group Convolutional Neural Network model with 19 Layers depth
VGGNet	Visual Geometry Group Convolutional Neural Network model
VQ	Vector Quantization
WBC	White Blood Cells
WHO	World Health Organization
WP	Wavelet Packet

Chapter 1
Stress Identification from Speech Using Clustering Techniques

Pradeep Tiwari and A. D. Darji

Abstract With the stressful environment of day to day life, pressure in the corporate world and challenges in the educational institutes, more and more children and adults alike are affected by lifestyle diseases. The Identification of the emotional state or stress level of a person has been accepted as an emerging research topic in the domain of Human Machine Interfacing (HMI) as well as psychiatry. The speech has received increased focus as a modality from which reliable information on emotion can be automatically detected. Stress causes variation in the speech produced, which can be measured as negative emotion. If this negative emotion continues for a longer period, it may bring havoc in the life of a person either physically or psychologically. The paper discusses the identification of stress by recognising the emotional state of a person. Herein, four approaches for automatic Emotion Recognition are implemented and their performances such as accuracy and computation time are compared. First approach is Stress/Emotion recognition based on Mel-Frequency Cepstral coefficients (MFCC) feature with Lib-SVM classifier. In other approaches, Vector Quantization (VQ) based clustering technique is used for feature extraction. Three algorithms based on VQ have been explored: (a) Linde-Buzo-Gray (LBG) algorithm, (b) Kekre's Fast Codebook Generation (KFCG) algorithm (c) Modified KFCG. The result obtained indicates that VQ based features perform better in comparison to MFCC, while KFCG modified algorithm gives further better results. The Surrey Audio-Visual Expressed Emotion (SAVEE) database of seven universal emotions and ENTERFACE database with six emotions is used to train and test the multiclass SVM.

P. Tiwari (✉) · A. D. Darji
Department of Electronics Engineering, Sardar Vallabhbhai Patel National Institute of Technology, Surat, Gujrat, India
e-mail: pradeep.tiwari@nmims.edu

A. D. Darji
e-mail: add@eced.svnit.ac.in

P. Tiwari
Department of Electronics and Telecommunication Engineering, Mukesh Patel School of Technology Management and Engineering, NMIMS University, Mumbai, India

M. Mehta et al. (eds.), *Tracking and Preventing Diseases with Artificial Intelligence*, Intelligent Systems Reference Library 206,
https://doi.org/10.1007/978-3-030-76732-7_1

1.1 Introduction

Mental stress is a serious problem nowadays that not only affects the capacity, performance and mood of an individual, but also induces physical and mental health issues [1]. Under several stressed circumstances or emotions, the attributes of speech signals vary [2]. Stressed speech is characterized as the speech generated under any situation that leads the speaker to vary the speech from the neutral condition in the production of speech [3]. If a speech generated is in a 'quiet place' with no work duties, therefore the generated speech is assumed to be neutral. Stress can be categorized as (a) Emotionally driven stress: Speech generated by a shift in the speaker's mental or psychological condition like angry speech, happy speech, etc. (b) External stress triggered by the atmosphere such as Lombard speech (c) Pathological stressed speech such as Cold influenced speech, Senior Citizens Speech. In this work, emotionally driven stressed speech and External stress triggered by the atmosphere are considered. Stress unlike physiological diseases does not show symptoms at an early stage, so it can be cured before it is a massacre. Stress can be identified with the help of seven universal human emotions like fear, anger, disgust, happiness, contempt, and surprise and sadness as suggested by Ekman and Friesen [4]. Speech being non-invasive, non-intrusive in nature attracts the majority of the researchers and deals with identification of stress or emotion. For the applications like Human Machine Interface to work on low-cost processors or mobile applications, it becomes challenging to obtain the accurate results in real time. Thus, this paper focuses on the investigation of feature extraction techniques which increases the recognition accuracy and decreases the computational time by suitable modification in features like clustering, thus arriving at a simpler approach to perform real time fast and efficient emotion classification. Clustering reduces the size of training vector by quantizing it into clusters. Hence, the major contribution of this paper is it investigates the techniques which reduce the complexity involved while training and testing of a classification model considerably which further decreases the computation time without compromising with the accuracy. Four approaches for automatic Emotion Recognition are implemented in this paper and their performances such as accuracy and computation time are compared. First approach is Stress/Emotion recognition based on Mel-Frequency Cepstral coefficients (MFCC) feature with Lib-SVM classifier. In other approaches, Vector Quantization (VQ) based clustering technique is used for feature extraction. Three algorithms based on VQ have been explored: (a) Linde-Buzo-Gray (LBG) algorithm, (b) Kekre's Fast Codebook Generation (KFCG) algorithm (c) Modified KFCG. The result obtained shows that the performance of VQ based features is better in comparison to MFCC, while KFCG modified algorithm shows further improvement in the results. Results also illustrates that the clustering technique reduces the possible overfitting, bias and variance issues along with reducing the dimensionality of the features thus improving the results.

The remaining paper is organized as follows.Related work is discussed in Sect. 1.2. Section 1.3 details the experiments conducted while the results and its analysis is discussed in Sect. 1.4. Section 1.5 describes the future scope and concludes this work.

1.2 Related Work

Lots of researchers have contributed to emotion and stress identification area in past one decade. Ramamohan et al. [4] have utilized Sinusoidal features which can be characterised by its Amplitude, Frequency and Phase as features. Its accuracy is calculated for four emotions such as Anger, Neutral, Happiness and Compassion with Vector Quantisation (VQ) and Hidden Markov model (HMM) classification algorithm which shows better results compared to cepstral features and the linear prediction algorithm. Shukla et al. [5] considered a database consisting of five emotions, namely angry, neutral, happy, sad and Lombard, using 33 words. VQ and Hidden Markov model (HMM) were used as classification models for 13 dimensional MFCC features. The result obtained was 54.65% for VQ and 56.02% for HMM, while a result of 59.44% was found to give human classification of stress. According to the survey conducted by Hasrul et al. [6] for emotion recognition with prosodic features such as pitch, MFCCs with Gaussian Mixture Model (GMM), formants with SVM and energy with GMM, energy with GMM gave the best result of 92.3%. Shilpi et al. in 2015 [7] proved that speech signals combined with textual information improves the accuracy of emotion recognition. MFCC and Mel-Energy Spectrum Dynamic Coefficients features with Lib-SVM classifier was used by Chavan et al. [8] for happiness, sad, anger, neutral, fear, from Berlin database with 93.75% accuracy. A comparative study for word level and sentence level utterances from the SUSE database was carried out by Sudhakar et al. in 2014 [9]. Linear Prediction Cepstral Coefficients (LPCC) and MFCC features were extracted. They conclude that word utterances performed better than sentences and 2nd, 3rd and 4th order coefficients also gave comparable results to 12/13 order coefficients. A novel technique was proposed by Amiya et al. in 2015 [10]. They combined prosody features, quality features, derived features and dynamic feature for robust emotion recognition. Anger, disgust, fear, happy, neutral, sad and surprise emotions were classified using SVM. Revathy et al. in 2015 [11] discusses the effectiveness of Hidden Markov Model tool kit (HTK) for speaker independent emotion recognition systems for EMO-DB database with 68% accuracy. A novel WP-based feature called Discriminative band WP power coefficients was introduced by Yongming et al. in 2015 [12] for emotion recognition. These features gave improved performance over MFCC. El Ayadi et al. [13] explains the features, formants, and energy-dependent properties related to pitch contribute to the recognition of speech emotion. For the SAVEE Database, Sanaul et al. proposed the speaker-dependent feature by case recommending feature selection on 106-dimensional audio features [14]. In addition, Davood et al. used Fast Correlation-Based Filter (FCBF) feature selection on MFCC, Formants, and related statistical features on the SAVEE database with an average accuracy of 53% for fuzzy ARTMAP neural networks in 2017 [15]. Significant changes were observed over spectral features when weighted MFCC features were combined with spectral and prosody features [16]. Deb et al. [17] suggested region flipping based classification strategy using vowel-like regions and non-vowel-like regions using the Extreme Learning Machine classification model on the EMO-DB database. Wissam et. al. build the SVM model by merging neurogram features and traditional speech features [18].

This work is targetted for low-cost processors or mobile applications where it becomes challenging to obtain the accurate results in real time. Thus, there was a need to investigate feature extraction techniques which increases the recognition accuracy and decreases the computational time. In the above mentioned literature, various approaches were considered but the computation time requirement in those approaches are more. Since, clustering reduces the size of training vector by quantizing it into clusters. Hence,this paper investigates the techniques which reduce the complexity involved while training and testing of a classification model considerably which further decreases the computation time without compromising with the accuracy. Experimental results show that the proposed feature considerably improves emotion recognition performance over MFCC feature.

1.3 Stress Identification System Setup

The Block diagram of stress identification system set-up is represented by Fig. 1.1.

The first step in stress identification system is speech signal acquisition which is obtained from two standard databases: (i) The Surrey Audio-Visual Expressed Emotion (SAVEE) database of seven universal emotion (neutral, fear, disgust, happy, anger, sad and surprise) (ii) ENTERFACE database with 6 emotion (neutral, fear, disgust, happy, anger, sad and surprise).

The acquired signal will have lots of unwanted part like silence, surrounding noise, dc offset values etc., and thus it is required to pre-process the speech signal. Pre-processing includes three steps: (a) Eliminating the redundant information in the signal (b) Removal of dc offset values which does not carry any information by the process called normalisation (c) Pre-emphasizing the speech signal by using a high pass filter since the speech produced is deemphasized at glottis. The next step is extracting the feature from speech signal. There are various features extraction techniques like Cepstral Co-efficients, MFCC and LPC coefficients which can be

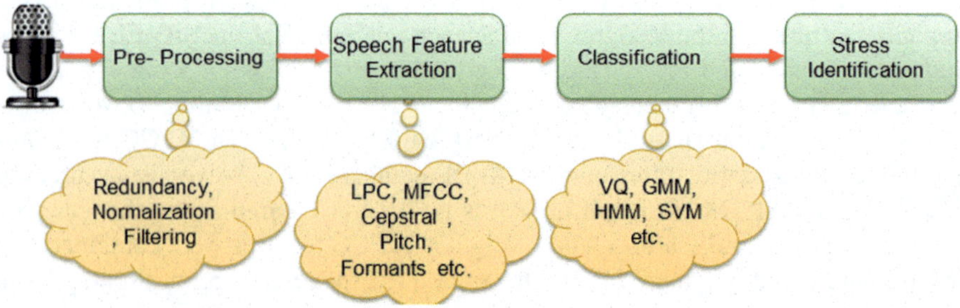

Fig. 1.1 Block diagram of stress identification system

applied to get feature vectors. There are two types of speech features which have been used by the researchers, (a) Prosodic speech features such as pitch and energy, also called local features. (b) Statistic or transform based features such as MFCC, Wavelet, also known as global features. MFCC in Fig. 1.2 and Vector Quantisation based features are considered in this research. Further, a pattern classifier called support vector machines (SVM) decides the emotion class of the utterance.

1.3.1 Signal Aquisition and Pre-processings

The first step is speech signal acquisition which is accomplished using standard database. The speech signal which is employed for AER is from standard SAVEE database of seven universal emotions (Anger, disgust, fear, happy, neutral, sad and surprise). The acquired signal would consist of unwanted part like silence, surrounding noise and dc offset values provided by microphone while recording, so it is required to pre-process the speech signal. The second step is extracting the feature from speech signal, wherein algorithms of various features extraction techniques like Mel frequency cepstral co-efficients from speech and facial landmarks from image are utilised to get feature vectors. These feature vectors will be used in third step where classifier models like Support vector machine (SVM) would classify the different emotion classes. The Pre-processing includes normalization and pre-emphasis.

1.3.2 Speech Feature Extraction

The second step is extracting the feature from speech signal, wherein algorithms which can provide intra-class resemblance and inter-class discrimination are applied to get feature vectors. The performance of any stress/ emotion identification system mainly depends on features extracted from speech emotion signal. Mel-frequency cepstral co-efficients (MFCC) is extracted for ASR and AER.

1.3.2.1 Mel-Frequency Cepstral Coefficients

This is a most widely used speech feature extraction technique found with the multiplication of Mel-filter bank with the frequency attribute of the signal called Power spectrum [7]. MFCC is based on human hearing perceptions, i.e. MFCC is calculated by considering the variation of the human ear's critical bandwidth with respect to frequency. The MFCC feature extraction technique is as shown in Fig. 1.2.

Mel-Scale or Melody-scale is computed if frequency f is given in Hz, with Eq. 1.1 is used.

$$Sk = Mel(f) = 2595 * (log_{10}(1 + \frac{f}{den})) \tag{1.1}$$

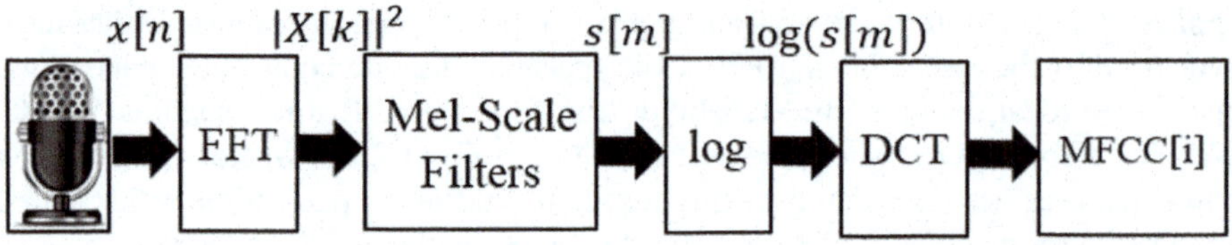

Fig. 1.2 MFCC Feature Extraction

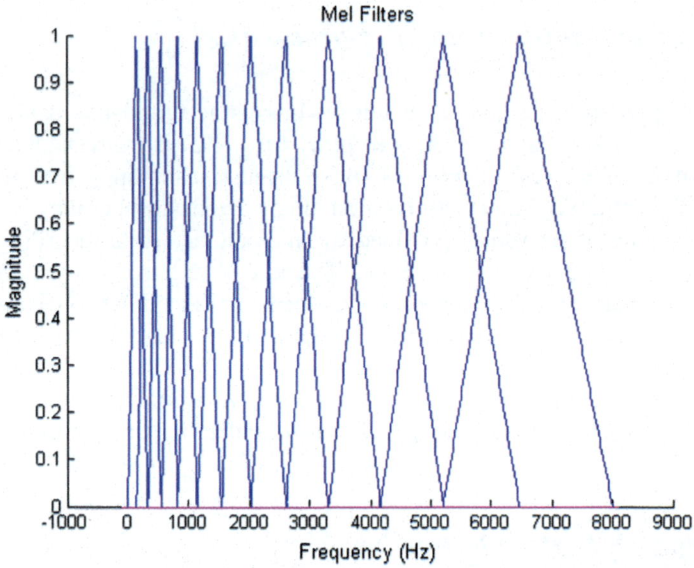

Fig. 1.3 Mel Filter Bank

The Mel filter bank obtained is given in Fig. 1.3.

First, the logarithm of the absolute value of the fast Fourier transform of input signal $x[n]$ is calculated whose inverse fast Fourier transform gives Cepstrum as shown in Eq. 1.2.

$$Cepstrum = IFT[abs(log(FT(x[n])))] \tag{1.2}$$

where, $FT(x[n])$ indicates to the fast Fourier transform of speech signal and $IFT(signal)$ means the inverse fast Fourier transform of the speech signal. The short time Fourier transform for a frame is illustrated in Eq. 1.3.

$$X_a[k] = \sum_{n=0}^{N-1} x[n]e^{\frac{-j2\pi nk}{N}}, \quad 0 \le k \le N \tag{1.3}$$

As $X_a[k]^2$ is known as Power spectrum and if it is applied to Mel frequency filter bank $H_m[k]$ consist of triangular filters, it results into Mel-frequency power spectrum as provided in Eq. 1.4.

$$S[n] = \sum_{n=0}^{N-1} X_a[k]^2 H_m[k], \quad 0 \leq m \leq M \tag{1.4}$$

Now, the log Mel-frequency power spectrum output is returned back to time domain by utilizing a compression algorithm called discrete cosine transform on $S[m]$. This gives MFCC calculated as shown in Eq. 1.5.

$$MFCC[i] = \sum_{m=1}^{M} log(S[m]) \cos[i(m - \frac{1}{2})\frac{\pi}{M}] \quad i = 1, 2, \ldots, L \tag{1.5}$$

The value of L is 13 i.e. it produces 13 MFCC coefficients for each frame and M indicates the length of the speech frames. The diagram shown in Fig. 1.4 represents the input speech signal and output as extracted MFCC features.

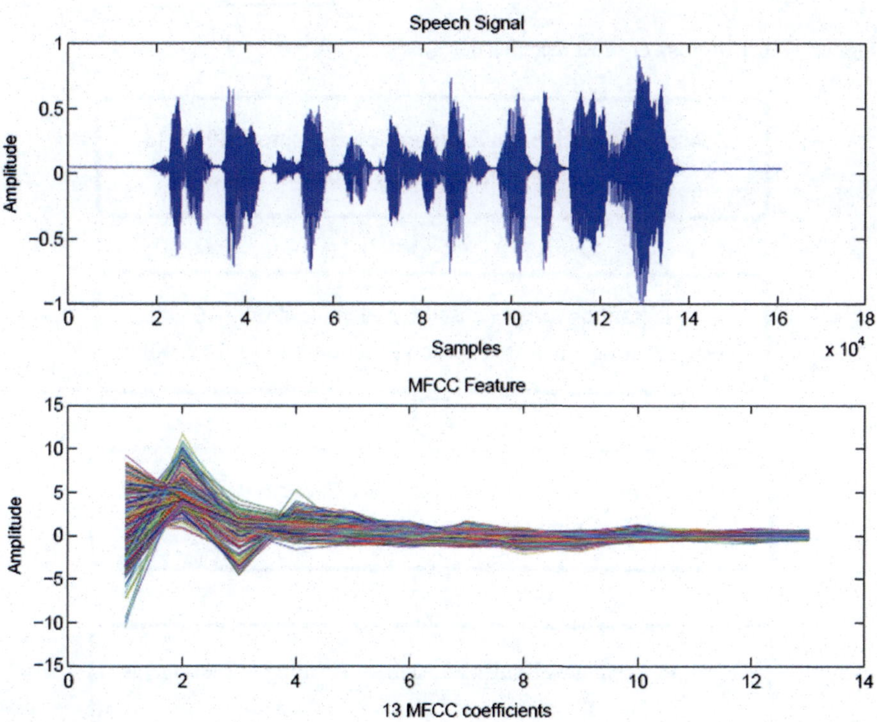

Fig. 1.4 MFCC Feature Extraction

1.3.2.2 VQ Based Features

Vector Quantization [19] reduces the size of training vector by quantizing it into clusters called codebook as shown in Fig. 1.5. It reduces the complexity involved while training and testing considerably as explained in Fig. 1.6.

Since this technique reduces the size of training vectors by quantizing it, it will be applied after extraction of MFCC feature. The complexity involved while training

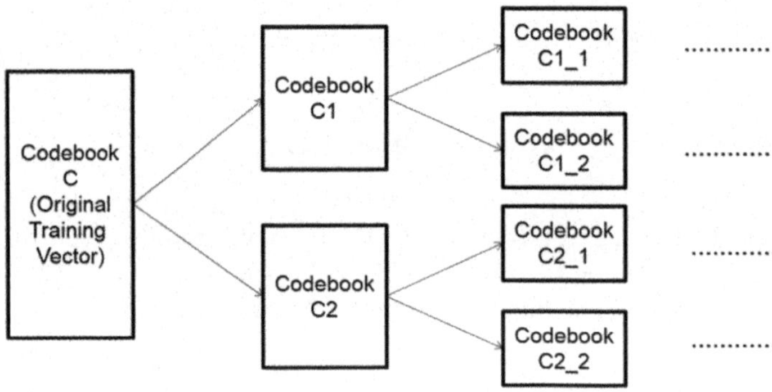

Fig. 1.5 Block diagram of stress identification system

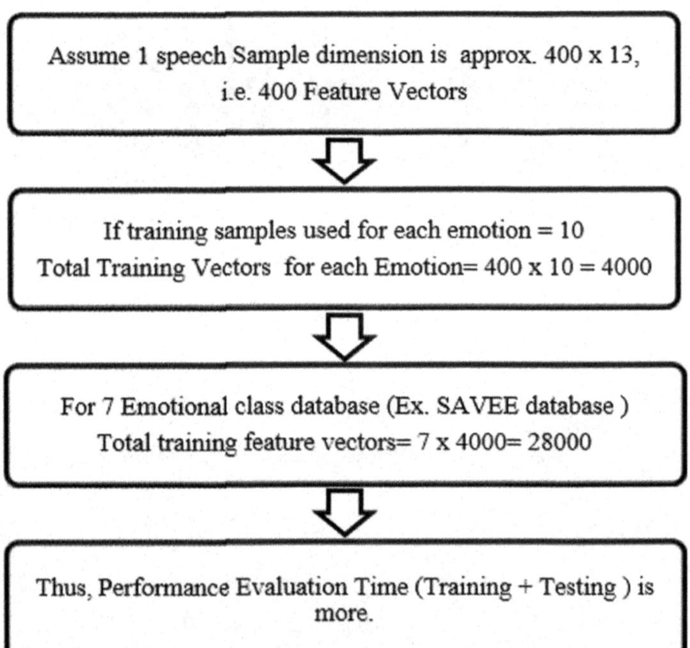

Fig. 1.6 Need of training data vector size reduction

decreases considerably, also the computational time decreases drastically. The VQ based algorithms given below are implemented:

- Linde-Buzo-Gray (LBG) algorithm
- Kekres Fast Codebook Generation (KFCG)
- Kekres Median Codebook Generation (KMCG)
- Modified Kekre Fast Codebook Generation (MKFCG)

These VQ algorithm starts from original data vector or cluster and then this cluster is converted into two clusters, further into four clusters, and so on, till N clusters are obtained, where N is the desired number of clusters or codebook size as shown in Fig. 1.7. **LBG** algorithm [19] is a divisive clustering approach and detailed steps are given as:

- 1-vector codebook is designed by calculating centroid C_i as shown in Eq. (1.6) where, T represents number of data vectors and X is the feature set called the main cluster.

$$C_i = \frac{1}{T}\sum_{j=1}^{T} X_j \tag{1.6}$$

- Each centroid C_i is divided into two close vectors C_{i_1} and C_{i_2} by splitting as shown in Eqs. (1.7) and (1.8) respectively. The 'δ' is a fixed perturbation value which was taken as 0.1 to split the centroid of the codebook.

$$C_{i_1} = C_i \times (1 + \delta) \tag{1.7}$$

$$C_{i_2} = C_i \times (1 - \delta) \tag{1.8}$$

- Identify the data vectors nearest to both new centroid by calculating Euclidean distance. Divide the data set X_j into two codebooks depending on their closeness with new centroids C_{i_1} and C_{i_2}.
- The splitting of the codebooks continues into 4, 8, 16 codebooks as shown in Fig. 1.7.
- Finally the centroids of the generated codebooks are used as feature vectors for training and testing.

KFCG algorithms [20] is fast because it does not calculate the Euclidean distance for splitting the cluster. The steps of KFCG are:

- 1-D centroid vector C_i represented as $C_i(1)$ is obtained by calculating centroid using Eq. (1.6) from the main cluster X.
- The first element of the centroid $C_i(1)$ and main cluster $X_j(1)$ is taken into consideration for splitting the codebook
- If $X_j(1) > C_i(1)$ then the vector X_j is assigned to codebook 1 else the vector X_j is assigned to codebook 2.
- Further calculate the centroid of two obtained codebooks and the splitting of the codebooks continues into 4, 8, 16 codebooks as shown in Fig. 1.7.

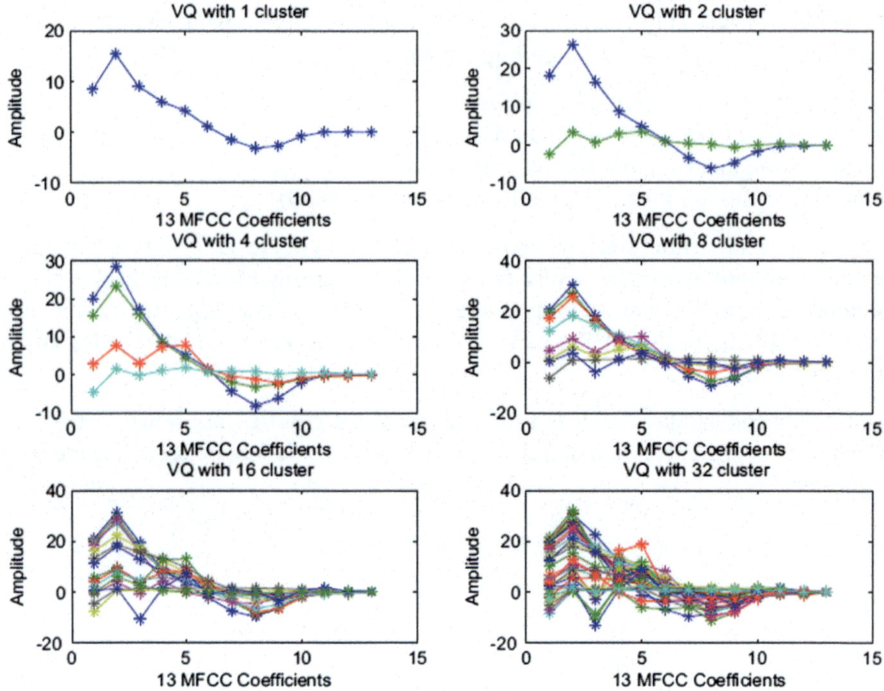

Fig. 1.7 4, 8, 16, 32 cluster centroid using LBG

- Finally the centroids of the generated codebooks are used as feature vectors during classification.

KMCG algorithm follows the same steps as in KFCG with the difference that instead of mean, it uses median for finding the centroid of the clusters [20]. By applying the above algorithms, 1 cluster is divided into 2, then 2 into 4 and so on till 32 clusters as shown in Fig. 1.7 are formed.

As it can be observed from Fig. 1.7, the first and second MFCC coefficients are more discriminative compare to other MFCC coefficients. Thus, the proposed modification in the algorithm is a vector quantization algorithm MKFCG based on either of these two co-efficient alone for comparison while dividing the clusters. The steps of the algorithm are same as KFCG. The only difference is that only first two coefficients are used in every iteration.

1.3.3 Support Vector Machine (SVM)

SVM is a binary classification algorithm which distinguishes two groups of clustered datapoints. Supervised training in SVM is carried out by putting a line a hyperplane

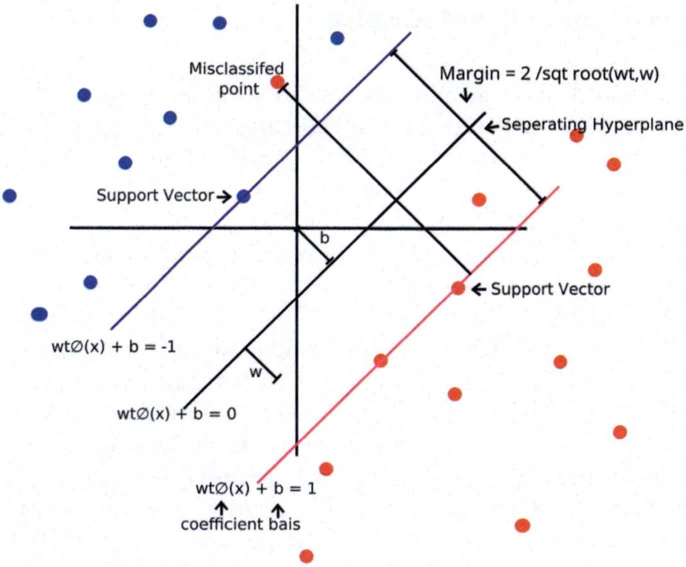

Fig. 1.8 Support Vector Machine [15]

between two separate groups by maximizing the gap between several data points. Figure 1.8 shows SVM classifier with hyperplane [15].

In an n-Dimensional feature space, a hyper plane can be expressed by the following Equation:

$$f(\mathbf{x}) = \mathbf{x}^T \times \mathbf{w} + b = \sum_{i=1}^{n} x_i \times w_i + b = 0$$

Dividing above equation by $||\mathbf{w}||$, we get

$$\frac{\mathbf{x}^T \times \mathbf{w}}{||\mathbf{w}||} = P_{\mathbf{w}}(\mathbf{x}) = -\frac{b}{||\mathbf{w}||}$$

representing that the projection of any point '**x**' on the plane onto the vector '**w**' is everytime '$-b/||\mathbf{w}||$', i.e., '**w**' is the normal direction of the plane, and '$|b|/||\mathbf{w}||$' is the length from the origin to the plane. Note that the equation found of the hyper plane can vary. $c\,f(\mathbf{x}) = 0$ represents the same plane for any c.

1.4 Implementation and Results

This section consists of the detail experimentation carried on the result obtained & supported by the tables and diagrams. The experimentation of Stress identification as shown in Fig. 1.7 includes two steps: 1. Training and 2. Testing. The speech samples utilized in the experimentation are obtained from standard databases SAVEE and enterface database which are used to train and test.

eNTERFACE database is a standard audio-visual emotion database which can be used for video, audio or audio-visual AER algorithms [21]. It contains six emotions: anger, disgust, fear, happy, sad, and surprise which are acted by 42 English speaking subjects. For each emotion, there are five videos and the total number of videos become $(42 \times 6 \times 5)$ i.e. 1260. The speech files extracted from video files have a sampling rate of 48 kHz. The abbreviation used here for various classes of emotion and the number of samples in each class are specified in Table 1.1.

SAVEE database comprises of 480 British English utterances recorded from 4 male actors with 7 emotions (anger, disgust, fear, happy, neutral, sad and surprise). The number of samples in each class of emotion are specified in Table 1.2. The various classes of emotion are abbreviated as represented in Table 1.2. The sampling frequency of the recorded samples 44,100 Hz (16 bit).

The performance results obtained for accuracy (%) and computation time (seconds) are shown below from Figs. 1.9, 1.10, 1.11, 1.12, 1.13, 1.14, 1.15, 1.16, 1.17, 1.18 and 1.19. Figure 1.19 shows the average overall results. The clustering technique reduces the possible overfitting and bias and variance issues along with reducing the

Table 1.1 eNTERFACE Database with 6 Emotion classes

S. No.	Emotion class	Abbreviation	No. of .wav files
1	Anger	A	210
2	Disgust	D	210
3	Fear	F	210
4	Happy	H	210
5	Sad	Sa	210
6	Surprise	Su	210

Table 1.2 SAVEE database with 7 emotion classes

S. No.	Emotion class	Abbreviation	No. of .wav files
1	Anger	A	60
2	Disgust	D	60
3	Fear	F	60
4	Happy	H	60
5	Neutral	N	120
6	Sad	Sa	60
7	Surprise	Su	60

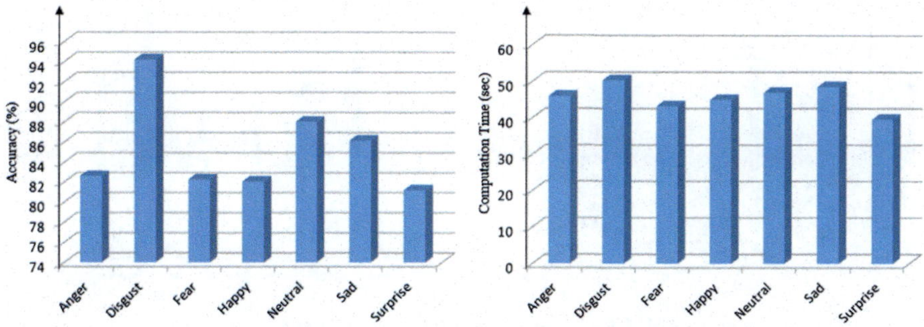

Fig. 1.9 Performance of MFCC on SAVEE database

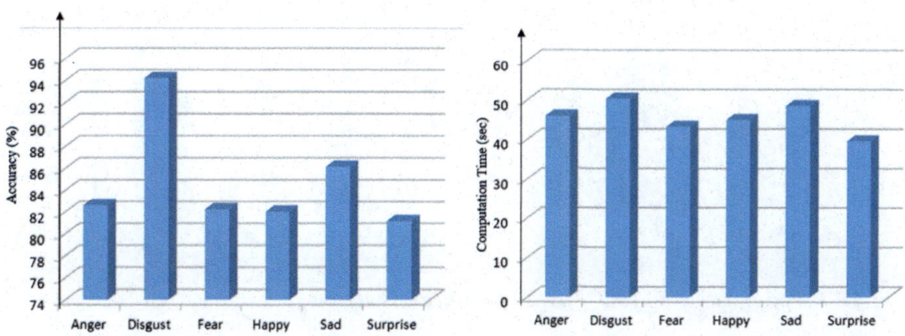

Fig. 1.10 Performance of MFCC on eNTERFACE Database

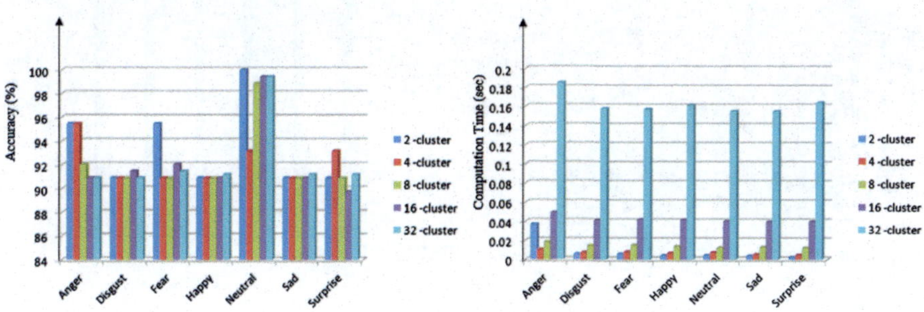

Fig. 1.11 Performance of MFCC+LBG on SAVEE Database

dimensionality of the features thus improving the results. The results that the proposed method gives better performance with 32 clusters compared to lower number of clusters i.e., 2, 4, 6, 8, 16. As clusters are increasing the average accuracy of emotion recognition is increasing but with a small value, however the computation time is increasing drastically specifically from 16 cluster to cluster. Thus, to provide real time fast computation on low-cost processors and mobile application, 32 clusters were selected as best bargain.

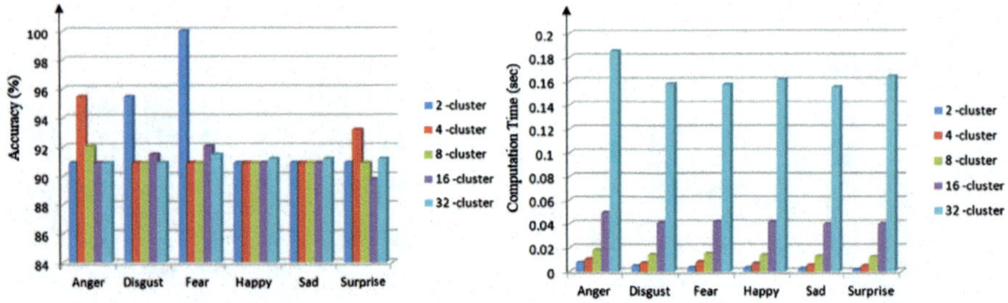

Fig. 1.12 Performance of MFCC+LBG on eNTERFACE Database

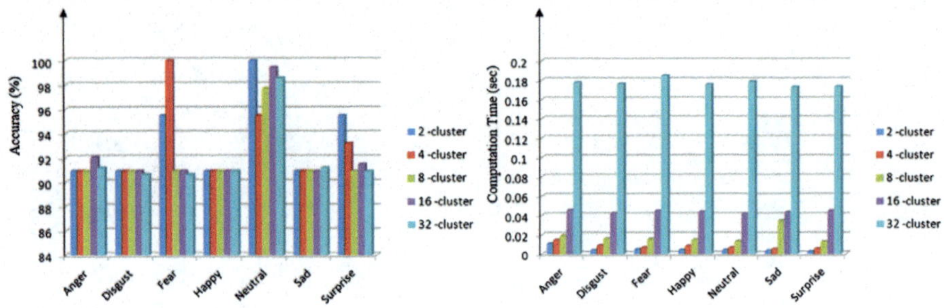

Fig. 1.13 Performance of MFCC+KFCG on SAVEE Database

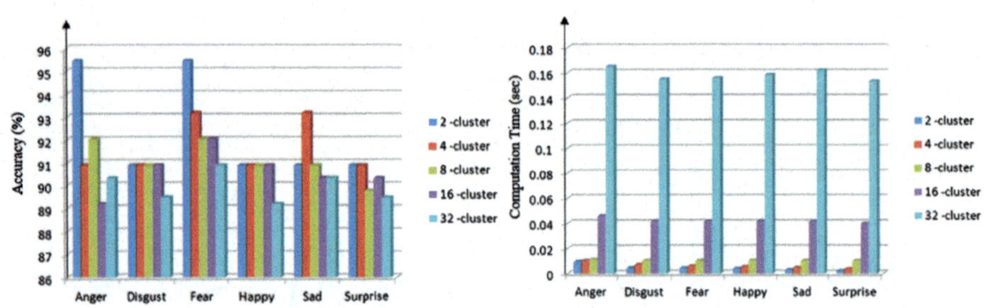

Fig. 1.14 Performance of MFCC+KFCG on eNTERFACE Database

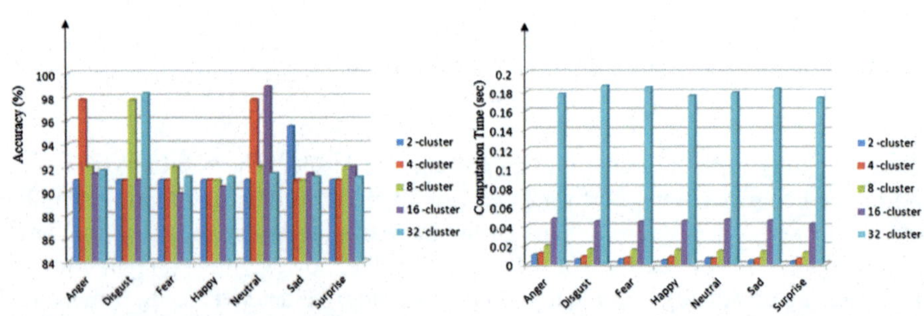

Fig. 1.15 Performance of MFCC+KMCG on SAVEE Database

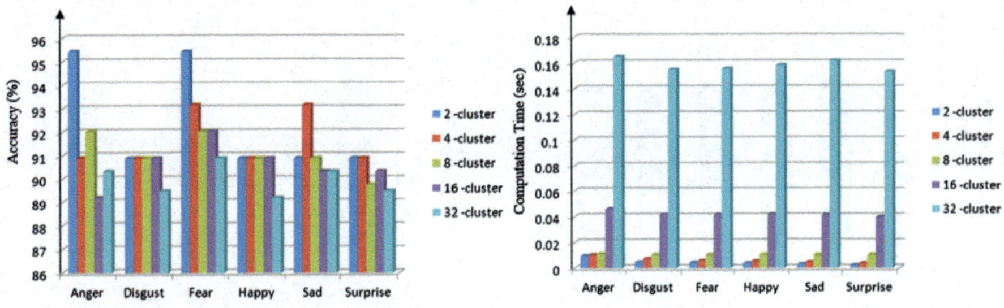

Fig. 1.16 Performance of MFCC+KMCG on eNTERFACE Database

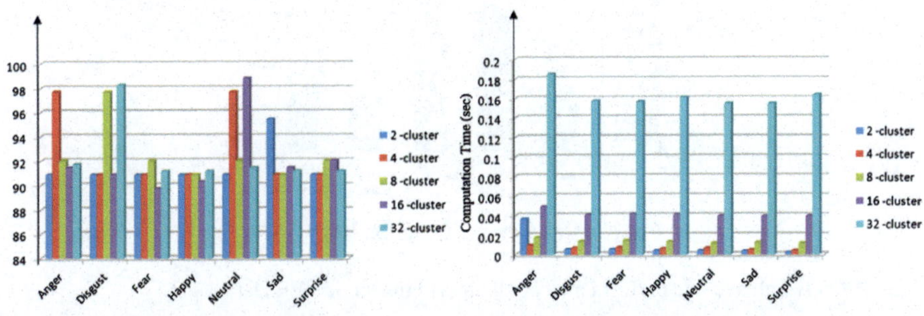

Fig. 1.17 Performance of MFCC+MKFCG (MF-1) on SAVEE Database

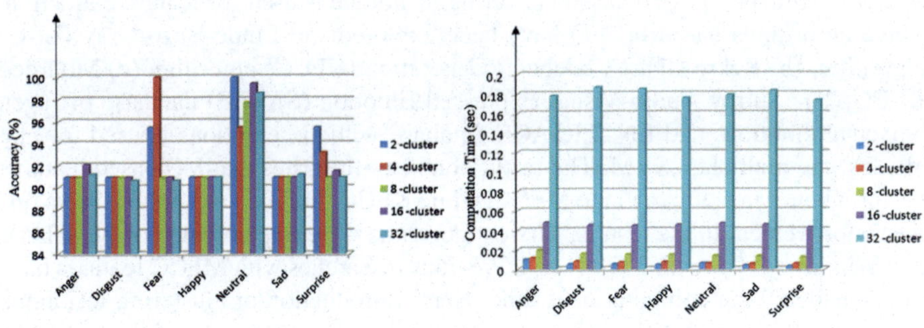

Fig. 1.18 Performance of MFCC+MKFCG (MF-2) on SAVEE Database

1.5 Conclusion

The human speech has received increased focus as a modality from which reliable information on emotion can be automatically detected. In this work, we have focussed the identification of stress by recognising the emotional state of a person. The paper proposed and implemented four approaches for automatic Emotion Recognition and their performances such as accuracy and computation time are compared. First approach is Stress/Emotion recognition based on Mel-Frequency Cepstral coefficients (MFCC) feature with Lib-SVM classifier. In other approaches,

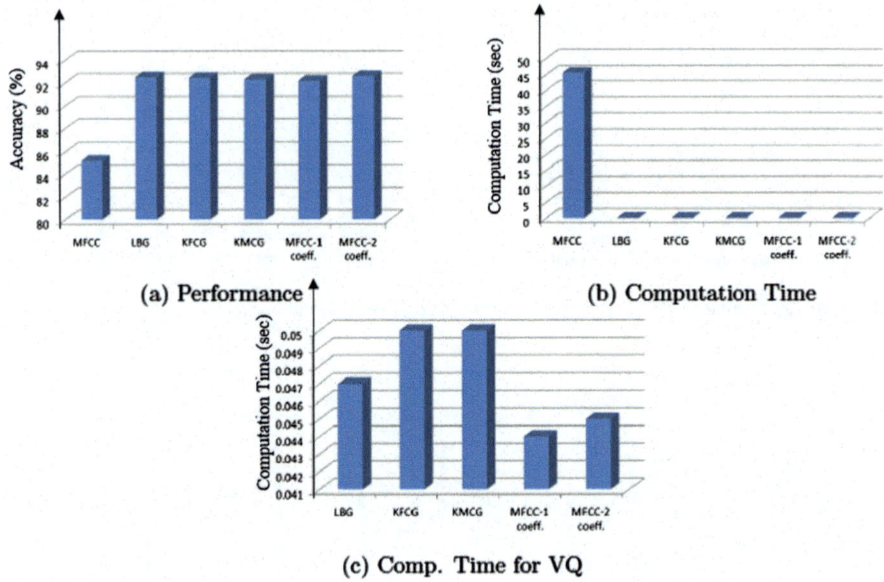

Fig. 1.19 Overall average performance of all algorithms on SAVEE Database

Vector Quantization (VQ) based clustering technique is used for feature extraction. Three algorithms based on VQ have been explored: (a) Linde-Buzo-Gray (LBG) algorithm, (b) Kekre's Fast Codebook Generation (KFCG) algorithm (c) Modified KFCG. The Surrey Audio-Visual Expressed Emotion (SAVEE) database of seven universal emotions and ENTERFACE database with six emotions is used to train and test the multiclass SVM. The result obtained illustrates that VQ based features perform better in comparison to MFCC, while KFCG modified algorithm shows further improvement in the results. These algorithms were implemented on MATLAB and Windows-7 OS with 3 GB RAM. Previous researches with MFCC features have not considered the computational time. Here, introduction of clustering technique has reduced the computational time by one-third of the computational time taken by MFCC technique.

Inclusion of features from other modalities like facial features, Electroencephalogram (EEG) features etc. can be added in future to increase the classification performance and decrease the computation time.

References

1. Vandyke, D.: Depression detection & emotion classification via data-driven glottal waveforms. In: Humaine Association Conference on Affective Computing and Intelligent Interaction (ACII), pp. 642–647. IEEE (2013)

2. John, H.L., Bou-Ghazale, S.E.: Robust speech recognition training via duration and spectral-based stress token generation. IEEE Trans. Speech Audio **3**, 415–421 (1995)
3. Ramamohan, S., Dandpat, S.: Sinusoidal Model based analysis and classification of stressed speech. IEEE Trans. Speech Audio Process. **14**(3), 737–746 (2006)
4. Ekman, P., Friesen, W.V.: Facial action coding system (1977)
5. Shukla, S., Prasanna, S.R.M., Dandapat, S.: Stressed speech processing: human vs automatic in non-professional speaker scenario. In: Proceedings of National Conference on Communications (NCC), Jan 2011, pp. 1–5
6. Hasrul, M.N., Hariharan, M., Yaacob, S.: Human affective (emotion) behaviour analysis using speech signals : a review. In: Proceedings of international conference on biomedical engineering, Feb 2012 (IEEE), pp. 217–222
7. Gupta, S., Mehra, A., Vinay: Speech emotion recognition using SVM with thresholding fusion. In: 2nd International Conference on Signal Processing and Integrated Networks (SPIN), pp. 570–574, 19–20 Feb 2015
8. Chavhan, Y.D., Yelure, B.S., Tayade, K.N.: Speech emotion recognition using RBF kernel of LIBSVM. In: 2nd International Conference on Electronics and Communication Systems (ICECS), pp. 1132–1135, 26–27 Feb 2015
9. Kumar, S., Das, T.K., Laskar, R.H.: Significance of acoustic features for designing an emotion classification system. In: International Conference on Electrical and Computer Engineering (ICECE), pp. 128–131, 20–22 Dec 2014
10. Samantaray, A.K., Mahapatra, K., Kabi, B., Routray, A.: A novel approach of speech emotion recognition with prosody, quality and derived features using SVM classifier for a class of North-Eastern Languages. In: IEEE 2nd International Conference on Recent Trends in Information Systems (ReTIS), pp. 372–377, 9–11 July 2015
11. Revathy, A., Shanmugapriya, P., Mohan, V.: Performance comparison of speaker and emotion recognition. In: 3rd International Conference on Signal Processing, Communication and Networking (ICSCN), pp. 1–6, 26–28 Mar 2015
12. Jing, H., Lun, X., Dan, L., Zhijie, H., Zhiliang, W.: Cognitive emotion model for eldercare robot in smart home. Communications, China **12**(4), 32–41 (2015)
13. El Ayadi, M., Kamel, M.S., Karray, F.: Survey on speech emotion recognition: features, classification schemes, and databases. Pattern Recogn. **44**(3), 572–587 (2011)
14. Jackson, P., Haq, S.: Surrey audio-visual expressed emotion (savee) database. University of Surrey, Guildford, UK (2014)
15. Kirchner, A., Signorino, C.S.: Using support vector machines for survey research. Surv. Pract. **11**(1) (2018)
16. Bozkurt, E., Erzin, E., Erdem, C.E., Erdem, A.T.: Formant position based weighted spectral features for emotion recognition. Speech Commun. **53**(9–10), 1186–1197 (2011)
17. Deb, S., Dandapat, S.: Emotion classification using segmentation of vowel-like and non-vowel-like regions. IEEE Trans. Affect. Comput. (2017)
18. Jassim, W.A., Paramesran, R., Harte, N.: Speech emotion classification using combined neurogram and INTERSPEECH 2010 paralinguistic challenge features. IET Signal Processing **11**(5), 587–595 (2017)
19. Soong, F.K., Rosenberg, A.E., Juang, B.-H., Rabiner, L.R.: Report: a vector quantization approach to speaker recognition. AT&T Tech. J. **66**(2), 14–26 (1987)
20. Kekre, H.B., Kulkarni, V.: Performance comparison of speaker recognition using vector quantization by lbg and kfcg. Int. J. Comput. Appl. **3**(10), 32–37 (2010)
21. Martin, O., Kotsia, I., Macq, B., et al.: The enterface' 05 audio-visual emotion database. In: 22nd International, p. 8. Atlanta, GA, USA, Conference on Data Engineering Workshops (2006)
22. Haq, S., Jan, T., Jehangir, A., Asif, M., Ali, A., Ahmad, N.: Bimodal human emotion classification in the speaker-dependent scenario. Pak. Acad. Sci. **52**(1), 27–38 (2015)

Chapter 2
Comparative Study and Detection of COVID-19 and Related Viral Pneumonia Using Fine-Tuned Deep Transfer Learning

Michael A. Fayemiwo, Toluwase A. Olowookere, Samson A. Arekete, Adewale O. Ogunde, Mba O. Odim, Bosede O. Oguntunde, Oluwabunmi O. Olaniyan, Theresa O. Ojewumi, and Idowu S. Oyetade

Abstract Coronavirus (or COVID-19), which came into existence in 2019, is a viral pandemic that causes illness and death in the lives of human. Relentless research efforts have been on to improve key performance indicators for detection, isolation and early treatment. The aim of this study is to conduct a comparative study on the detection of COVID-19 and develop a Deep Transfer Learning Convolutional Neural Network (DTL-CNN) Model to classify chest X-ray images in a binary classification task (as either COVID-19 or Normal classes) and a three-class classification scenario (as either COVID-19, Viral-Pneumonia or Normal categories). Dataset was collected from Kaggle website containing a total of 600 images, out of which 375

M. A. Fayemiwo (✉) · T. A. Olowookere · S. A. Arekete · A. O. Ogunde · M. O. Odim · B. O. Oguntunde · O. O. Olaniyan · T. O. Ojewumi · I. S. Oyetade
Department of Computer Science, Redeemer's University, Ede, Osun, Nigeria
e-mail: fayemiwom@run.edu.ng

T. A. Olowookere
e-mail: olowookereta@run.edu.ng

S. A. Arekete
e-mail: areketes@run.edu.ng

A. O. Ogunde
e-mail: ogundea@run.edu.ng

M. O. Odim
e-mail: odimm@run.edu.ng

B. O. Oguntunde
e-mail: oguntunden@run.edu.ng

O. O. Olaniyan
e-mail: olaniyano@run.edu.ng

T. O. Ojewumi
e-mail: ojewunmit@run.edu.ng

I. S. Oyetade
e-mail: oyetadei@run.edu.ng

M. Mehta et al. (eds.), *Tracking and Preventing Diseases with Artificial Intelligence*, Intelligent Systems Reference Library 206,
https://doi.org/10.1007/978-3-030-76732-7_2

were selected for model training, validation and testing (125 COVID-19, 125 Viral Pneumonia and 125 Normal). In order to ensure that the model generalizes well, data augmentation was performed by setting the random image rotation to 15 degrees clockwise. Two experiments were performed where a fine-tuned VGG-16 CNN and a fine-tuned VGG-19 CNN with Deep Transfer Learning (DTL) were implemented in Jupyter Notebook using Python programming language. The system was trained with sample datasets for the model to detect coronavirus in chest X-ray images. The fine-tuned VGG-16 and VGG-19 DTL models were trained for 40 epochs with batch size of 10, using Adam optimizer for weight updates and categorical cross entropy loss function. A learning rate of $1e^{-2}$ was used in fine-tuned VGG-16 while $1e^{-1}$ was used in fine-tuned VGG-19, and was evaluated on the 25% of the X-ray images. It was discovered that the validation and training losses were significantly high in the earlier epochs and then noticeably decreases as the training occurs in more subsequent epochs. Result showed that the fine-tuned VGG-16 and VGG-19 models, in this work, produced a classification accuracy of 99.00% for binary classes, and 97.33% and 89.33% for multi-class cases respectively. Hence, it was discovered that the VGG-16 based DTL model classified COVID-19 better than the VGG-19 based DTL model. Using the best performing fine-tuned VGG-16 DTL model, tests were carried out on 75 unlabeled images that did not participate in the model training and validation processes. The proposed models, in this work, provided accurate diagnostics for binary classification (COVID-19 and Normal) and multi-class classification (COVID-19, Viral Pneumonia and Normal), as it outperformed other existing models in the literature in terms of accuracy.

Keywords Convolutional neural networks · Coronavirus · COVID-19 test results · Deep transfer learning · Machine learning

2.1 Introduction

Viral pandemics have been known to be a serious threat to the world, and coronavirus disease 2019 (COVID-19) is not an exception. The World Health Organization (WHO) in 2020 reported that coronavirus is from a huge body of viruses that are responsible for ailments in humans or animals. Several coronaviruses are known in humans as the basis for varieties of respiratory disease such as common cold to more severe illnesses such as Middle East Respiratory Syndrome (MERS) and Severe Acute Respiratory Syndrome (SARS) [63]. The recently discovered coronavirus is the purported cause of COVID-19. This virus contaminates people and multiplies effortlessly from person-to-person. On March 11, 2020, the COVID-19 epidemic was described as a deadly disease by the World Health Organization (WHO). According to WHO, one million, seven hundred and seventy-five thousand and seven hundred and seventy-six (1,775,776) humans have lost their lives as a result of the COVID-19 epidemic as at the 30th of December, 2020, 9:26 AM CET.

As at the time of writing this chapter, the number of confirmed cases stands at 82,502,518 the world over. The fatality rate is still being assessed [66]. Numerous researchers globally are putting their efforts to collect data and develop solutions. The constant emphasis has been on advancing key performance metrics, for example, gradually increasing the speed of case identification, segregation and early cure. The introduction of these containment procedures was maintained and made possible by the innovative and effective use of cutting-edge technology. This work is, therefore, aimed at showing how Artificial Intelligence (AI), specifically machine learning, can help in identifying who is most at risk and diagnose patients early. According to BBC [8], a superhuman attempt is required towards the ease of the global epidemic killing. AI may have been overestimated—but in the case of medicine, it already has established evidence. According to Arora et al. [6], the role of AI is going to be crucial for predicting the outcome based on symptoms, CT-Scan, X-ray reports, etc.

Laboratory checking of suspected cases are described with extended periods and an exponential rise in request for tests [25]. Speedy prognosis with shorter turnaround times within a range of 10–30 mins have been to ameliorate the problem, although the majorities are presently going through clinical validation and therefore not in regular use [16]. In the process of result expectation, there is a need to continue to self-isolate. Once results are received, there is a need to remain on self-isolation until the symptoms resolve after being in seclusion for at least 14 days. In situations where the symptoms worsen during the seclusion time or continued after the 14 days, the patient would be asked to contact the accredited healthcare providers. Even the Rapid Test Kits deliver results after hours. Therefore, there is much interest to develop computer algorithms and methods to optimize screening and early detection, which is the primary purpose of this research where deep learning, most especially Convolutional Neural Network (CNN) is deployed. Deep learning provides the chance to increase the accuracy of the early discovery by automating the primary diagnosis of medical scans [31]. A CNN is a type of a DNN consisting of many hidden layers such as a convolution layer, a Rectified Linear Unit (ReLU) layer, a pooling layer and a fully connected layer.

CNN divides weights in the convolutional layer, thus reducing memory footprint and increasing network efficiency [44]. There are existing deep learning approaches to the detection of COVID-19 in clinical images (especially chest X-ray) in literature. However, it is the opinion of the authors of this chapter that the detection accuracies obtained in these approaches could be improved upon. The aim of this chapter therefore is to deploy Deep Transfer Learning Convolutional Neural Network Model to classify real-life COVID-19 dataset consisting of X-ray images. X-ray data are used since most hospitals have X-ray equipment and now the COVID-19 X-ray dataset is available on the Internet.

The rest of this chapter is organized thus: Sect. 2.2 gives an overview of literature where some current related works that are relevant to this chapter are explored, while Sect. 2.3 details the methodology used in this chapter for the detection of COVID-19 and related Viral Pneumonia. Section 2.4 of this chapter presents the results obtained from the experiments carried out by the authors of this chapter, and performance comparison drawn with other related works, while Sect. 2.5 concludes the chapter.

2.2 Literature Review

This section provides a brief review of literature tied to this work.

2.2.1 The COVID-19 Coronavirus

The COVID-19 coronavirus was initially observed in the Wuhan province of China and is fast spreading to all parts of the world [34]. The attention of the WHO was drawn to a collection of cases of the novel pneumonia discovered in the City of Wuhan, Hubei Province, China. The coronavirus disease (COVID-2019) was recognized as the contributory virus by Chinese authorities on January 7, 2020. On January 30, 2020, the WHO Director-General stated that the outbreak constitutes a Public Health Emergency of International Concern (PHEIC), by following the recommendations of the Emergency Committee. WHO has activated the R&D Blueprint in reaction to the occurrence to speed up diagnostics, vaccines and therapeutics for this new coronavirus [64]. The spread was so frightening that it became a source of concern globally, such that as at 28/04/2020, over two million cases (2,954,222), and 202,597 deaths have been recorded. Africa has over 22,239 cases, 881 death [36]. Nigeria, as of April 29, 2020, has about 1532 cases with 44 deaths and 255 recoveries with 12,004 tested samples [36]. The coronavirus is transmitted through droplets from coughing or sneezing and on close contacts with infected persons. The incubation period of COVID-19 is about 14 days; it attacks the lung, and it is highly contagious.

Nadeem [34] presented a short review of coronavirus that was made available by various journals and companies around the world. The review reported that the following journals/publishers had decided to make their COVID-19, coronavirus-related articles and the obtainable supporting data, accessible in PubMed Central (PMC), and authorized it for reuse: "American Chemical Society (ACS), The British Medical Journal (BMJ), American Society for Microbiology, Bulletin of the World Health Organization, Annals of Internal Medicine, Chinese Journal of Lung Cancer, Cambridge University Press (CUP), Cell Press, eLife, Elsevier, EMBO Press, Emerald Publishing, European Respiratory Society, Frontiers, Future Science Group (FSG), Global Virus Network Healthcare Infection Society, Hindawi, IOP Publishing, JMIR Publications, Karger Publishers, F1000 Research Limited, The Lancet, Life Science Alliance, MDPI, Microbiology Society, New England Journal of Medicine (NEJM), Oxford University Press, PLOS, PNAS—Proceedings of the National Academy of Sciences of the USA, Rockefeller University Press (RUP), The Royal Society, SAGE Publishing, Science Journals—American Association for the Advancement of Science, Springer Nature, Taylor & Francis, WikiJournal User Group, Wiley, Wolters Kluwer." The author noted that the review was open to updates.

2.2.2 *COVID-19 Clinical Features*

COVID-19 is predominantly a respiratory illness and pulmonary appearances which constitute main presentations of the disease. COVID-19 infects the respiratory system but could affect other organs as reported in some studies. Renal dysfunction [14, 69], gastrointestinal complications [39], liver dysfunction [23], cardiac manifestations [75], mediastinal findings [58], neurological abnormalities, and haematological manifestations [51], are among the reported extrapulmonary features. Some of the clinical symptoms of COVID-19 are cough, expectoration, asthenia, dyspnoea, muscle soreness, dry throat, pharyngeal dryness and pharyngalgia, fever, poor appetite, shortness of breath, nausea, vomiting, nasal obstruction, and rhinorrhoea. According to a WHO report on COVID-19, the disease has no specific manifestation to note, and the patients' presentation can range from completely asymptomatic to severe pneumonia and death [63–65].

Some symptoms presented by patients include sore throat, fever, dry cough, aches, nasal congestion and in severe cases patients build up respiratory breakdown, various organ malfunction, acute respiratory distress syndrome and in the end death [9].

The symptoms of COVID-19 infection begin to appear 5–6 days subsequent to contracting it either from the droplet or close contact with an infected person. According to Wang et al. [60], the period between the beginning of symptom and demise ranges from 6 to 41 days depending on the age and immune system of the patient. The period is shorter among older people. General symptoms at the beginning of the disease include fatigue, sore throat, fever, dry cough, aches, nasal congestion, sputum production, diarrhea and in severe cases patients show respiratory dysfunctions, chronic breathing distress syndrome, manifold organ failure and eventual demise [9, 18]. Chest CT scan shows pneumonia-like and some other viral pneumonia.

Furthermore, it presents other symptoms like severe respiratory distress syndrome, acute cardiac injury, RNA anemia and incidence of grand-glass opacities which lead to death [18]. COVID-19 shows some unique clinical symptoms like targeting the lower airway that manifests in the form of sore throat, sneezing and rhinorrhea. In addition, the chest radiographs result in some cases possesses "infiltrate in the upper lobe of the lung" related to growing dyspnea with hypoxemia [18]. COVID-19 has become pandemic and rapid diagnosis is imperative to identify patients and carriers for possible isolation and treatment in order to curb the spread of the disease. Attempts have been made to diagnose the disease, but a number of them are slow and not accurate in that they often give false-negative and false-positive results.

2.2.3 Related Works on the Detection of COVID-19

Li et al. [27, 28] examined chest images for the diagnosis of COVID-19. High-Resolution Computed Tomography (HRCT) was implemented for the primary diagnosis of the virus infection. HRCT objectively evaluates lung lesions giving a better understanding of the pathogenesis of the disease. CT scans were taken with the following parameters: collimation of 5 mm; 512 × 512 matrix; 100–250 mAs; collimation of 5 mm; pitch of 1–1.5; and 120 kV. The images were reconstructed by high resolution and conventional algorithms. The experiments were repeated several times, running into days for each patient.

Furthermore, COVID-19 have overlapping imaging manifestation with other cases of pneumonia such as SARS, other tests like the nucleic acid tests of samples from the respiratory tract are necessary for an accurate diagnosis. Long et al. [30] evaluated the suitability of Computed Tomography (CT) and "real-time reverse-transcriptase-polymerase Chain Reaction (rRT-PCR)." Clinical experiment with life data was executed, and the results presented that CT examination outperforms that of rRT-PCR at 97.2% and 84.6% respectively. In [57], seven important AI applications for the novel COVID-19 were recognized that can perform vital roles in screening, analyzing, tracking, and prediction of patients. Application areas identified comprise quick discovery and prognosis of the infection, treatment monitoring, individuals contact tracing, projection of cases and mortality, treatments and vaccines engineering, lessening of healthcare workers assignment and deterrence of the disease by providing updated supportive information.

Raajan et al. [42] suggested the use of image-based detection approach for COVID-19 detection, noting observations on the accuracy, speed and reliability of Chest CT imaging approach in the diagnosing and testing of COVID-19, in contrast with the Real-time Reverse Transcription Polymerase Chain Reaction (RT-PCR) detection approach. The work has the primary objective of proposing a high-speed, accurate and very sensitive CT scan approach for the diagnosis of COVID-19. The study further gives credence to novel computer-aided diagnosis system in the detection of COVID-19 using Deep Convolution Neural Network. The ResNet-16 network architecture was adopted in the training and labelling the CT dataset. The trained ResNet-16 model was then used to effectively diagnose COVID-19 patients. In evaluating the model, the accuracy obtained was 95.09%, specificity obtained was 81.89%, while the sensitivity obtained was 100%. On speed of test, it was reported that after four tests, the CT scan using ResNet CNN model required a maximum of one hour for test as compared with the RT-PCR that would require 2–3 hours for each test.

In considering the fact that the results of COVID-19 tests take long time, within a range of 2 hours and two days and the limited number of RT-PCR test kits, Altan and Karasu [3] argued on the importance of applying another diagnostic approach and thus proposed a hybrid model that consists of two-dimensional (2D) curvelet transformation, Chaotic Salp Swarm Algorithm (CSSA) and deep learning scheme to detect patients that contracted COVID-19 pneumonia from X-ray images. The model developed applied 2D curvelet transformation on the images acquired from the patient's chest X-ray radiographs and then formed a feature vector based on the coefficients realized. The Chaotic Salp Swarm Algorithm was employed to optimize the coefficients in the feature matrix. Diagnosis of COVID-19 infections in patients was then done by using a trained deep learning model,EfficientNet-B0 model. The propose hybrid model for COVID-19 diagnosis, in the study therefore consists of a framework of data synthesis using image processing method, revolution of RGB into grayscale images, application of the two-dimensional curvelet transformation to every image, the training and testing segments of the EfficientNet-B0 deep learning model, along with the model evaluation stage. The accuracy of the EfficientNet-B0 model alone is 95.24%, its specificity is 96.05%, while the accuracy of the model developed by applying only the 2D curvelet transformation is 96.87%, and its specificity is 97.46%. Also, the hybrid model, in which the feature matrix is formed using optimal coefficients obtained from CSSA optimization technique has an accuracy of 99.69%, while its specificity is 99.81%.

In a study, Pu et al. [40] aimed to develop and test feasibility of software for the detection, quantification, and monitoring of progression of pneumonia cases that are accompanying COVID-19 disease from chest CT scans. To achieve this, two datasets were collected and used. The first dataset in the study contained 120 chest CT scans which was used in the training and testing of deep learning algorithms for the segmentation of the lung boundaries and main lung vessels. The second dataset consisted of 72 serial chest CT scans that were obtained from 24 patients diagnosed with confirmed COVID-19, which was used to develop and test the deep learning algorithms for the detection and quantification of the presence and progression of infiltrates that accompany COVID-19, in the pneumonic regions. The computerized scheme flowchart of the algorithm used captured four important aspects of the detection process, first, an automated segmentation of the lung boundary and vessel based on the U-Net framework deep learning technique; secondly, an elastic lung registration stage for registering lung boundary between two serial CT scans at different time points using a bidirectional elastic registration algorithm; thirdly, a computerized automated identification of COVID-19 disease of the pneumonitis regions, and lastly, a quantifiable valuation of disease progression subjectively rated by radiologists. Radiologists rated 95% accuracy of heatmaps at least "acceptable" for representing disease progression. This suggests the feasibility of using computer software to detect and quantify pneumonic regions associated with COVID-19 and to generate heatmaps that can be used to visualize and assess progression.

Li et al. [27, 28] in a study on using Artificial Intelligence for the detection of COVID-19 and Community-acquired Pneumonia developed a fully automatic three-dimensional deep learning framework for the discovery of COVID-19 disease from chest CT scan images and evaluated the performance of the framework. A deep learning model based on RestNet50 was developed and christened COVID-19 detection neural network (COVNet) for the extraction of visual characteristics from volumetric chest CT scan images in order to detect COVID-19 in such CT scan images. The CT scan images of community-acquired or viral pneumonia and some other non-pneumonia conditions were involved in order to evaluate how effective the model would be in the differentiation of these conditions from COVID-19 disease proper.

The chest CT scan datasets that was used were those collected from six medical centers from the month of August 2016 to the month of February 2020. The results of evaluation of the developed model show that the deep learning neural network was able to detect COVID-19 distinctly having to differentiate it from the community-acquired or viral pneumonia and some other non-pneumonia lung diseases from chest CT scan images. It detected COVID-19 on CT scan images with an AUC value of 0.96, and viral pneumonia on chest CT scan images with an AUC value of 0.95.

Dipayan et al. [15] proposed a deep learning-based Convolutional Neural Network (CNN) architecture called Truncated Inception Net. The posited model classified COVID-19 positive cases from combined Pneumonia and normal cases with an accuracy of 99.96% (AUC of 1.0). They employed six different types of datasets to validate their proposal by taking the chest X-rays (CXRs) from COVID-19 positive, Pneumonia positive, Tuberculosis positive, and normal cases into considration. They obtained an accuracy of 99.92% (AUC of 0.99) in predicting COVID-19 positive cases from combined Pneumonia, Tuberculosis, and healthy CXRs. They therefore concluded that the posited Truncated Inception Net would be effective in predicting COVID-19 positive cases using CXRs.

Alazab et al. [2] proposed VGG16, a deep convolutional neural network model to detect COVID-19 from the chest X-ray. Their dataset contained 128 images of both COVID-19 healthy and non-healthy persons. The dataset was augmented to become 1000 images, 500 of which are for healthy and the other half for non-healthy persons. The proposed system achieved a weighted average F-measure of 95% on non-augmented dataset and a weighted average F-measure of 99% when trained on an augmented dataset. Additionally, three forecasting methods namely: the prophet algorithm (PA), autoregressive integrated moving average (ARIMA) model, and long short-term memory neural network (LSTM) were adopted to predict the numbers of COVID-19 confirmations, recoveries, and deaths over the next 7 days. The prediction results exhibit promising performance and offer an average accuracy of 94.80 and 88.43% in Australia and Jordan, respectively. They therefore concluded that the proposed system can significantly help identify the most infected cities, and also revealed that coastal areas are heavily impacted by the COVID-19 spread as the number of cases is significantly higher in those areas than in non-coastal areas.

The work of Sharma [46] proposed lung Computed Tomography (CT) scan as the first screening test and an alternative to real-time reverse transcriptase-polymerase chain reaction (RT-PCR) for diagnosis COVID-19 patients. About 2200 CT scan images consisting or 800 COVID-19 patients, 600 other viral pneumonia patients and 800 healthy persons were collected and used to train the machine learning model. A fresh set of 600 CT images with 200 COVID-19, 150 other viral pneumonia and 250 healthy persons were used to test the model for the three cases. Both training and testing used the custom vision software based on Residual Neural Network (ResNet) architecture of Microsoft azure. The model achieved a high accuracy and it is fast as it requires no blood sample collection, no tests kits and the diagnosis could be done on the spot of the scan. The results obtained compared favourably with other models in literature with about 91% accuracy, 92.1% sensitivity, 90.29% specificity in classifying the CT scan images into COVID-19, viral pneumonia and healthy cases. The author also recommends inclusion of polymerase chain reaction (PCR) for final diagnosis for images that were wrongly classified.

In [50], a multi-objective differential evolution-based convolutional neural network was proposed for COVID-19 diagnosis from the chest CT scan images of suspected patients. The model was built to classify images as COVID-19 positive or COVID-19 negative, i.e. the images were classified into two, positive and negative cases of COVID-19. Convolution operator with 3 Kernel/filter and 2 strides was used to extract potential features, max pooling layer of size 2 kernel and 1 stride was used to minimize the spatial size of the convolved features and Rectified linear unit (ReLU) activation function was employed to study the complex functional mappings between inputs and output parameters. The model was implemented with MATLAB 2019a software with deep learning toolbox. The population size of 40 was varied at different ratio for training and testing, 20:80, 30:70, 40:60, 60:40, 70:30, 80:20 and 90:10 ratios were used for the experiment. The model's performance was compared with those of competing models CNN, ANFIS and ANN. The model outperformed other models with accuracy > 92% for all the data ratio divisions, it presented improved and consistent true positive and true negative values as well as had lower false negative and false positive values when compared with other models. The results also showed that the model outperformed other models with 2.0928% F-measure, about 1.8262% more in sensitivity, 1.6827% more specificity and 1.9276% more in Kappa statistics. It could be deduced from the experiment that multi-objective differential evolution-based convolutional neural network was suitable for real-time COVID-19 diagnosis from chest CT scan images.

A deep learning-based software called "uAI Intelligent Assistant Analysis System" was proposed by Zhang et al. [71, 72] to diagnose COVID-19. This AI was developed to assess COVID-19 by United Imaging Medical Technology Company Limited in China. The software consists of an adapted 3D convolutional Neural Network and a combined VB-Net with bottle neck structure. The system has the capability to quickly and accurately localize and quantify COVID-19 infection, comprising the volume of the infection in the lung, lung lobes and bronchopulmonary segments. The percentage of infections can be calculated to determine the seriousness of the disease and to define the anatomical pattern within the lung. The software used the values Hounsfield Unit histogram within the infection region to evaluate the ground glass opacity (GGO), solid and sub-solid components in the affected region and to classify them. A data set of 2460 chest CT scan images were used, 90% (2215) showed COVID-19 pneumonia in both lungs, 7% (167) showed unilateral pulmonary infection with 81 and in the right and left lung respectively, 84 had negative chest CT scan. In terms of lung appearance, 298 (12%) showed pure GGO, 778 (32%) showed GGO with sub-solid lesion and 1300 (53%) with GGO with solid and sub-solid lesion. Their results showed that elderly patients ($\geq$ 60 years) were more susceptible to COVID-19, in addition, the dorsal segment of the right lobe of the lung is the preferred site for the pneumonia, this could be attributed to the distinctive anatomic features of the lobar bronchus, the bronchus of the right lower lobe of the lung is straight and steep. It was reported that the model would aid in prognosis of COVID-19 in addition to determining the seriousness of the disease to guide in the treatment plans.

Statistical analysis using Linear regression, Multilayer perceptron and Vector autoregression techniques were explored in [53] to predict the occurrence and spread of COVID-19 infection. The impact of COVID-19 in India resulting from statistics of confirmed, death and recovered cases were used to predict the rates of infection, death and recovered cases on COVID-19 for the subsequent 69 days. Linear regression (LR) predicted death cases with 95% Confidence Interval indicating an increase in death rates and recovery rates in the future based on existing case data. Multilayer perceptron (MLP) predicted a reduction in the confirmed cases with a slow rate and fluctuation in the death and recovered cases with 95% confidence Interval. The Vector autoregression (VAR) model of order 10, with AIC optimize information criteria with constant and linear trend vector and 95% confidence interval for the confirmed, recovered and death cases also predicted for the period under consideration perfectly.

Tuncer et al. [56], presented an automated residual exemplar local binary pattern and Iterative RelieF-based COVID-19 detection method with chest X-ray images. The scheme involved 87 X-ray images with COVID-19 and 234 healthy X-ray images, preprocessing, feature extraction using residual exemplar local binary pattern (ResExLBP), feature selection with the aid of iterative ReliefF (IRF), and the classification stage. A Grayscale transformation was carried out on the input X-ray images while resizing them into 512×512 size. These input images were subsequently divided into 128×128 sized exemplars with the aid of the ResExLBP and features extracted from these input images and their exemplars using local binary pattern. Discriminative features were selected using IRF after the generated features were concatenated. The chosen features served as input to different classifiers implemented with varying modes of validation. Evaluation of the system was carried out with accuracy, sensitivity, specificity, confusion matrix, AUC value, and geometric mean. Classification result showed that the SVM gave an accuracy of 100% by outperforming other classifiers, while the decision tree gave the worst accuracy of 91%.

Xu et al. [67, 68] proposed an early screening model to differentiate COVID-19 pneumonia from IAVP and healthy cases via pulmonary computed tomography (CT) images using deep learning procedures. Initially, the CT images were preprocessed to mine the effective pulmonary regions, and multiple candidate image cubes were partitioned using a three-dimensional (3D) CNN segmentation model. This was achieved after the central image, gathered for further processing in accordance with the two adjacent partners of each cube. The study used 618 CT samples and had 86.7% overall accuracy.

A model for detection and classification of COVID-19, using convolutional neural network as feature extractor with Bayesian optimization and 3 machine learning algorithms as classifiers was proposed by Nour et al. [37]. A total of 2905 X-ray images with three classes (COVID-19, viral pneumonia and normal) from an open-access dataset that covers the posterior-to-anterior chest X-ray images was employed for investigation in the study. Implementation of the model was actualized by applying data augmentation like flipping and rotation on the COVID-19 database and Bayesian optimization algorithm to enhance the CNN hyperparameters. The proposed CNN model was trained from scratch and validated and confusion matrix with other metrics derivable from it, like accuracy, sensitivity, specificity, as evaluation parameters. The model yielded an accuracy of 98.97%, F-1 score of 95.75%, a sensitivity of 89.39%, and specificity of 99.75% with the SVM classifier, whereas KNN and decision tree resulted in 95.76% and 96.10% accuracies respectively.

Toğaçar et al. [55] developed a deep learning model for detection of COVID-19 utilizing fuzzy colour and social mimic optimization techniques for pre-processing and SVM as classifier. The study utilized COVID-19 chest images, normal chest images and pneumonia chest images. The dataset was reconstructed by way of preprocessing using fuzzy and stacking techniques. Training and validation of the model were accomplished on the three datasets using the MobileNetV2 and SqueezeNet deep learning models and classification was actualized using the support vector classifier. For this investigation, two openly available databases comprising COVID-19 images were consolidated upon, since COVID-19 is a novel infection and the quantity of images associated with the virus is low. A total of 458 images were used to implement the model. This total comprises 295 images in COVID-19 category, 65 images in the normal class and 98 X-ray images in the pneumonia category. The experiments were implemented in python and confusion matrix and other parameters derivable from it were used as performance metrics. It was observed that the SVM classifier obtained an accuracy of 99.27% with the proposed method.

Amrane et al. [4] adopted a genetic approach using a rapid virological diagnosis on sputum and nasopharyngeal samples from suspect patients. Two real-time RT-PCR systems by means of a "hydrolysis probe and the LightCycler Multiplex RNA Virus Master Kit" were used. The primary technique probes the envelope protein (E)-encoding gene and uses a synthetic RNA positive control. The subsequent system targets the spike protein-encoding gene (forward priming, reverse priming, and probe) and uses optimistic control methods for synthetic RNA.

Bai et al. [9] proposed the use of medical technology through the internet of things (IoT) to develop an intelligent diagnosis and treatment assistance program (nCapp). The proposed cloud-based IoT platform includes the basic IoT roles and has a core graphics processing unit (GPU). Cloud computing systems link existing electronic medical records, image archiving, image archiving and communication to assist in profound mining and smart diagnosis.

A review of different machine learning methods with their attendant challenges for predicting the number of confirmed cases of COVID-19 was performed by Ahmad et al. [1]. The study involved a comprehensive review of different articles that employed machine learning strategies to forecast number of confirmed cases of COVID-19 and came up with a nomenclature system that categorized these techniques into four broad groups. The research characterized four titles or classes to which each methodology can be associated, taking care to ensure each method was placed in the most related category. However, it was discovered that certain methods belonged to more than one category. The four themes are: data-based search queries, traditional regression of machine learning, network and social media analysis, and methods of deep learning regression. It was noted that the traditional regression method of machine learning is a supervised technique of machine learning, which approximates the relationship between a dependent variable and independent variables. The study also revealed that two approaches have been used in the prediction of confirmed cases of COVID-19 using regression methods and that Random Forests and Neural Networks fall under this group. The first approach is the time series analysis and the other in which the associations between confirmed instances of COVID-19 and the other factors such as dampness, temperature etc. are mined from the data and the relationships were utilized to forecast the number of confirmed cases with the new values of the factors. The study further reviewed that deep learning essentially is related to ANNs which simulated human brain and used its many hidden layers to make accurate predictions, whereas analysis comprises networks or graphs which consisted of nodes and edges utilized for social networks analysis, community detection, web mining, etc. Ultimately, the work pointed out that social media and internet queries have huge data embedded in them, which can be utilized to estimate the number of confirmed cases of COVID-19. The study was concluded by highlighting the challenges associated with the methodologies and suggestions to address them were as well presented.

A feasibility study was performed by Brinati et al. [11] on the detection of COVID-19 infections using machine learning methods with blood as sample. The essence of the study was to find a sound and cost effective substitute to the costly and scarce rRT-PCR reagent for discovering COVID-19 positive cases. The feasibility study was carried out using 279 cases randomly selected from the patients admitted into the IRCCS Ospedale San Raffaele clinic between February ending 2020 and the middle of March 2020. The study employed descriptive statistics like mean, median, standard deviation, skewness and curtosis as descriptive statistics for the features considered. Implementation was actualized with the standard Python data analysis framework, comprising pandas for data loading and pre-processing, scikit-learn for both pre-processing and the classifiers analysis and matplotlib for visualization purposes. Different machine learning classifiers like KNN, decision tree, random forest, SVMs, Naïve Bayes, Logistic regression and extremely randomized trees were used for classification and their results compared. The resulting models were evaluated on balanced accuracy, positive predictive value (PPV), sensitivity, accuracy and specificity. The highest accuracy obtained from the study was 86%, and as such, the resulting model presented a good substitute to the gold standard method which requires highly sophisticated laboratory and the costly and scarce rRT-PCR reagent for detection of COVID-19.

A review of the use of Artificial Intelligence techniques amidst the COVID-19 Pandemic was presented by Bansal et al. [10]. The study demonstrated that the role of machine learning strategies, which heavily rely on artificial intelligence, in the prediction and management of COVID-19 pandemic. Different aspects of management explored were outbreak detection, spread prediction, preventive strategies and vaccine development, early case detection and tracking, prognosis prediction and treatment development. Further, the study also highlighted the role of big data generation, data cleaning and standardization in building reliable prediction models. The review also emphasized that noise and other anomalies in data, which could result in outliers, leading to prejudiced outcome must be avoided by all means. After a comprehensive review of different machine learning tools in combating the scourge of COVID-19, it was concluded that machine learning approaches are fast, more flexible than the traditional methods, and are free from human intervention and prejudice in detecting positive cases of COVID-19.

Lokuge et al. [29] proposed an effective, rapid, and scalable surveillance approach to spot all residual COVID-19 community transmissions over exhaustive identification of each energetic spread chain. They combined efficiency and sensitivity to identify community transmission chains by monitoring of patients with primary care fever and cough, hospital cases or asymptomatic community members, using surveillance evaluation approaches in addition to mathematical modeling, varying testing capacities and prevalence of COVID-19 and non-COVID-19 fever and cough, and duplication quantity. The study results showed that screening all demonstrations of syndromic fever and cough primary care, in combination with thorough and meticulous case and contact identification and management, allows for the proper primary discovery and removal of COVID-19 community transmission. If the test capability is limited, interventions such as combining allow for increased case discovery, even given the sensitivity of the concentrated test.

Hyafil and Moriña [24] carried out an impact analysis of the lockdown on the evolution of COVID-19 epidemics in Spain. The study was to assess the effect of the measures that started in Spain to deal with the epidemic. The amount of cases and the influence of the imposed lockdown on the multiplicative quantity resulting in hospitalization reports were estimated. The projected instances displayed acute rise till the lockdown, followed by deceleration and then a reduction after the full lockdown was imposed. The basic reproduction ratio reduced meaningfully from 5.89 (95% CI: 5.46–7.09) before the lockdown to 0.48 (95% CI: 0.15–1.17) thereafter. The study opined that managing a pandemic in the magnitude of COVID-19 was very intricate and required timely decisions. The great modifications found in the rate of infestation displayed that being able to employ inclusive participations in the first phase was vital to reducing the effect of a possible transferrable threat. This study likewise stressed the significance of dependable up-to-date epidemiological facts to precisely measure the influence of Public Health guidelines on the virus-related outburst.

"A hybrid deterministic and stochastic formalism that allows for time-variable transmission rates and discovery probabilities modelling for COVID-19" was presented by Romero-Severson et al. [43]. The model was fitted using iterative particle filtering to case and death counts time series analysis of data obtained from 51 countries. The study established the fact of a declining spread rate in 42 among the 51 countries studied. Out of the 42 countries, 34 had a major proof for subcritical transmission rates, though the turndown in novel cases was moderately slow in contrast to the early development rates. The study recommended that the adoption of social distancing efforts to reduce the incidence of COVID-19 were efficient, although they could be strengthened and maintained in various regions to prevent more renaissance of the disease. The study also proposed other approaches to manage the virus prior to the relaxation of social distancing efforts.

Previous related works have achieved good performance results in terms of accuracy, however detection accuracies obtained in the previous approaches could be improved upon. The aim of this chapter therefore is to improve the accuracy of COVID-19 detection model by deploying Deep Transfer Learning Convolutional Neural Network to classify real-life COVID-19 dataset consisting of X-ray images. Further studies on efficient diagnosis and detection of the virus and vaccine development are continuing owing partly to the fact that the disease is new and available research efforts have not been able to effectively address the concerns. Therefore, in this chapter, a deep transfer learning framework for efficient identification, classification and provision of new insights for the diagnosis is presented. Also, the prediction of probable patients of the novel COVID-19 and related Viral Pneumonia using radiology scanned images of suspected patients are presented.

2.3 Methodology

This research is centered on classifying COVID-19 chest X-ray dataset using a Convolutional Neural Network (CNN). The CNN consisted of one or more convolution layers and one or more fully connected layers as in a standard multilayer neural network. COVID-19 Radiology Dataset (chest X-ray) for Annotation and Collaboration was collected from the Kaggle website. The collected data was pre-processed, where the median filter was used to restore the image under examination by reducing the effects of the acquisition degradations. In [33], various pre-processing and segmentation techniques were discussed. The median filter replaces every pixel value, including itself, with the mean value of its neighbours. Therefore the pixel values that are very different from those of their neighbors have been eliminated. Following the preprocessing of the image dataset, the images were sectioned by using a simulated annealing algorithm. Feature extraction and classification were done using CNN. The neural network-based convolutional segmentation was implemented in Jupyter Notebook using Python programming language, and the system was trained with sample datasets for the model to recognize and classify the coronavirus. The model generated could be used to develop a simple web-based application, where medical personnel handling COVID-19 tests can input new cases and quickly predict the presence of the coronavirus, with a very high level of accuracy.

2.3.1 Dataset Description

Researchers from Qatar University, Doha, Qatar and the University of Dhaka, Bangladesh together with collaborators from Pakistan and Malaysia and some medical doctors have collated a database of chest X-ray images for COVID-19 positive cases along with normal and viral pneumonia images. There were 219 COVID-19 healthy images in their latest publication, 1341 normal images and 1345 images with viral pneumonia [13]. The study selected 125 images from each category. Data augmentation was performed by setting the random image rotation to 15° in clockwise direction to ensure the model generalizes.

2.3.2 The VGGNet Architecture

The VGGNet that Simonyan and Zisserman [49] proposed is a convolutional neural network that performed very well in the 2014 ImageNet Large Scale Visual Recognition Challenge (ILSVRC). The VGG-16 and VGG-19 are variants of the VGGNet Architecture. The network of the VGG-16 CNN has 13 convolutional layers (that is, 3 × 3 convolutional layers in blocks that are stacked on top of one another with growing depth). Two blocks house two 3 × 3 convolutional layers of the same setup in a sequential arrangement, while three blocks have three 3 × 3 convolutional layers of the same configuration in a sequential arrangement. Max pooling handles the reduction of the volume size of inputs at each layer. It further has two fully-connected layers, each with 4096 nodes and one fully-connected layer with 1000 nodes, and is followed by the SoftMax classifier, as shown in Fig. 2.1.

The network of the VGG-19 CNN has 16 convolutional layers (that is, 3 × 3 convolutional layers in blocks that are stacked on top of one another with growing depth). Two blocks house two 3 × 3 convolutional layers of the same setup in a sequential arrangement, while three blocks have four 3 × 3 convolutional layers of the same configuration in a sequential arrangement. Max pooling handles the reduction of the volume size of inputs at each layer. It further has two fully-connected

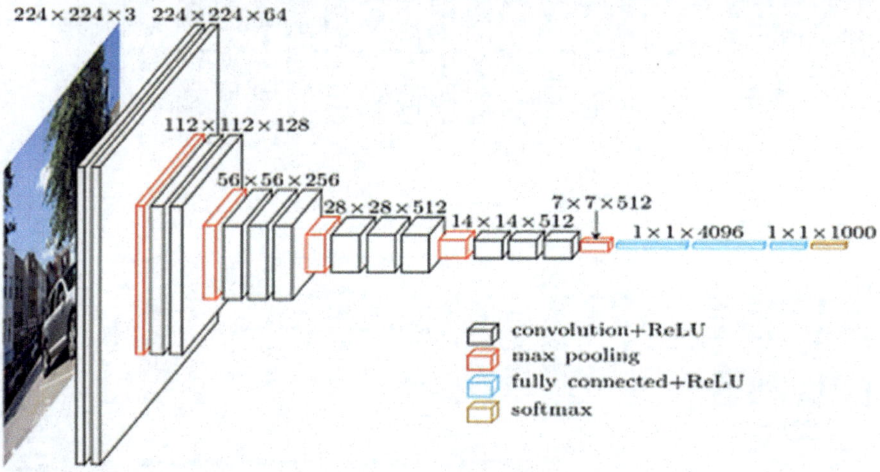

Fig. 2.1 A visualization of the VGG architecture. *Source* [17]

layers, each with 4096 nodes and one fully-connected layer with 1000 nodes, and is followed by the SoftMax classifier.

This chapter employed the VGG-16 and VGG-19 Convolutional Neural Networks (CNN) with Deep Transfer Learning (DTL) approach for COVID-19 detection. The Deep Transfer Learning (DTL) approach focused on the storage of weights that have been grown while unravelling some image classification tasks and then engaging it on a related task. Several DTL networks have been proposed, some of which include VGGNet [49], GoogleNet [54], ResNet [20], DenseNet [22] and Xception [12]. In this chapter, VGG-16 CNN and VGG-19 CNN, variants of the VGGNet were trained on the popular ImageNet images dataset. The VGG-16 and VGG-19 CNN were pre-trained deep neural networks for computer vision having 16 weight layers and 19 weight layers, respectively. They can be used as pre-trained models to help learn the distinguishing features in COVID-19 X-ray images with the aid of transfer learning approach and thus train DTL models for the detection of COVID-19 infection from X-ray images of patients.

As shown in the workflow in Figs. 2.2 and 2.3, to train the VGG-16 based Deep Transfer Learning model and VGG-19 based Deep Transfer Learning model for the detection of COVID-19, the VGG-16 CNN and VGG-19 CNN were used as pre-trained models respectively and were fine-tuned for COVID-19 detection based on the principles of transfer learning. In order to implement transfer learning with fine-tuning, the weights of the lower layers of the network, which learn very generic features from the pre-trained model served as feature extractors. The pre-trained model's lower layers weights were frozen and were therefore not updated through

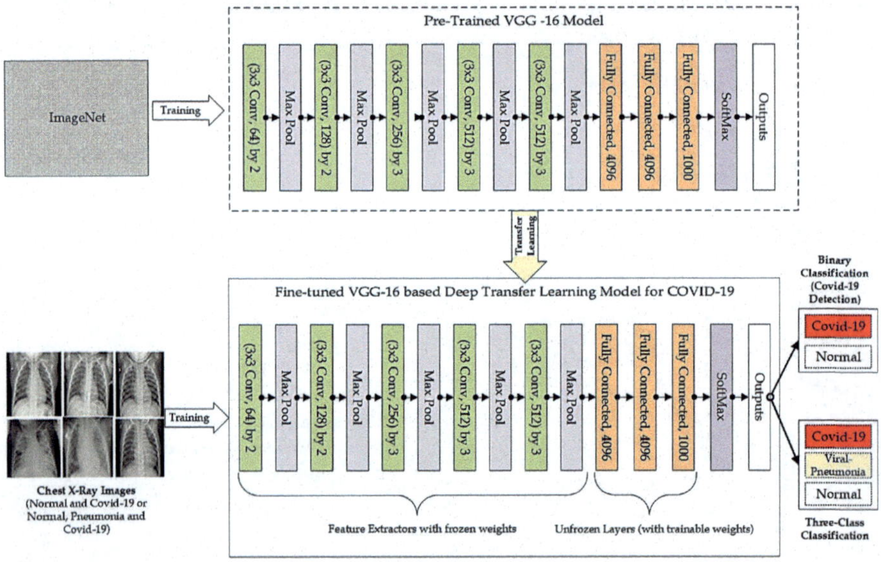

Fig. 2.2 The architectural workflow of the proposed fine-tuned VGG-16 based deep transfer learning model for COVID-19 detection

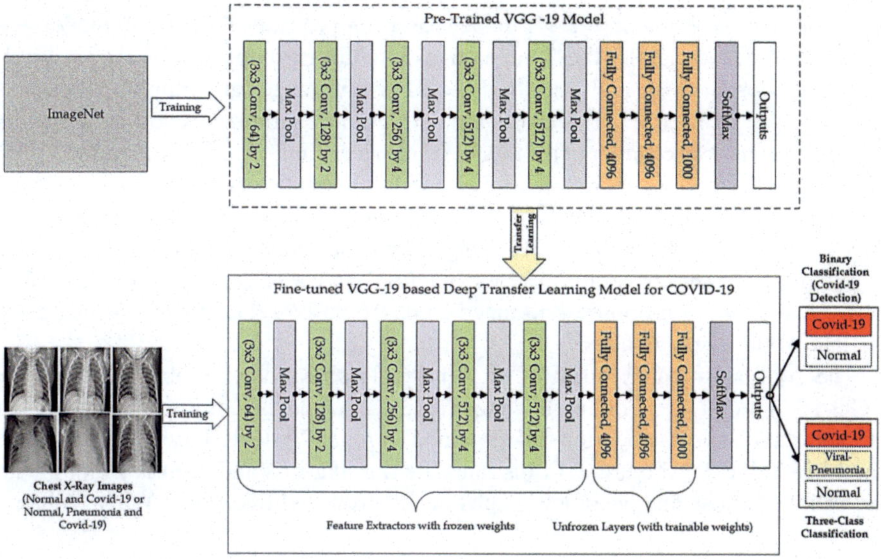

Fig. 2.3 The architectural workflow of the proposed fine-tuned VGG-19 based deep transfer learning model for COVID-19 Detection

the training process, thus not participating in the transfer-learning process. The higher layers of the pre-trained model were used for learning task-specific features from the COVID-19 images dataset. In this case, the higher layers of the pre-trained model were relaxed and thus made trainable or fine-tuned in which the weights of the layers were updated. Therefore, the layers were allowed to participate in the transfer-learning process. Each of these models ends with the SoftMax layer, which produces the outputs. Both the binary classification task and three-class classification scenarios were considered in the workflow, in which the DTL model determined the class of the chest X-ray images either as the COVID-19 category or Normal category in the binary classification and either as any of the COVID-19 category, Viral-Pneumonia category, or Normal category in the three-class classification.

2.4 Experimentation and Results

Two different experiments were performed to classify radiological X-ray images using Deep Transfer learning approaches. For the experiments, out of a total of 375 images that were used, 225 images were used to train the models, 75 images were used to perform validation and hyper-parameter tuning, and 75 images were used for testing in order to provide an unprejudiced assessment of a final model fit on the training dataset.

In the first experiment, a DTL model based on pre-trained VGG-16 model was trained in order to classify the X-ray images into three groups of COVID-19, Viral-Pneumonia or Normal; and also, to detect if X-ray images are simply of the class COVID-19 or Normal. The VGG-16 based DTL model summary, detailing the layers and parameters in each layer of the model is shown in Table 2.1. The fine-tuned VGG-16 based DTL model consists of 14,747,715 total parameters, with 33,027 of them made trainable while 14,714,688 are non-trainable. The model was trained on 40 epochs and with a batch size of 10, using Adams optimizer specifically for updates of weights, certain cross-entropy loss function with a learning rate of $1e^{-2}$. The performance of the proposed fine-tuned VGG-16 based DTL model was evaluated on 25% of the X-ray images.

The confusion matrix result of the binary classification task obtained from the VGG-16 based DTL model is shown in Table 2.2, while the confusion matrix result of the three-class classification task obtained from the VGG-16 based DTL model is shown in Table 2.3. Figure 2.4 illustrates the training loss and accuracy along with the validation loss and accuracy graphs of the proposed fine-tuned VGG-16 based

Table 2.1 The layers and layer parameters of the proposed fine-tuned VGG-16 based DTL model

	Layers	Layer's type	Shape of output	Number of trainable parameters
1	Convolution_1 of Block_1	Convolution 2D	[64, 224, 224]	1,792
2	Convolution_2 of Block_1	Convolution 2D	[64, 224, 224]	36,928
3	Convolution_1 of Block_2	Convolution 2D	[128, 112, 112]	73,856
4	Convolution_2 of Block_2	Convolution 2D	[128, 112, 112]	147,584
5	Convolution_1 of Block_3	Convolution 2D	[256, 56, 56]	295,168
6	Convolution_2 of Block_3	Convolution 2D	[256, 56, 56]	590,080
7	Convolution_3 of Block_3	Convolution 2D	[256, 56, 56]	590,080
8	Convolution_1 of Block_4	Convolution 2D	[512, 28, 28]	1,180,160
9	Convolution_2 of Block_4	Convolution 2D	[512, 28, 28]	2,359,808
10	Convolution_3 of Block_4	Convolution 2D	[512, 28, 28]	2,359,808
11	Convolution_1 of Block_5	Convolution 2D	[512, 14, 14]	2,359,808
12	Convolution_2 of Block_5	Convolution 2D	[512, 14, 14]	2,359,808
13	Convolution_3 of Block_5	Convolution 2D	[512, 14, 14]	2,359,808
14	Flatten	Flatten	[512]	0
15	Dense	Dense	[64]	32,832
16	Dense_1	Dense	[3]	195

Table 2.2 The confusion matrix of the binary classification task obtained from the fine-tuned VGG-16 based DTL

	COVID-19	Normal
COVID-19	22	1
Normal	0	27

Table 2.3 The confusion matrix of the three-class classification task obtained from the fine-tuned VGG-16 based DTL

	COVID-19	Normal	Viral pneumonia
COVID-19	23	0	0
Normal	0	25	2
Viral pneumonia	0	0	25

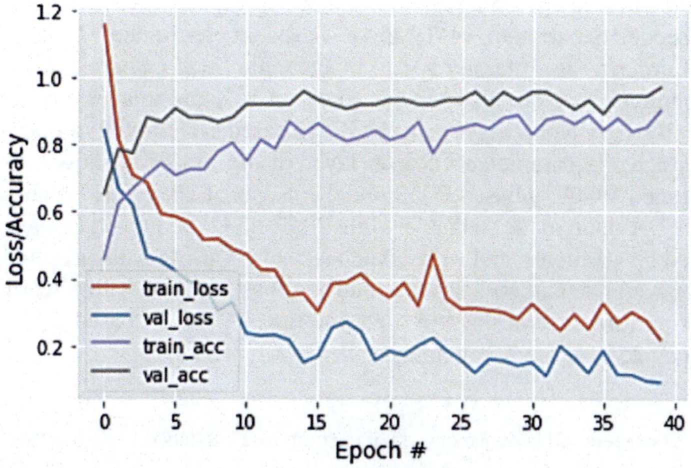

Fig. 2.4 The Training loss and accuracy with the validation loss and accuracy curves obtained for the fine-tuned VGG-16 based deep transfer learning model

DTL model. The precision, recall and F1-Score of the proposed fine-tuned VGG-16 based DTL model were also obtained for both binary and three-class classifications and shown in Tables 2.4 and 2.5.

It was observed from Fig. 2.4 that the validation and training losses were significantly high in the earlier epochs and then noticeably decreased as the training occured in more subsequent epochs. This sharp decrease in the loss values at the 40th epoch

Table 2.4 The precision, recall and F1-score obtained for the binary classification task using the fine-tuned VGG-16 based DTL model

	Precision	Recall	F1-score
COVID-19	1.00	0.96	0.98
Normal	0.96	1.00	0.98

Table 2.5 The precision, recall and F1-Score obtained for the three-class classification task using the fine-tuned VGG-16 based DTL model

	Precision	Recall	F1-score
COVID-19	1.00	1.00	1.00
Normal	1.00	0.93	0.96
Viral pneumonia	0.93	1.00	0.96

can be attributed to the fact that the fine-tuned VGG-16 based DTL model has been exposed to all the available X-ray images time and again during each of the epoch considered during training.

The accuracy obtained for the fine-tuned VGG-16 based DTL model was 97.33%, its sensitivity was 100% while its specificity stood at 92.59%. The obtained precision, recall and F1-Score metrics for the binary classification task and three-class classification task are given in Tables 2.4 and 2.5 respectively.

In the second experiment, a DTL model based on pre-trained VGG-19 model was trained in order to also classify X-ray images into three categories of COVID-19, Viral-Pneumonia or Normal; and also, to detect if X-ray images are simply of the class COVID-19 or Normal. Also, the VGG-19 based DTL model summary, detailing the layers and the parameters in each layer of the model is shown in Table 2.6. The fine-tuned VGG-19 based DTL model consists of 20,057,411 total parameters, with 33,027 of them made trainable while 20,024,384 are non-trainable. The model was trained on 40 epochs and with a batch size of 10, using Adams optimizer for the updates of weights, categorical cross-entropy loss function with a learning rate of $1e^{-1}$. The performance of the proposed fine-tuned VGG-19 based DTL model was evaluated on 25% of the X-ray images.

Table 2.6 The layers and layer parameters of the proposed fine-tuned VGG-19 based DTL model

	Layers	Layer's type	Shape of output	Number of trainable parameters
1	Convolution_1 of Block_1	Convolution 2D	[64, 224, 224]	1,792
2	Convolution_2 of Block_1	Convolution 2D	[64, 224, 224]	36,928
3	Convolution_1 of Block_2	Convolution 2D	[128, 112, 112]	73,856
4	Convolution_2 of Block_2	Convolution 2D	[128, 112, 112]	147,584
5	Convolution_1 of Block_3	Convolution 2D	[256, 56, 56]	295,168
6	Convolution_2 of Block_3	Convolution 2D	[256, 56, 56]	590,080
7	Convolution_3 of Block_3	Convolution 2D	[256, 56, 56]	590,080
8	Convolution_4 of Block_3	Convolution 2D	[256, 56, 56]	590,080
9	Convolution_1 of Block_4	Convolution 2D	[512, 28, 28]	1,180,160
10	Convolution_2 of Block_4	Convolution 2D	[512, 28, 28]	2,359,808
11	Convolution_3 of Block_4	Convolution 2D	[512, 28, 28]	2,359,808
12	Convolution_4 of Block_4	Convolution 2D	[512, 28, 28]	2,359,808
13	Convolution_1 of Block_5	Convolution 2D	[512, 14, 14]	2,359,808
14	Convolution_2 of Block_5	Convolution 2D	[512, 14, 14]	2,359,808
15	Convolution_3 of Block_5	Convolution 2D	[512, 14, 14]	2,359,808
16	Convolution_4 of Block_5	Convolution 2D	[512, 14, 14]	2,359,808
17	Flatten	Flatten	[512]	0
18	Dense	Dense	[64]	32,832
19	Dense_1	Dense	[3]	195

Table 2.7 The Confusion Matrix of the Binary classification task obtained from the fine-tuned VGG-19 based DTL

	COVID-19	Normal
COVID-19	21	2
Normal	0	27

The confusion matrix result of the binary classification task obtained from the fine-tuned VGG-19 based DTL model is shown in Table 2.7, while the confusion matrix result of the three-class classification task obtained from the fine-tuned VGG-19 based DTL model is shown in Table 2.8.

Figure 2.5 illustrates the training loss and accuracy along with the validation loss and accuracy graphs of the proposed fine-tuned VGG-19 based DTL model. The precision, recall and F1-Score of the proposed fine-tuned VGG-19 based DTL model were also obtained for both binary and three-class classifications as shown in Tables 2.9 and 2.10.

The same trend as of the fine-tuned VGG-16 based DTL model could be observed from Fig. 2.5 in that the validation and training losses were significantly high in the earlier epochs and then noticeably decreased as the training occured in more subsequent epochs. These sharp decreases in the loss values at the 40th epoch can

Table 2.8 The confusion matrix of the three-class classification task obtained from the fine-tuned VGG-19 based DTL

	COVID-19	Normal	Viral pneumonia
COVID-19	22	0	1
Normal	0	26	1
Viral pneumonia	5	1	19

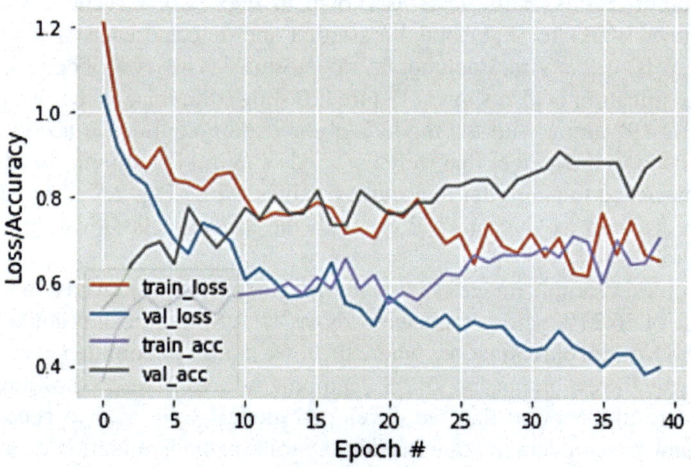

Fig. 2.5 The training loss and accuracy with the validation loss and accuracy curves obtained for the fine-tuned VGG-19 based deep transfer learning model

Table 2.9 The precision, recall and F1-Score obtained for the Binary classification task using the fine-tuned VGG-19 based DTL model

	Precision	Recall	F1-score
COVID-19	1.00	0.91	0.95
Normal	0.93	1.00	0.96

Table 2.10 The precision, recall and F1-score obtained for the three-class classification task using the fine-tuned VGG-19 based DTL model

	Precision	Recall	F1-score
COVID-19	0.81	0.96	0.88
Normal	0.96	0.96	0.96
Viral Pneumonia	0.90	0.76	0.83

be attributed to the fact that the fine-tuned VGG-19 based DTL model has been exposed to all the available X-ray images over and over again during each of the epoch considered during training.

The accuracy obtained for the fine-tuned VGG-19 based DTL model was 89.33%, its sensitivity was 95.65% while its specificity stood at 96.30%. The obtained precision, recall and F1-Score metrics for the binary classification task and the three-class classification task are given in Tables 2.9 and 2.10 respectively.

It could be noted from the obtained confusion matrices and the computed performance evaluation metrics of the three-class classification task that the fine-tuned VGG-16 based deep transfer learning model classified COVID-19 well than the fine-tuned VGG-19 based deep transfer learning model. Based on this, tests were carried out on unlabeled images using the developed fine-tuned VGG-16 multi-classification model. This is in order to yield an unprejudiced evaluation of the final model fit on the training dataset. Some results of the tests are shown in Figs. 2.6, 2.7 and 2.8.

The output results of the tests as shown in Figs. 2.6, 2.7 and 2.8 show how the fine-tuned VGG-16 DTL model classified and detected each of the images as either "COVID-19", "Viral Pneumonia" or "Normal". The level of confidence in the model classification is also shown. Figure 2.6 shows the images that were detected as "COVID-19" along with the model's classification confidence accuracy values. Figure 2.7 shows the images that were detected as "Viral Pneumonia" along with the model's classification confidence accuracy values, while Fig. 2.8 shows the images that were detected as "Normal" along with the model's classification confidence accuracy values.

Out of the six sample images shown in Fig. 2.6, only one showed a lower level of confidence of 76.37% while others were above 94%. Similar results could be seen in Fig. 2.8 for Normal classification, where the lowest level of confidence was 78.92%. However, the lowest output for Viral Pneumonia was 96.21%, as shown in Fig. 2.7. These test results showed that the developed models were able to generalize and adapt to new unseen data that are outside the training and validation dataset. These test results are necessary to show the adaptability of the developed models when given any related data.

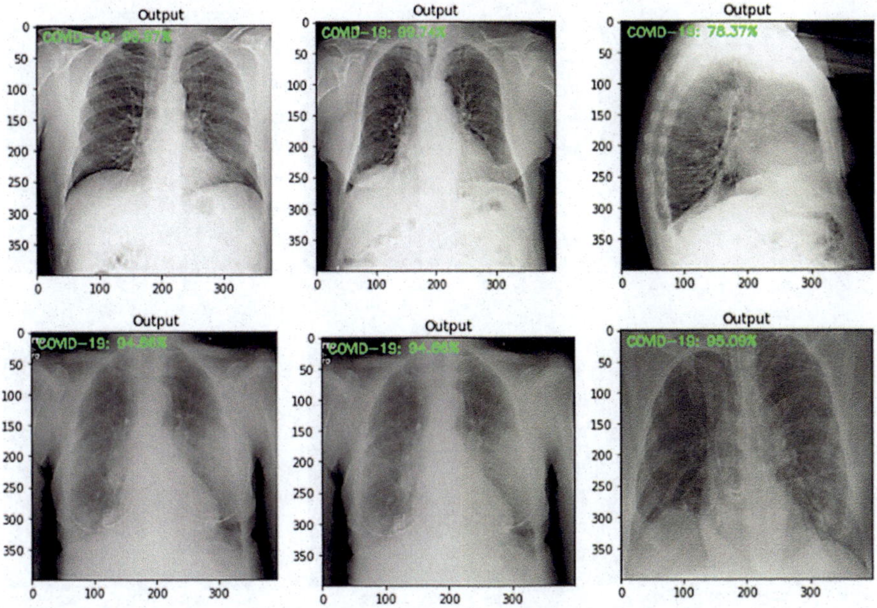

Fig. 2.6 COVID-19 sample test results with predicted level of confidence value

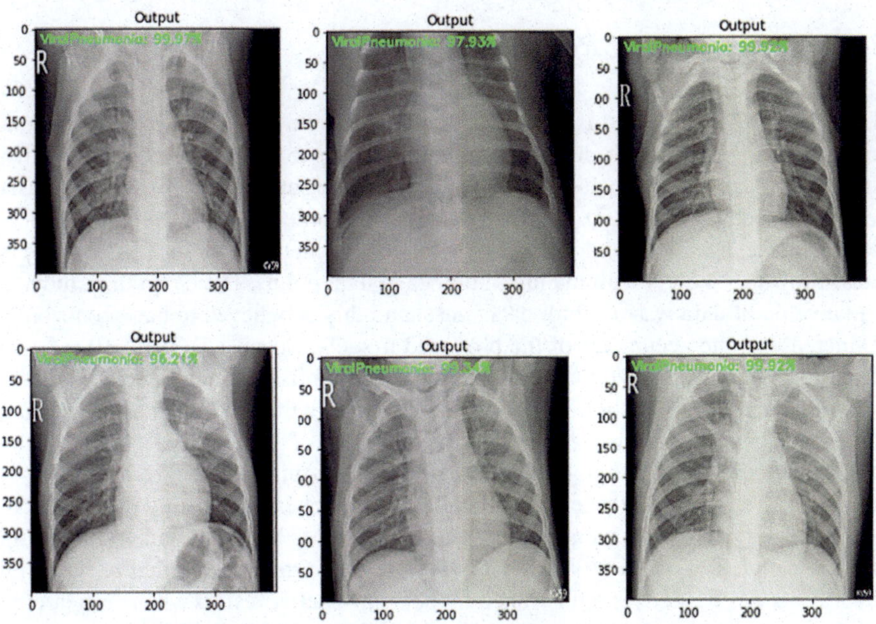

Fig. 2.7 Viral pneumonia sample test results with a predicted level of confidence value

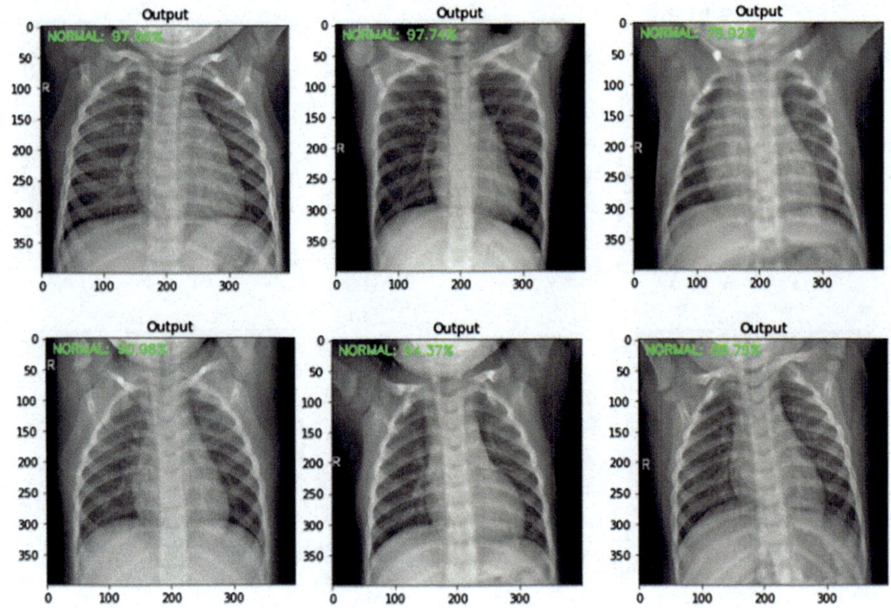

Fig. 2.8 Normal sample test results with predicted level of confidence value

2.4.1 Evaluation of Results

The findings obtained in this study were compared with twenty other current literature surveys. Few number of studies conducted prior to this study used twenty-five and fifty images in each class [21, 35, 45], while eight comparative reviews used imbalanced data (Table 2.7). Generally, there is a problem in modelling imbalanced data, which can lead to the inability of the model to generalize, or the model could be biased towards a class with the high number of data points. Hence, in this study, an equal value of data was used for each class, and this is believed to have contributed to increase in the accuracies of the proposed models. Because COVID-19 is a new disease, there are few X-ray images available to establish the automated diagnostic program. Data augmentation was performed by setting the random image rotation to 15° in clockwise direction to ensure the generalization of the models developed in this work. The proposed new models were based on fine-tuning VGG-16 and VGG-19 methods by constructing a new fully-connected layer head consisting of POOLING $\Rightarrow$ FC $\Rightarrow$ SOFTMAX layers and append it on top of VGG-16 and VGG-19, the CONV weights of VGG-16 and VGG-19 were then frozen, such that *only* the FC layer head was trained. The fine-tuned models gave better results when compared to other models that used ordinary pre-trained VGG-16 and VGG-19 [5], Asnaoui and Chawki [7]. Full results of the comparison with twenty other existing results from the literature are presented in Table 2.11. The proposed model outperformed all the twenty existing results in terms of accuracy.

Table 2.11 Comparison of the proposed COVID-19 diagnostic methods with other deep learning methods developed using radiology images

S. No.	Study	Type of images	Number of cases	Method used	Accuracy (%)
1	Apostolopoulos and Mpesiana [5]	Chest X-ray images	224 COVID-19 700 Pneumonia 504 Normal	VGG-19	93.48
2	Wang and Wong [61]	Chest X-ray images	53 COVID-19 5526 Normal 8066 Healthy	COVID-Net	92.4
3	Sethy and Behra [45]	Chest X-ray images	25 COVID-19 25 Normal	ResNet50 + SVM	95.38
4	Hemdan et al. [21]	Chest X-ray images	25 COVID-19(+) 25 Normal	COVIDX-Net	90.0
5	Narin et al. [35]	Chest X-ray images	50 COVID-19 50 Normal	Deep CNN ResNet-50	98.0
6	Ying et al. [52]	Chest CT scan images	777 COVID-19 708 Normal	DRE-Net	86.0
7	Wang et al. [62]	Chest CT scan images	195 COVID-19 258 Normal	M-Inception	82.9
8	Zheng et al. [74]	Chest CT scan images	313 COVID-19(+) 229 COVID-19(-)	UNet + 3D Deep Network	90.8
9	Xu et al. [67, 68]	Chest CT scan images	219 COVID-19 224 Viral pneumonia 75 Normal	ResNet	86.7
10	Ozturk et al. [38]	Chest X-ray images	125 COVID-19 500 Normal	DarkCovidNet	98.08
			125 COVID-19 500 Pneumonia 500 Normal		87.02

(continued)

Table 2.11 (continued)

S. No.	Study	Type of images	Number of cases	Method used	Accuracy (%)
11	Asnaoui and Chawki [7]	Chest X-ray and CT Scan	2780 bacterial pneumonia 1493 corona-virus 231 COVID-19 1583 Normal	Inception_Resnet_V2	92.18
				DensNet201	88.09
				Resnet50	87.54
				Mobilenet_V2	85.47
				Inception_V3	88.03
				VGG-16	74.84
				VGG-19	72.52
12	Raajan et al. [42]	Chest CT scan images	349 COVID-19 CT images of 216 patients and 463 non-COVID-19 CT images	ResNet-16	95.09
13	Pu et al. [40]	Chest CT scan images	120 chest CT scans and 72 serial chest CT scans from 24 COVID-19 patients	U-Net framework deep-learning technique	95
14	Li et al. [27, 28]	Chest CT scan images	4563 3D chest CT scan images from 3506 patients	A three-dimensional deep-learning framework based on RestNet50 (COVID-19 detection neural network)	0.96 AUC
15	Yoo et al. [70]	Chest X-ray images	240 images were used for the augmentation of 1216 images	Resnet18 model with PyTorch frame	95
16	Qianqian et al. [41]	Chest X-ray images	14,435 participants 2154 COVID-19 5874 Pneumonia 6434 Normal	Deep learning algorithm	95.0
17	Harsh et al. [19]	Kaggle's Chest X-Ray Images	192 COVID-19	VGG-16	88.0
				nCOVnet	97.0

(continued)

Table 2.11 (continued)

S. No.	Study	Type of images	Number of cases	Method used	Accuracy (%)
18	Shashank et al. [47]	https://github.com/ieee8023/covid-chestxray-dataset	181 COVID-19 with 364 chest X-ray images	VGG-19	96.3
19	Xu et al. [67, 68]	Chest X-ray images	618 Cases	CNN segmentation with Bayesian function	86.7%
20	Nour et al. [37]	Chest X-ray images	2905 Cases	CNN with Bayesian Optimization	98.97%
21	Proposed Study	Chest X-ray images	100 COVID-19(+) 100 Normal	VGG-16	**99.0**
				VGG-19	**99.0**
			100 COVID-19(+) 100 Viral Pneumonia 100 Normal	VGG-16	**97.3**
				VGG-19	**89.3**

2.5 Conclusion

Many researchers around the world are coordinating their efforts to gather data and develop solutions for COVID-19. Laboratory testing of suspicious cases characterized by long waiting times and an increasing rise in testing demand has been a major global bottleneck. To ameliorate this, rapid diagnostic test kits are being developed; most of which are currently undergoing clinical validation and therefore, are yet to be adopted for routine use. While waiting for results, this chapter proposes a solution of using Deep Learning Convolutional Neural Network Architecture to classify real-life COVID-19 dataset of chest X-ray images into a three-class classification scenario of COVID-19, Viral-Pneumonia or Normal categories. In this study, a Convolutional Neural Network was fine-tuned to instinctively prognose or detect COVID-19 using deep learning. A total of 300 images (100 COVID-19, 100 Viral Pneumonia and 100 Normal) were used to develop the models of which 225 was used to train the model, 75 was used for validation and to perform hyper-parameter tuning. Also, a total of 75 new images were used for testing the models (25 COVID-19, 25 Viral Pneumonia and 25 Normal). The proposed models were developed to provide accurate diagnostics for binary classification (COVID-19 and Normal) and multi-class classification (COVID-19, Viral Pneumonia and Normal). The fine-tuned VGG-16 and VGG-19 models both produced a classification accuracy of 99.00% for binary classes and 97.33% and 89.33% for multi-class cases respectively. Two experiments were performed where the VGG-16 and VGG-19 Convolutional Neural Networks (CNN) with Deep Transfer Learning (DTL) was implemented in Jupyter Notebook using Python programming language, and the result showed that the DTL model based on pre-trained VGG-16 model classified COVID-19 better than the VGG-19 based deep transfer learning model. The proposed model, in this work, also outperformed existing methods in terms of accuracy, although different publicly available dataset other than those used in the various related works have been used in this work. The findings have a high potential of increasing the prediction accuracy for coronavirus disease, which would be of immense benefit to the medical field and the entire human populace as it could help save many lives from untimely death. Finally, the publicly available image datasets of COVID-19 are limited at the moment, and this is a limitation in this research work. Future works would, therefore, consider increasing the volume of data used for the study and further hyper-parameter tweaking to get more accurate results.

References

1. Ahmad, A., Garhwal, S., Ray, S.K., Kumar, G., Malebary, S.J., Barukab, O.M.: The number of confirmed cases of COVID-19 by using machine learning: methods and challenges. Arch. Comput. Methods Eng. 1–9 (2020)
2. Alazab, M., Awajan, A., Mesleh, A., Abraham, A., Jatana, V., Alhyari, S.: COVID-19 prediction and detection using deep learning. Int. J. Comput. Inf. Syst. Ind. Manag. Appl. **12**, 168–181 (2020)
3. Altan, A., Karasu, S.: Recognition of COVID-19 disease from X-ray images by hybrid model consisting of 2D curvelet transform, chaotic salp swarm algorithm and deep learning technique. Chaos, Solitons Fractals **140**, 1–10 (2020). https://doi.org/10.1016/j.chaos.2020.110071
4. Amrane, S., Tissot, D., Doudier, H., Eldin, B., Hocquart, C., Mailhe, M., Colson, M.: Rapid viral diagnosis and ambulatory management of suspected COVID-19 cases presenting at the infectious diseases referral hospital in Marseille, France, - January 31 to March 1, 2020: A respiratory virus snapshot. Travel Med. Infect. Dis. (2020). https://doi.org/10.1016/j.tmaid.2020.101632
5. Apostolopoulos, I.D., Mpesiana, T.A.: COVID-19: automatic detection from X-ray images utilizing transfer learning with convolutional neural networks. Phys. Eng. Sci. Med. **43**, 635–640 (2020). https://doi.org/10.1007/s13246-020-00865-4
6. Arora, K., Bist, A., Chaurasia, S., Prakash, R.: Analysis of deep learning techniques for COVID-19 detection. Int. J. Sci. Res. Eng. Manag. (IJSREM) **4**(4), 1–5 (2020)
7. Asnaoui, K., Chawki, Y.: Using X-ray images and deep learning for automated detection of coronavirus disease. J. Biomol. Struct. Dyn., 1–12 (2020)
8. BBC: British Broadcasting Corporation (2020). Retrieved from https://www.bbc.com/news/technology-52120747
9. Bai, L., Dawei, Y., Wang, X., Tong, L., Zhu, X., Zhong, N., et al.: Chinese experts' consensus on the Internet of Things-aided diagnosis and treatment of coronavirus disease 2019 (COVID-19). Clin. eHealth **3**, 7–15 (2020). https://doi.org/10.1016/j.ceh.2020.03.001
10. Bansal, A., Padappayil, R.P., Garg, C., Singal, A., Gupta, M., Klein, A.: Utility of artificial intelligence amidst the COVID-19 pandemic: a review. J. Med. Syst. **44**(9), 1–6 (2020)

11. Brinati, D., Campagner, A., Ferrari, D., Locatelli, M., Banfi, G., Cabitza, F.: Detection of COVID-19 infection from routine blood exams with machine learning: a feasibility study. medRxiv (2020)
12. Chollet, F.: Xception: deep learning with depthwise separable convolutions. In: 2017 IEEE Conference on Computer Vision and Pattern Recognition (CVPR), Honolulu, HI, pp. 1800–1807 (2017)
13. Chowdhury, M.E., Rahman, T., Khandakar, A., Mazhar, R., Kadir, M.A., Mahbub, Z.B., Reaz, M.B.: Can AI help in screening Viral and COVID-19 pneumonia?, 29 Mar 2020. Retrieved from https://arxiv.org/abs/2003.13145; https://www.kaggle.com/tawsifurrahman/covid19-radiography-database
14. Chu, K.H., Tsang, W.K., Tang, C.S.: Acute renal impairment in coronavirus-associated severe acute respiratory syndrome. Kidney Int. **67**(2), 698–705 (2005)
15. Dipayan, D., Santosh, K.C., Umapada, P.: Truncated inception net: COVID-19 outbreak screening using chest X-rays. Phys. Eng. Sci. Med (2020). https://doi.org/10.1007/s13246-020-00888-x
16. ECDC.: ECDC: an overview of the rapid test situation for COVID-19 diagnosis in the EU/EEA (2020). https://doi.org/10.1101/2020.03.18.20038059
17. Frossard, D.: VGG in TensorFlow, 17 June 2016. Retrieved 24 May 2020, from https://www.cs.toronto.edu/~frossard/post/vgg16/
18. Hamid, S., Mir, M.Y., Rohela, G.K.: Novel coronavirus disease (COVID-19): a pandemic (epidemiology, pathogenesis and potential therapeutics). New Microbes New Infect. **35** (2020). https://doi.org/10.1016/j_nmni.2020.100679
19. Harsh, P., Gupta, P.K., Mohammad, K.S., Morales-Menendez, R., Vaishnavi, S.: Application of deep learning for fast detection of COVID-19 in X-Rays using nCOVnet. Chaos, Solitons Fractals **138**, 1–8 (2020). https://doi.org/10.1016/j.chaos.2020.109944
20. He, K., Zhang, X., Ren, S., Sun, J.: Deep residual learning for image recognition. In: 2016 IEEE conference on computer vision and pattern recognition (CVPR), Las Vegas, NV, pp. 770–778 (2016)
21. Hemdan, E.E.D., Shouman, M.A., Karar, M.E.: COVIDX-Net: a framework of deep learning classifiers to diagnose COVID-19 in X-ray images (2020). arXiv preprint arXiv:2003.11055
22. Huang, G., Liu, Z., Van Der Maaten, L., Wein, K.Q.: Densely connected convolutional networks. In: 2017 IEEE Conference on Computer Vision and Pattern Recognition (CVPR), pp. 2261–2269. Honolulu, HI (2017)
23. Huang, C., Wang, Y., Li, X., Ren, L., Zhao, J., Hu, Y., Zhang, L., Fan, G., Xu, J., Gu, X., Cheng, Z., Yu, T., Xia, J., Wei, Y., Wu, W., Xie, X., Yin, W., Li, H., Liu, M., Xiao, Y., Gao, H., Guo, L., Xie, J., Wang, G., Jiang, R., Gao, Z., Jin, Q., Wang, J., Cao, B.: Clinical features of patients infected with 2019 novel coronavirus in Wuhan, China. Lancet **395**(10223), 497–506 (2020)
24. Hyafil, A., Moriña, D.: Analysis of the impact of lockdown on the evolution of COVID-19 epidemics in Spain. medRxiv preprint, pp. 1–20 (2020). https://doi.org/10.1101/2020.04.18.20070862
25. Kobia, F., Gitaka, J.: COVID-19: are Africa's diagnostic challenges blunting response effectiveness? AAS Open Res. 1–11 (2020)
26. Kumar, S.V., Damodar, G., Ravikanth, S., Vijayakumar, G.: An overview on infectious disease. Indian J. Pharm. Sci. Res. **2**(2), 63–74 (2012)
27. Li, L., Qin, L., Xu, Z., Yin, Y., Wang, X., Kong, B., Xia, J., et al.: Using Artificial Intelligence to detect COVID-19 and community-acquired pneumonia based on pulmonary CT: evaluation of the diagnostic accuracy. Radiology (2020). https://doi.org/10.1148/radiol.2020200905
28. Li, M., Lei, P., Zeng, B., Li, Z., Yu, P., Fan, B., Liu, H., et al.: Coronavirus disease (COVID-19): Spectrum of CT findings and temporal progression of the disease. Acad Radiol. **27**(5), 603–608 (2020). https://doi.org/10.1016/j.acra.2020.03.003
29. Lokuge, K., Banks, E., Davies, S., Roberts, L., Street, T., Glass, K., et al.: Exit strategies: optimizing feasible surveillance for detection, elimination and ongoing prevention of COVID-19 community transmission. medRxiv preprint (2020). https://doi.org/10.1101/2020.04.19.20071217

30. Long, C., Xu, H., Shen, Q., Zhang, X., Fan, B., Wang, C., Li, H., et al.: Diagnosis of the coronavirus disease (COVID-19): rRT-PCR or CT? Eur. J. Radiol. (2020). https://doi.org/10.1016/j.ejrad.2020.108961
31. Madan, B., Panchal, A., & Chavan, D.: Lung cancer detection using deep learning. In: 2nd International Conference on Advances in Science & Technology (ICAST-2019) (2019)
32. Makhoul, M., Ayoub, H.H., C. H., Seedat, S., Mumtaz, G., Sarah, A.-O., Abu-Raddad, L. J.: Epidemiological impact of SARS-CoV-2 vaccination: mathematical modeling analyses (2020). medRxiv preprint. https://doi.org/10.1101/2020.04.19.20070805
33. Manikandarajan, A., Sasikala, S.: Detection and segmentation of lymph nodes for lung cancer diagnosis. In: National Conference on System Design and Information Processing (2013)
34. Nadeem, S.: Coronavirus COVID-19: Available free literature provided by various companies, Journals and Organizations around the World. J. Ongoing Chem. Res. **5**(1), 7–13 (2020). https://doi.org/10.5281/zenodo.3722904
35. Narin, A., Kaya, C., Pamuk, Z.: Automatic detection of coronavirus disease (COVID-19) using X-ray images and deep convolutional neural networks (2020). arXiv preprint arXiv:2003.10849
36. Nigeria Centre for Disease Control [NCDC]: Coronavirus (COVID-19) highlights. Nigeria Centre for Disease Control (NCDC), Abuja (2020). Retrieved 29 Apr 2020, from https://covid19.ncdc.gov.ng/index.php
37. Nour, M., Cömert, Z., Polat, K.: A novel medical diagnosis model for COVID-19 infection detection based on deep features and bayesian optimization. Appl. Soft Comput. 106580 (2020)
38. Ozturk, T., Talo, M., Yildirim, E.A., Baloglu, U.B., Yildirim, O., Acharya, U.R.: Automated detection of COVID-19 cases using deep neural networks with X-ray images. Comput. Biol. Med. (2020). https://doi.org/10.1016/j.compbiomed.2020.103792
39. Pan, L., Mu, M., Ren, H.G.: Clinical characteristics of COVID-19 patients with digestive symptoms in Hubei, China: a descriptive, cross-sectional, multicenter study. Am. J. Gastroenterol. **115**(5), 766–773 (2020)
40. Pu, J., Leader, J.K., Bandos, A., Ke, S., Wang, J., Shi, J., Jin, C., et al.: Automated quantification of COVID-19 severity and progression using chest CT images. Eur. Radiol. 1–11 (2020). https://doi.org/10.1007/s00330-020-07156-2
41. Qianqian, N., Zhi, Y.S., Li, Q., Wen, C., Yi, Y., Li, W., Xinyuan, Z., Liu, Y., Yi, F., Zijian, X., Zhen, Z., Yizhou, Y., Guang, M.L, Long, J.Z.: A deep learning approach to characterize 2019 coronavirus disease (COVID-19) pneumonia in chest CT images. Eur. Radiol. 1–11 (2020). https://doi.org/10.1007/s00330-020-07044-9
42. Raajan, N. R., Ramya Lakshmi, V. S., & Prabaharan, N. (2020, July). Non-Invasive Technique-Based Novel Corona (COVID-19) Virus Detection Using CNN. *National Academy of Sciences Letters*, 1–4. doi:https://doi.org/10.1007/s40009-020-01009-8
43. Romero-Severson, E., Hengartner, N., Meadors, G., Ke, R.: A decline in global transmission rates of COVID-19. medRxiv preprint (2020). https://doi.org/10.1101/2020.04.18.20070771
44. Sasikala, S., Bharathi, M., Sowmiya, B.R.: Lung cancer detection and classification using deep CNN. Int. J. Innov. Technol. Explor. Eng. (IJITEE) **8**(25), 259–262 (2018)
45. Sethy, P.K., Behera, S.K.: Detection of coronavirus disease (COVID-19) based on deep features (2020)
46. Sharma, S.: Drawing insights from COVID-19-infected patients using CT scan images and machine learning techniques: a study on 200 patients. Environ. Sci. Pollut. Res. 1–9 (2020). https://doi.org/10.1007/s11356-020-10133-3
47. Shashank, V., Reza, K., Mohit, B.: Deep learning COVID-19 detection bias: accuracy through artificial intelligence. Int. Orthop. **44**, 1539–1542 (2020). https://doi.org/10.1007/s00264-020-04609-7
48. Shinde, G.R., Kalamkar, A.B., Mahalle, P.N., Dey, N., Chaki, J., Hassanien, A.E.: Forecasting models for coronavirus disease (COVID 19): a survey of the state-of-the-art. SN Comput. Sci. 1–15 (2020). https://doi.org/10.1007/s42979-020-00209-9
49. Simonyan, K., Zisserman, A.: Very deep convolutional for large-scale image recognition. In: International Conference on Learning Representations. San Diego (2015)

50. Singh, D., Kumar, V., Vaishali, Kaur, M.: Classification of COVID-19 patients from chest CT images using multi-objective differential evolution–based convolutional neural network. Eur. J. Clin. Microbiol. Infect. Dis. 1–11 (2020). https://doi.org/10.1007/s10096-020-03901-z
51. Song, Y.G., Shin, H.-S.: COVID-19, a clinical syndrome manifesting as hypersensitivity pneumonitis. Infect. Chemother. **52**, 110–112 (2020)
52. Song, Y., Zheng, S., Li, L., Zhang, X., Zhang, X., Huang, Z., Chen, J., Zhao, H., Zha, Y., Shen, J., Wang, R., et al.: Deep learning enables accurate diagnosis of novel coronavirus (COVID-19) with CT images. medRxiv (2020)
53. Sujath, R., Chatterjee, J.M., Hassanien, A.E.: A machine learning forecasting model for COVID-19 pandemic in India. Stochast. Environ. Res. Risk Assess. 1–14 (2020). https://doi.org/10.1007/s00477-020-01827-8
54. Szegedy, C., Liu, W., Jia, Y., Sermanet, P., Reed, S., Anguelov, D., Rabinovich, A.: Going deeper with convolutions. In: 2015 IEEE Conference on Computer Vision and Pattern Recognition (CVPR), Boston, MA, pp. 1–9 (2015)
55. Toğaçar, M., Ergen, B., Cömert, Z.: COVID-19 detection using deep learning models to exploit Social Mimic Optimization and structured chest X-ray images using fuzzy color and stacking approaches. Comput. Biol. Med 103805 (2020)
56. Tuncer, T., Dogan, S., Ozyurt, F.: An automated residual exemplar local binary pattern and iterative ReliefF based corona detection method using lung X-ray image. Chemometrics and Intelligent Laboratory Systems, 104054 (2020)
57. Vaishya, R., Javaid, M., Khan, I.H., Haleem, A.: Artificial intelligence (AI) applications for COVID-19 pandemic. Diabetes and metabolic syndrome. Clin. Res. Rev. **14**, 337–339 (2020). https://doi.org/10.1016/j.dsx.2020.04.012
58. Valette, X., du Cheyron, D., Goursaud, S.: Mediastinal lymphadenopathy in patients with severe COVID-19. Lancet Infect Dis. pii: S1473-3099(20)30310-8 (2020)
59. Vasilarou, M., Alachiotis, N., Garefalaki, J., Beloukas, A.: Population genomics insights into the recent. bioRxiv (2020). https://doi.org/10.1101/2020.04.21.054122
60. Wang, W., Tang, J., Wei, F.: Updated understanding of the outbreak of 2019 novel coronavirus (2019-nCoV) in Wuhan, China. J. Med. Virol. **92**(4), 441–447 (2020). https://doi.org/10.1002/jmv.25689
61. Wang, L., Wong, A.: COVID-Net: a tailored deep convolutional neural network design for detection of COVID-19 cases from chest radiography images (2020). 2020 arXiv preprint arXiv:2003.09871
62. Wang, S., Kang, B., Ma, J., Zeng, X., Xiao, M., Guo, J., Xu, B.: A deep learning algorithm using CT images to screen for Corona Virus Disease (COVID-19). medRxiv (2020)
63. World Health Organization [WHO]: Report of the WHO-China Joint Mission on Coronavirus Disease 2019 (COVID-19). World Health Organization, Geneva (2020)
64. World Health Organization [WHO]: Coronavirus disease (COVID-2019) R&D. R&D, Geneva (2020). Retrieved 24 Apr 2020, from https://www.who.int/blueprint/priority-diseases/key-action/novel-coronavirus/en/
65. World Health Organization [WHO]: Coronavirus disease 2019 (COVID-19). World Health Organization, R&D. World Health Organization, Geneva (2020). Retrieved 29 Apr 2020
66. Worldometer (2020). Retrieved from https://www.worldometers.info/coronavirus/coronavirus-death-toll/
67. Xu, X., Jiang, X., Ma, C., Du, P., Li, X., Lv, S.: Deep learning system to screen coronavirus disease 2019 pneumonia (2020). arXiv preprint arXiv:200209334
68. Xu, X., Jiang, X., Ma, C., Du, P., Li, X., Lv, S., Su, J., et al.: A deep learning system to screen novel coronavirus disease 2019 pneumonia. Engineering (2020)
69. Xu, Z., Shi, L., Wang, Y.: Pathological findings of COVID-19 associated with acute respiratory distress syndrome. Lancet Respir Med 420–422 (2020)
70. Yoo, S.H., Geng, H., Chiu, T.L., Yu, S., Cho, D.C., Heo, J., Lee, H., et al.: Deep learning-based decision-tree classifier for COVID-19 diagnosis from chest X-ray Imaging. Front. Med. (2020). https://doi.org/10.3389/fmed.2020.00427

71. Zhang, H., Zhang, J., Zhang, H., Nan, Y., Zhao, Y., Fu, E., Zhang, T., et al.: Automated detection and quantification of COVID-19 pneumonia: CT imaging analysis by a deep learning-based software. Eur. J. Nucl. Med. Mol. Imaging 1–8 (2020). https://doi.org/10.1007/s00259-020-04953-1
72. Zhang, L., Zheng, Z., Yang, L., Tianyu, Z., Liangxin, G., Dakai, J., Yuling, T., et al.: (2020). From community-acquired pneumonia to COVID-19: a deep learning–based method for quantitative analysis of COVID-19 on thick-section CT scans. Eur. Radiol.
73. Zhao, B., Wei, Y., Sun, W., Qin, C., Zhou, X., Wang, Z., Wang, Y., et al.: Distinguish coronavirus disease 2019 patients in general surgery emergency by CIAAD scale: development and validation of a prediction model based on 822 cases in China. medRxiv preprint (2020). https://doi.org/10.1101/2020.04.18.20071019
74. Zheng, C., Deng, X., Fu, Q., Zhou, Q., Feng, J., Ma, H., Wang, X.: Deep learning-based detection for COVID-19 from chest CT using a weak label. medRxiv (2020). https://doi.org/10.1101/2020.03.12.20027185
75. Zhou, B., She, J., Wang, Y., Ma, X.: The clinical characteristics of myocardial injury 1 in severe and very severe patients with 2019 novel coronavirus disease. J. Infect. (2020). pii: S0163-4453(20)30149-3

Chapter 3
Predicting Glaucoma Diagnosis Using AI

Falguni Ranadive, Akil Z. Surti, and Hemant Patel

Abstract This chapter is based on a research carried out with primary focus to develop an intelligent diagnostic system for Glaucoma—an eye related disease, from the data obtained through clinician by various examination devices or equipment used in ophthalmology. The data is used as training set to multi-classifier, developed using hybridization of various techniques of Artificial Intelligence. The classification is done by a hybrid approach using Artificial Neural Network, Naïve Bayes Algorithms and Decision Tree Algorithms. A design/development of a new technique or algorithm is required for such diagnosis and it is tested for its efficacy. Using the algorithms and techniques of Neural Network, Naïve Bayes Algorithm and Decision Tree based classifiers, the proposed hybrid technique is anticipated to intelligently analyze and perform diagnosis for patient's visionary predicaments, thus lessening the intervention of medical practitioners in terms of decision making. The proposed ensemble FGLAUC-99 takes 18 medical examination parameters for a patient and predicts if patient is suffering from glaucoma or not. If patient is detected with glaucoma, FGLAUC-99 also predict the type of glaucoma. Ensemble of classifiers shown improved accuracy compared to single individual classifiers. The classifiers were selected from group of classifiers. First classifier was included in ensemble from Probability based classifier such as, Naïve Bayes, As the probabilistic classifiers gives more accuracy. Second classifier was selected from Decision Tree based classifier, such as, Random Forest, J48. Decision Tree based classifier provides good interpretability. Third classifier was selected from neural Network based classifier, such as, MLP. Neural network based classifiers provide better prediction. FGLAUC-99 is developed with J48, Naïve Bayes and MLP classifiers with accuracy of 99.18%. The accuracy cannot be compared one to one with other classifiers as the dataset is exclusively developed. However, accuracy obtained by FGLAUC-99 classifier is better than the accuracy obtained from other classifiers available in literature.

F. Ranadive
Rishabh Software, Vadodara, Gujarat, India

A. Z. Surti (✉)
Enlighten Infosystems, Vadodara, Gujarat, India

H. Patel
Sumandeep Vidyapeeth, Vadodara, Gujarat, India

M. Mehta et al. (eds.), *Tracking and Preventing Diseases with Artificial Intelligence*, Intelligent Systems Reference Library 206,
https://doi.org/10.1007/978-3-030-76732-7_3

Keywords Glaucoma · Artificial Intelligence · Classification · FGLAUC-99 · Ophthalmology · Neural Network · Ensemble classifier

3.1 AI in Medical Diagnosis

In recent years, artificial intelligence (AI) has shown a great commitment to integrating different computer paradigms such as fuzzy systems, genetic algorithms (genetic algorithms) and ANNs in order to create more efficient hybrid systems. The goal is the provision of flexible information processing systems that can use tolerance for imprecision, uncertainty, reasoning and partial information in order to achieve tractability, robustness, low cost and near-like decision-making [1]. This chapter focuses on the active machine assisted diagnostics of neuro-fuzzy genetic hybrid and presents various steps in soft computing strategies for computer-aided medical diagnosis. The cause of this study is discussed as well as different research stages and contributions.

Medical computer-assisted diagnostics combines computer science, processing of images, pattern recognition and AI techniques and their accuracy and reliability depend upon a range of factors such as segmentation, functional selection, the reference size of the database, computational efficiency etc. Computer-Aided Diagnosis (CAD) is a diagnostic technology that helps pharmacists view medical images. Methods of imaging for mammography, CT, X-ray, MRI, and ultra-sound (US) diagnostics carry a lot of information to radiologists that they need to be thoroughly interpreted and checked in a limited period of time. The radiologist has been able to do so. The main application field of this research was computer-based algorithms like linear programming, tree decision [2, 3] and neural networks [4, 5] (Fig. 3.1).

3.2 AI in Ophthalmology Diagnosis

One of the key uses for retinal image processing is disease identification. The entire planning process is focused upon the results of these detection techniques. This area is underlined. There must also be substantially high precision of disease detection techniques, as misidentification can lead to fatal outcomes. In addition to being cost-effective, convergence times for results must also be good enough. The main cause of vision loss is, for instance, diabetes retinopathy (DR) disease, which will continue to increase its prevalence. The lesion of diabetic retinopathy imitates bright lesions from macular degeneration linked to age. The differentiation of lesion types is also

Fig. 3.1 Block diagram of computer aided diagnosis

very significant as they have diverse diagnostic and management consequences. The potential risk for blindness in these patients can be minimized by the diabetic patient screening for diabetic retinopathy. Early detection currently has allowed laser therapy to prevent or delay visual loss and may be used to facilitate better diabetic control. Therefore the diagnosis and assessment approach of current diabetic retinopathy are manual, expensive and require skilled eye doctors.

Thus, the efficient classification of retinal pathologies includes novel techniques with the above-mentioned characteristics. Four distinct yet closely linked eye diseases such as Choroidal Neo-Vascular Membrane (CNVM), Central Serous Retinopathy (CSR), Central Retinal Vein Occlusion (CRVO) yet Non-Proliferative Diabetic Retinopathy (NPDR) are automatically detected and categorized using Artificial Intelligence (AI) techniques. Also here is a short overview of these diseases.

3.3 Artificial Intelligent Techniques in Glaucoma Diagnosis

a. An Ophthalmic Condition-Glaucoma

Glaucoma is a retinal ganglion cells and axons without eye. Medically, this loss is suggestive of cupping, also known as OD digging and the concomitant loss of visual field. There are several subgroups of glaucoma, characterized by causes, biology or morphology and there may exist in any category tens of different belief forms. Open-angle and corner-close glaucoma are two primary types. These are characterized by increased intraocular pressure (IOP) or eye pressure. Figure 3.2a, b depict the vision with a normal and Glaucomatous eye, respectively.

Glaucoma is an eye nerve injury condition. It gets worse over time if not treated. Glaucoma calls for many reasons. The key pressure accumulation in the eye (IOP-Intra Ocular Pressure). Glaucoma is inherited and happens in life only later. Intraocular pressure growth can damage the optic nerve severely. Optic Nerve conveys images of the brain. The optic nerve is compromised, leading to glaucoma when the eye begins to exert high pressure. Faith can cause loss of vision permanently. Unless treated early, glaucoma will cause total blindness within a few years.

a b

Fig. 3.2 **a** Vision with normal eye, **b** vision with abnormal eye

Daily consultation with an eye doctor or an ophthalmologist is important, as most people with glaucoma are without early signs or pain from this additional strain. Daily appointments are important until long-term vision loss for the diagnosis and treatment of glaucoma. Every one or two years a person over 40 with a Glaucoma family history should invariably be examined by an ophthalmologist fully. If the person is at risk for other eye disorders like diabetes or a family history of glaucoma, the ophthalmologist needs to see the person more often.

When eye pressure increases, glaucoma generally develops. This can occur if the eye fluid is not in the front part of the eye naturally. This fluid normally flows through a mesh-like tunnel, called "Aqueous Humor." Blocked, the fluid levels are produced inside this channel, which cause glaucoma. The specific reason for this obstruction is yet unknown, but doctors know it can be passed on, so parents can pass it to children.

Gluteal or chemical eye injury, severe blurred ocular inflammation, blockage of the arteries in the eye, inflammatory conditions in the eye, and sometimes eye surgery to fix other conditions are the least common causes of glaucoma. Glaucoma is commonly present in both eyes, but at different levels each eye can be elevated.

The signs of glaucoma are generally low or no in most people. The first symptom of belief is the loss of peripheral or side vision several months before the later stages of the illness. One explanation why a patient should have a full-eye examination every one to 2 years is to detect glaucoma early. High intraocular pressure will sporadically increase. Sudden eye pain, headache, blurred vision, or even halos around lights can be seen as symptoms.

i. **Types of Glaucoma**

There are two main types of Glaucoma:

Open-angle Glaucoma: Often known as Glaucoma, the most common form of Glaucoma. The eye structures look fine, but fluid in the eye does not flow properly through the eye's drainage canals, called Trabecular Meshwork.

Angle-closure Glaucoma: Often called acute angle-closure or narrow angle glaucoma. This form of glaucoma is less common but can cause a sudden eye pressure buildup. Drainage can be low, as the angle between Iris and Cornea (where an eye drainage channel is located) is extremely narrow.

Other types of Glaucoma include

- Primary Angle-closure Glaucoma.
- Secondary Glaucoma.
- Primary Normal Tension Glaucoma.
- Primary Ocular Hypertension Glaucoma.
- Symptoms of Glaucoma

ii. **Symptoms of Glaucoma**

A person likely to develop Glaucoma may exhibit any or all of the following symptoms:

- Seeing halos around lights

- Vision loss
- Redness in the eye
- Eye that looks hazy (particularly in infants)
- Nausea or vomiting
- Pain in the eye
- Narrowing of vision (tunnel vision)

Glaucoma cannot be stopped completely but it can be fairly handled if it is diagnosed and treated early. The vision loss of Glaucoma becomes permanent at its later stages and cannot be regained. However, effective decrease in the eye pressure may help prevent further glaucoma vision loss. Most people with glaucoma are not blind, as long as their treatment scheme and regular assessments are followed.

iii. **Glaucoma Statistics**

The loss of visual vision known as blindness leads to physiological or neurological causes. Glaucoma is one of the world's leading causes of permanent blindness and affects about 70 million people after cataracts [6]. It is the second-biggest cause of global blindness after cataract, particularly because of Primary Open Angle Glaucoma (POAG) [7]. Data in 2002 shows a worldwide blindness of nearly 161 million people and a loss of sight of 37 million. Glaucoma caused about 12.3% of the world's blindness, and 47.8% of Cataract.

Glaucoma's visual mutilation is more extreme in some of the backward countries, with adults more affected than children and women than men [7] 57. Around 60.5 million people around the world will be affected by Open Angle Glaucoma (OAG) and Angle Closure Glaucoma (ACG) by 2010. It is estimated that this number will hit 79.6 million by 2020. OAG is likely to suffer from the mainstream (74%). Two-sided blindness associated with glaucoma is expected to reach 11 million by 2020. Glaucoma is a significant foundation of global vision loss that affects women and Asians unexplainedly [8]. A new methodical analysis of all population-based blindness and vision surveys from around 55 countries in 2002 by the World Health Organization (WHO) was recently completed and expanded to 17 epidemiological Sub-regions of the WHO [7]. The numbers shown in Tables 3.1 and 3.2 are blind and visually impaired [7].

Tables 3.1 and 3.2 show that approximately 39 million people are fully blind and 246 million have limited vision in the world. The overall number of people with visual disability is therefore 285 million globally, including 39 million people who are blind and 246 million with poor vision; 65% of the visually impaired and 80% of all blind people are 50% or older. Table 3.1, with the percentage of global impairment, indicates a different distribution of visually impaired individuals among the six WHO regions with both India and China.

Blindness pervasiveness ranges from 1% in Europe and America to 7.0% in Africa. Of 37 million blind people, 1.4 million are aged 0–14 years, 5.2 million are aged 15–49 years, and 30.3 million are over 50 years, with women more affected than men. The male-to-male blindness ratio ranges from 1.5 to 2.2. The world's chief causes of blindness are cataract, glaucoma, corneal scarring, including trachoma, age-related

Table 3.1 Global estimates by WHO of visual impairments, 2010

WHO region	Total population (millions)	Total population%	Blindness %	Low vision%	Visual impairment%
Africa	804.9	11.9	15	8.3	9.2
America	915.4	13.6	8	9.5	9.3
Eastern Mediterranean	580.2	8.6	12.5	7.6	8.2
Europe	889.2	13.2	7	10.4	9.9
South-East Asia	579.1	8.6	10.1	9.7	9.8
Western Pacific	442.3	6.6	6	5	5.2
India	1181.4	17.5	20.5	22.2	21.9
China	1344.9	20	20.9	27.3	26.5
World	6737.5	100	100	100	100

Table 3.2 Global estimates of visual impairment in people by age, 2010

Age (years)	Population (millions)	Blind (millions)	Low vision (millions)	Visually impaired (millions)
0–14	1848.5	1.42	17.52	18.93
15–49	3548.2	5.78	74.46	80.24
50 and older	1340.8	32.16	154.04	186.20
All ages	6737.5	39.36	246.02	285.38

macular degeneration, and diabetic retinopathy. Figure 3.3 summarizes 2010 global blindness triggers.

Visual deficiency and blindness occurred in six separate WHO regions across three age groups, viz. 0–14 years, 15–49 years and older. Figures 3.3 and 3.4 show the 2010 global vision disability and blindness figures [9]. Internationally, the main causes of visual impairment are uncorrected refractive errors and cataracts of around 43 and 33%. Other factors are glaucoma up to 2%, whereas age-related macular degeneration (AMD), diabetic retinopathy, trachoma and corneal opacities, all around 1%. A large percentage of triggers is undetermined (Fig. 3.4). Blindness is caused by Cataract—51%, Glaucoma—8%, AMD—5%, Childhood Blindness and Corneal Opacity—4%, Uncorrected Refractive Errors and Trachoma—3%, and Diabetic Retinopathy—1%, up to 21% (Fig. 3.5).

A 100% centrally funded scheme was introduced in 1976 (currently 60:40 in all states and 90:10 in the Northeast states), with the goal of reducing blindness to 0.3% by 2020. By 2020, it was established in 1976, the National Program for Management of Blindness and Visual Effects. The fast-packed NPCB survey for prevention of blindness during 2006–07 showed a decline from 1.1% (2001–02) to 1.0% of blindness occurrences (2006–07).

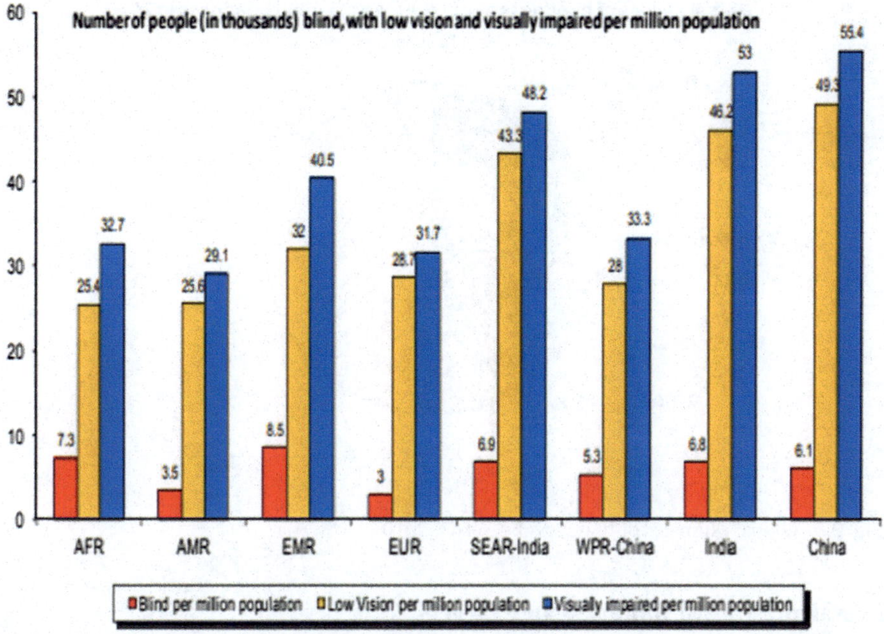

Fig. 3.3 Causes of blindness in millions of people

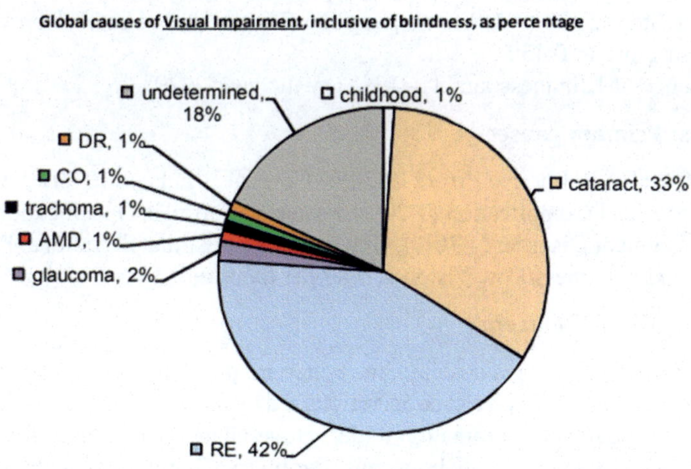

Fig. 3.4 Global estimates of visual impairment by WHO, 2010

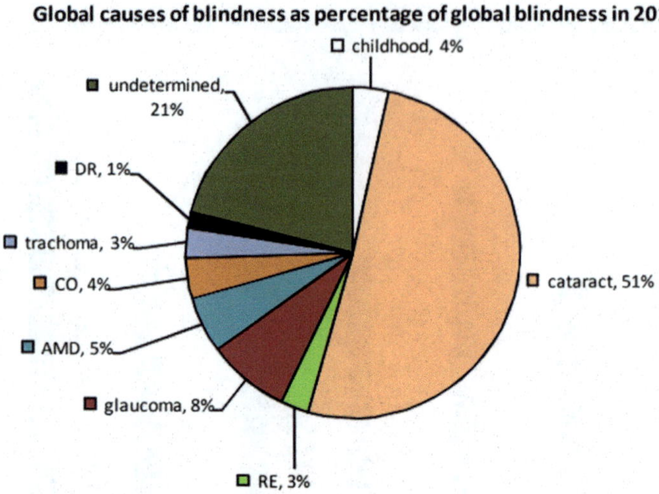

Fig. 3.5 Global estimates of blindness by WHO, 2010

Prevalence rate of Blindness and Targets

Prevalence of Blindness—1.1%. (Survey 2001–02).

Prevalence of Blindness—1.0%. (Survey 2006–07).

Current Survey (2015–18) is in progress. The estimated rate of prevalence of blindness is close to 0.45%.

Prevalence of Blindness target—0.3% (by the year 2020).

Blindness: Primary causes

Cataract (62.6%) Refractive Error (19.70%) Corneal Blindness (0.90%), Glaucoma (5.80%), Surgical Complication (1.20%) Posterior Capsular Opacification (0.90%) Posterior Segment Disorder (4.70%), Others (4.19%) Estimated National Prevalence of Childhood Blindness/Low Vision is 0.80 per thousand.

iv. **Diagnosis of Glaucoma**

The earlier the Glaucoma is detected, the better are the chances to protect the vision from getting damaged which can be accomplished by regular and complete eye examinations. A through eye examination to determine Glaucoma involves five common tests such as Tonometry, Ophthalmoscopy, Perimetry, Gonioscopy, and Pachymetry.

3.4 Ensemble Method for Classification

Combined methods boost the outcomes of classification. It consists of many models. This approach will increase predictive performance compared to single classifiers.

Prediction modeling in a single model, based on a single data sample, can lead to biases, great variability or inaccuracies which affect the reliability of the results. By integrating different models, this constraint effect can be reduced.

Combination approaches are techniques that create and combine many models to obtain better performance. The basic models are described by two or more models combined in assembly processes. Meta-algorithms integrate various machine learning techniques into a predictable model that minimizes variations (bagging), bias (improvement) and predictions (stacking).

Ensemble methods can be divided into two groups:

a. Sequential Ensemble methods:
   The basic learners are generated sequentially in these methods (e.g. AdaBoost)
b. Parallel Ensemble Methods:

The basic learners are generated in parallel in these methods (e.g. Random Forest)

In addition, the base learners must be as exact and diverse as possible in order for the ensemble methods to be more exact than all of its individual participants.

Voting is a group tool used to identify data sets. Voting system employs various prediction approaches:

i. Majority Voting
ii. Average of Probabilities
iii. Maximum Probability
iv. Minimum Probability
v. Product of Probability
vi. Median

3.5 Ensemble FGLAUC-99

The chapter focuses on multidisciplinary research involving both computer science and the medical field, which is diagnostic in nature. It is multidisciplinary, as a branch of medicine it maps those areas of computer science and ophthalmology. It is also diagnostic in conjunction with a case-based approach to achieve simple causal interrelationships by using a profound approach with different classification techniques. Even the sample size is limited, and data collection instruments for ophthalmic testing are required.

Ophthalmic disorders lack fatality, but appear over time to develop, leaving life-long disability-morbidity, which has greater repercussion on patients' daily life [10]. In this respect, because of their genetic predisposition, change in lifestyle, and prompt diagnosis, the Indian population is more vulnerable to these diseases. Due to the chronic and subsequent untreatable illness, it puts a major economic and social burden in the form of working hours and the cost of care on developing economy.

One of the well-known problems of machine learning is that of the balance between prediction capacity and interpretability. The black box models, such as, neural networks show strong prediction ability. They are therefore not completely

Fig. 3.6 Flow of model

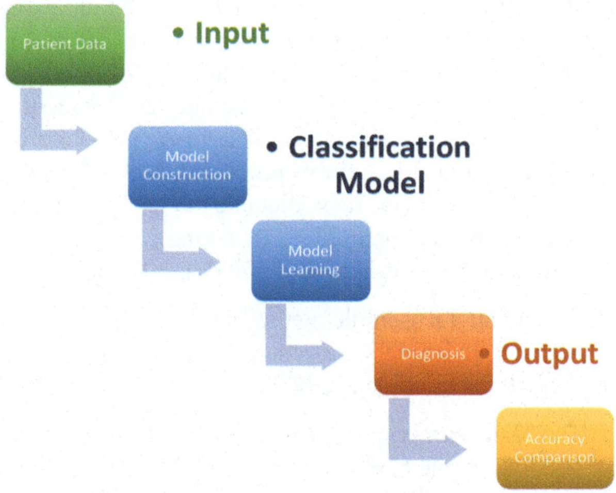

appropriate for medical diagnosis as physicians are interested in understanding both the prediction and why. Decision tree models like C5.0 demonstrate good interpretability and poor predictability. Logistic regression and Naïve Bayes are probabilistic classification algorithms [11].

The proposed ensemble model FGLAUC-99 exploits prediction power of neural network and interpretability power of decision tree based models in order to achieve maximum accuracy of prediction. Here F stands of the first letter in the first author name. GLAUC stands for Glaucoma—an eye disease, the area which the proposed algorithm addresses and 99 shows the accuracy of obtained using the proposed algorithm towards Glaucoma diagnosis.

The sample size is 1082. Out of these 1082 samples 709 were used for training set and 373 were used for testing set (Fig. 3.6).

The Glaucoma dataset used in this research was obtained from practitioner. The data was validated by the practitioner. It was normalized and standardized. The training set was used for construction of a classification model. The model was constructed using tenfold cross validation. The model learning was carried out using training set. It was used for classification of the testing set. The model provided diagnosis of glaucoma disease. The accuracy obtained by the model was compared with other models found in the literature review. The glaucoma dataset had 2 classes, 'Glaucoma' and 'No Glaucoma'. Figure 3.7 shows the configuration of ANN classifier with 2 classes.

The classifier was then updated to include 4 classes, 'Glaucoma', 'No Glaucoma', 'Angle Closure Glaucoma' (Suspect for closed angle glaucoma) and 'Open Angle Glaucoma' (Suspect for open angle glaucoma) to diagnose more specific condition of glaucoma from the glaucoma dataset. Figure 3.8 shows the configuration of ANN classifier with 4 classes.

The classes further specified as 7 classes, 'No Glaucoma', 'Primary Open Angle Glaucoma', 'Primary Normal Tension Glaucoma', 'Primary Ocular Hypertension',

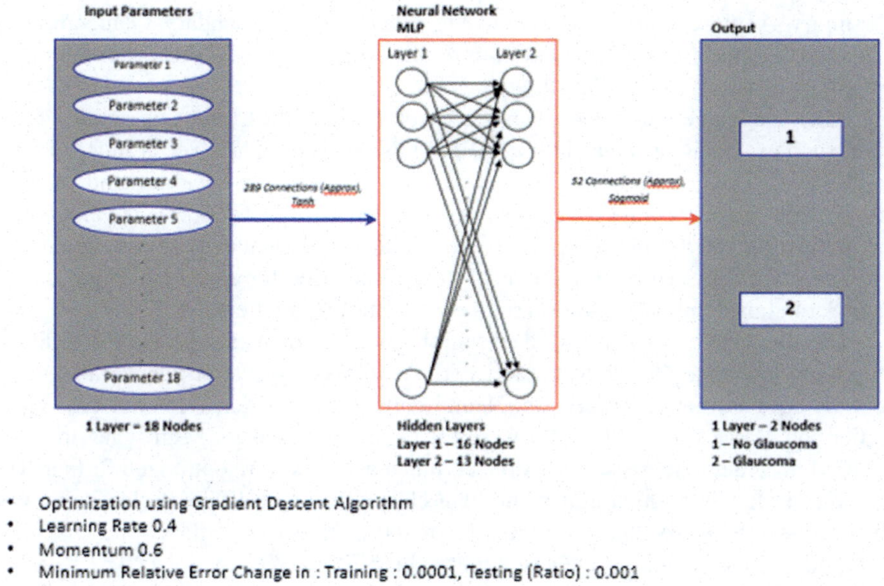

Fig. 3.7 Configuration of ANN for classification with 2 classes

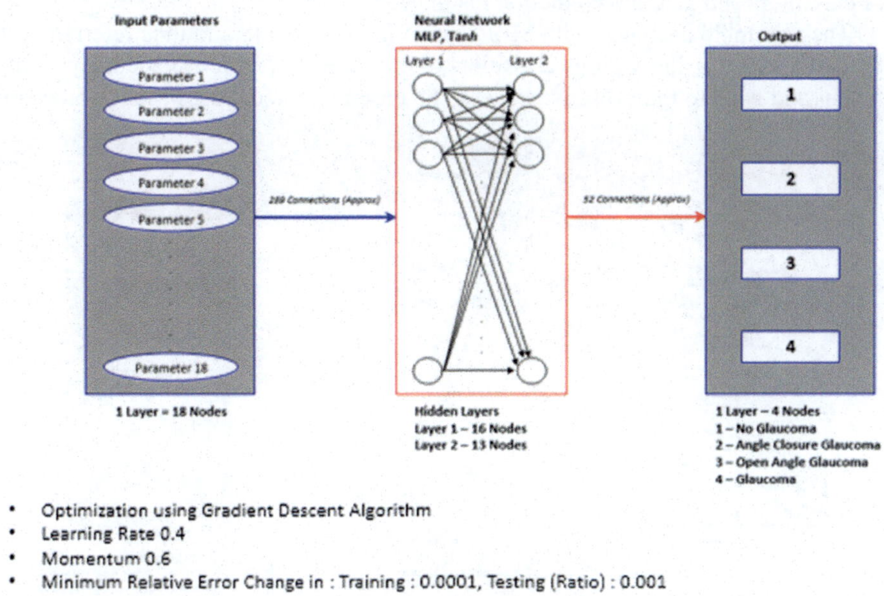

Fig. 3.8 Configuration of ANN for classification with 4 classes

'Primary Angle closure', 'Angle closure Glaucoma', 'Secondary Glaucoma' for detailed diagnosis of various conditions of glaucoma. Figure 3.7 shows configuration of ANN classifier with 7 classes (Fig. 3.9).

Later, the glaucoma dataset was also used for classification using single classifier, such as, Decision Tree (J48 in WEKA), Naïve Bays Classifier, Random Forest and SVM (SMO in WEKA). To improve the classification accuracy and in order to provide more accurate diagnosis, an ensemble FGLAUC-99 was developed for classification and prediction from glaucoma dataset-clinical eye examination data.

There are different combinations of algorithms used for ensemble. Figures 3.10 and 3.11 shows the configuration of these ensemble classifiers.

The diagnostic classifier used ensemble of more than one group for different types of glaucoma. The Multi-classifier (Ensemble) configuration was Parallel to the combined classifier configuration. Neural Network Classifier, Decision Tree Classifier and Naive Bayes Classifier were the categories used for Multi-Classification (MCS). In order to increase precision, certain classifiers were optimized further. The logic of judgment, which makes the final classification of different classifiers, was also optimized to resolve the partiality of the different classifiers. The ensemble method used was Vote, which is a meta classifier in WEKA 3.8.1. The Average of Probabilities was used as an evaluation operator in Vote ensemble classifier. This ensemble is a heterogeneous classifier. It can combine classifiers from different group of classifiers and give the prediction result.

The ensemble designed with MLP, Naïve Bayes and J48 and thus referred to as FGLAUC-99 is using various classifiers to enhance the predictive accuracy of this model and adds to their disadvantages. The ensemble model uses MLP's precision

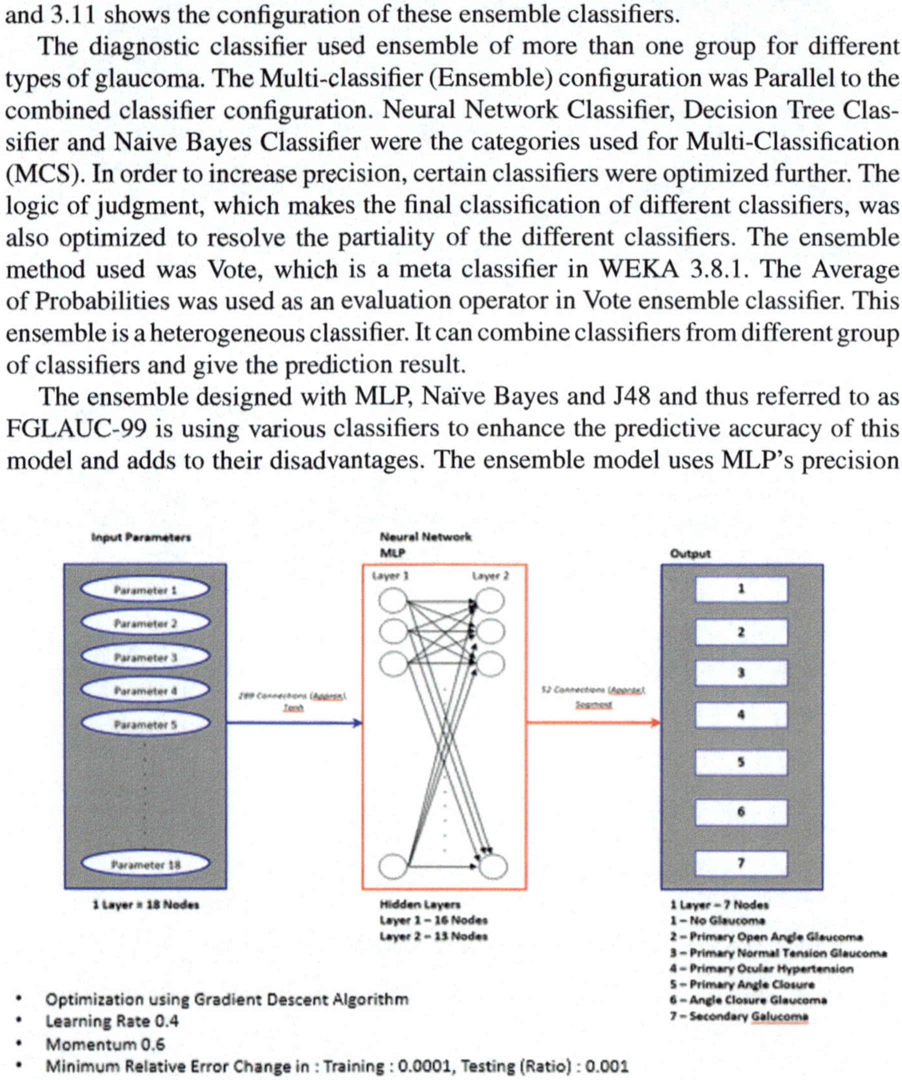

Fig. 3.9 Configuration of ANN for classification with 7 classes

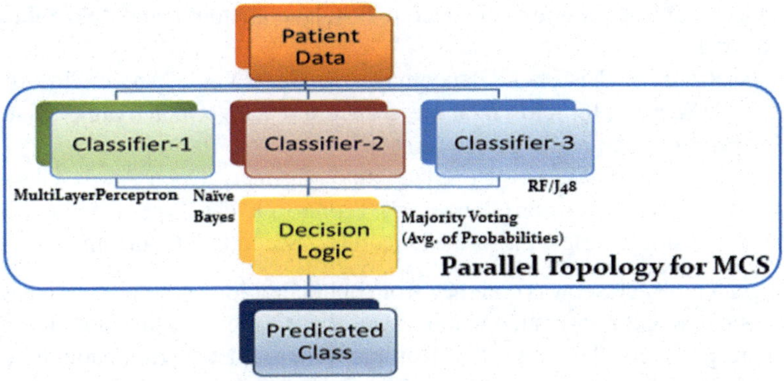

Fig. 3.10 Configuration of initial ensemble classifier

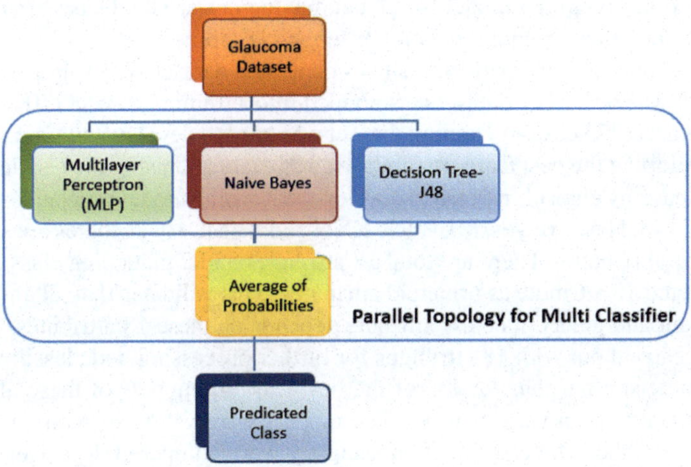

Fig. 3.11 Configuration of ensemble classifier

and overcomes ANN's ineffective interpretability. Decision Trees are responsive to data values, and minor data changes might lead to different decision tree construction. But decision-making trees are useful for analysis. Naïve Bayes works on measuring probabilities, but almost every time attributes are independent, they perform well. In the event of significant data changes the accuracy of Naïve Bayes is not affected.

3.6 Results and Discussions

This research has used three types of Glaucoma datasets and developed the learning models:

1. Dataset with 2 Classes—'Glaucoma' (Glaucomatous) and 'No Glaucoma' (Normal)
2. Dataset with 4 Classes—'Glaucoma', 'No Glaucoma', 'Angle Closure Glaucoma' (Suspect for closed angle glaucoma) and 'Open Angle Glaucoma' (Suspect for open angle glaucoma)
3. Dataset with 7 Classes—'No Glaucoma', 'Primary Open Angle Glaucoma', 'Primary Normal Tension Glaucoma', 'Primary Ocular Hypertension', 'Primary Angle closure', 'Angle closure Glaucoma', 'Secondary Glaucoma'

The glaucoma dataset was obtained from ophthalmology practitioner. It had glaucomatous cases and non-glaucomatous cases that formed our diseased and healthy controls respectively. The important features were extracted from patients' clinical examinations and arranged as a data table. The records with missing values were removed.

The glaucoma dataset with 2 classes, 4 classes and 7 classes were supplied to an ANN classifier to build a model for glaucoma diagnosis. The dataset was also used for classification and prediction using different classifiers.

There were total 1082 patients' data in glaucoma dataset. The split percentage of the dataset was 66%. The dataset was divided into a training dataset (709 cases) and testing dataset (373 cases). Training dataset was used to develop and train the model. After developing the best learning model, it was evaluated using the test dataset. The machine learning algorithms used to develop the learning model for glaucoma prediction were: J48, Naïve Bayes, MLP, RF, KNN and SVM. All attributes are selected as principal components. There are total 18 attributes and 2 glaucoma class attributes. PCA selected 18 attribute as principal attributes. This validates that all attributes are independent and glaucoma class attribute depends on these 18 attributes. Therefore the study carried out with 18 attributes for further processing and classification.

The models, after construction of the best learning models of these algorithms, were evaluated the in various ways. The models were validated with tenfold cross validation method. The classification accuracy were compared. Receiver operating characteristics, (ROC) curves and areas under the curve (AUC) value were also analyzed. All the classification algorithms were implemented in java using WEKA APIs. Figure 3.12 Shows the input screen and diagnosis of glaucoma diagnosis system.

Further to improve the accuracy of glaucoma diagnosis, an ensemble of classifiers FGLAUC-99 was used. The ensemble used here was heterogeneous ensemble of multiple classifiers. The ensemble Vote method uses combination of classifiers, such as, Naïve Bayes, Decision Tree, Artificial Neural Network, Support Vector Machine, K-Nearest neighbor.

a. **Demographic Profile of Patients**

The data gathered in this research was of the patients showing symptoms related to diverse eye ailments. The data of the ophthalmic patients was collected from one of the practitioner's clinic in the city of Vadodara, Gujarat. The probability of

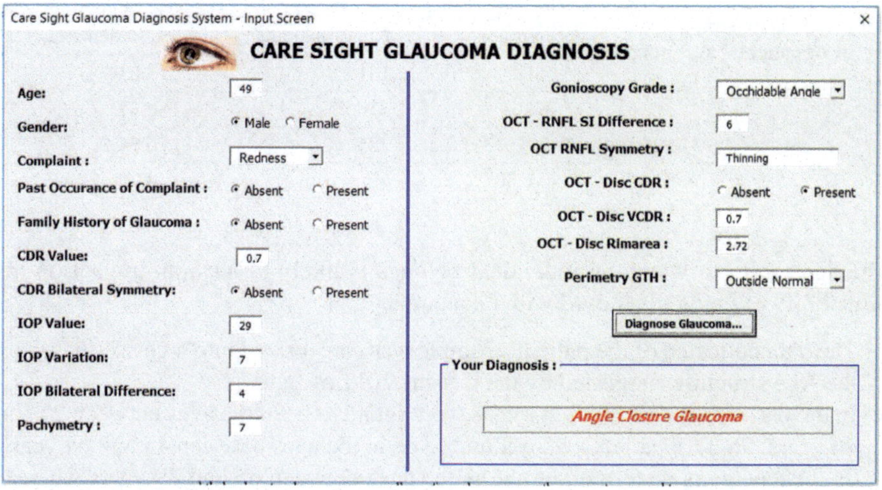

Fig. 3.12 Glaucoma diagnosis system

eye patients likely to suffer from eye disease such as Glaucoma is also influenced by demographic factors such as Age and Gender. In this type of a research work, there is a requisite to underscore the demographic characteristics of the sample. Taking into account this sample as a representative sample, one can surely develop or can extrapolate and judge the demographic characteristics of the patients at large. The analysis of Demographic profiles especially—Age and Gender of patient would definitely reveal relationships among examination data and such demographic factors of the patients.

In this study, the data of total 1082 patients has been collected and used for further analysis. All the required information pertaining to relevant aspects of the ophthalmic disease—Glaucoma under the study was extensively covered. The information collected from the practitioner was about the patients' examination data on different parameters such as, symptoms (complaints), past history of symptoms, family history, posterior segment CDR, CDR asymmetry, IOP, IOP Bilateral Difference, IOP min max difference, pachymetry, goniosopy grades, OCT-RNFL superior inferior value difference, OCT-RNFL Avg, OCT-RNFL symmetry, OCT-disc vertical CDR, OCT-Disc Rimarea and Perimetry Glaucoma Hemifield Test. In addition, the two demographic factors viz. patient's age, patient's gender were also recorded for the purpose of this research, the characteristics of which have been detailed in Table 3.3.

AGE

Since the proliferation of Glaucoma and related eye disorders are more prominently observed in specific age groups such as Adults in the age groups of 40 years and above, the major age groups of population are middle aged to old aged individuals.

Table 3.3 Demographic profile of patients' age groups

Age groups	No. of patients	% Patients
25–44	141	13.03
45–64	610	56.37
65–79	285	26.34
Above 80	46	4.25

The age group to which an individual belongs is likely to have an impact on his probability of being diagnosed with Glaucoma.

- The data collected of the patient's sample was categorized into 4 classes based on the Age structure suggested by the Census 2011 of India.
- In the patient's data 13.03% patients were found to be in the ages between 25 and 44 years. 56.37% patients were found to be in the ages between 45 and 64 years. 26.34% patients were found to be in the ages between 65 and 79 years whereas 4.25% patients were above the age of 80 years.

GENDER

- The data collected of the patient's sample was categorized into 2 classes based on the Age structure suggested by the Census 2011 of India.
- In the patient's data 65.06% patients were found to be Males and 34.93% were found to be Females (Table 3.4).

Table 3.5 shows the distribution of the two genders of patient sample population across four age groups. 9.25% male and 3.78% females were found to be in the ages between 25 and 44 years. Also 37.07% males and 19.31% female patients were found to be in the ages between 45 and 64 years. Further, 17.20% males and 9.14% female patients were found to be in the ages between 65 and 79 years whereas 1.21% male and 3.04% female patients were found to be in the ages above 80 years.

Table 3.4 Demographic profile of patients gender groups

Gender	Nos.	%
Male	704	65.06
Female	378	34.94

Table 3.5 Demographic profile of patients' age group wise gender groups

Age groups	No. of patients	% Patients	Male patients	% Male patients	Female patients	% Female patients
25–44	141	13.03	100	9.25	41	3.78
45–64	610	56.38	401	37.07	209	19.31
65–79	285	26.34	186	17.20	99	9.14
Above 80	46	4.25	13	1.21	33	3.04

b. **Result from Artificial Neural Network Classifier**

An Artificial Neural Network Classifier model is developed to predict the glaucoma diagnosis for new data supplied to the model. The model was implemented in SPSS software. The model had 1 input layer with 18 nodes to input 18 attribute values, 2 hidden layers with 16 and 13 nodes respectively and 1 output layer. The learning rate of the model was set to 0.4 and momentum was set to 0.6. Sigmoid function was used as activation function for this model.

A structure of ANN (Multilayer Perceptron) with 2 classes—'Glaucoma' and 'No Glaucoma' is shown in Fig. 3.13:

The model accuracy was 79%. The Area Under the Curve (AUC) of ROC Curve are as shown in Table 3.6. The AUC were approximately 0.8, which shows that the classifier is excellent classifier.

A structure of ANN (Multilayer Perceptron) with 4 classes—'Glaucoma', 'Angle Closure Glaucoma' (Suspect for closed angle glaucoma), 'Open Angle Glaucoma' (Suspect for open angle glaucoma) and 'No Glaucoma' is shown in Fig. 3.14:

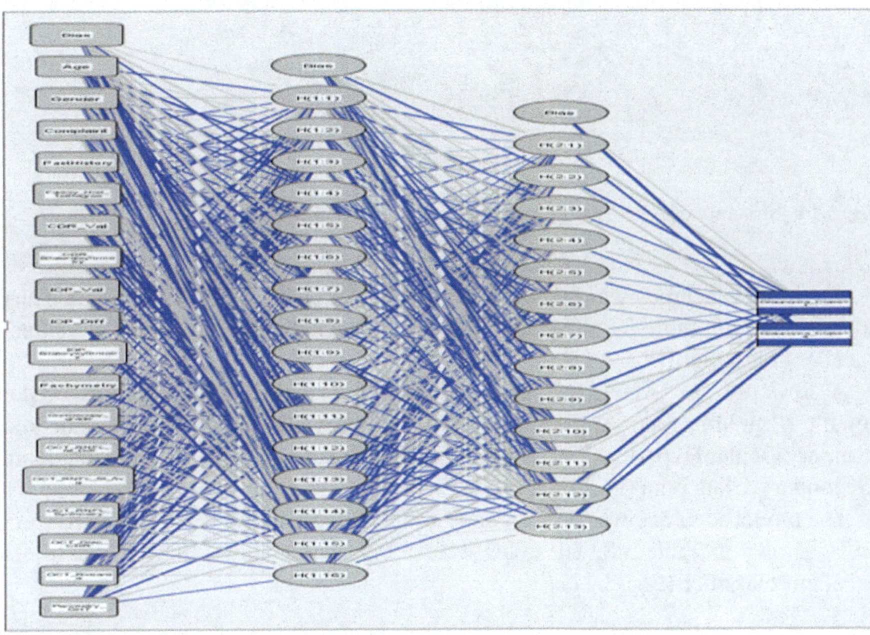

Fig. 3.13 Structure of ANN (multilayer perceptron) with 2 classes

Table 3.6 Area under the curve for 2-classes

Glaucoma class	AUC
1. No glaucoma	0.83
2. Glaucoma	0.83

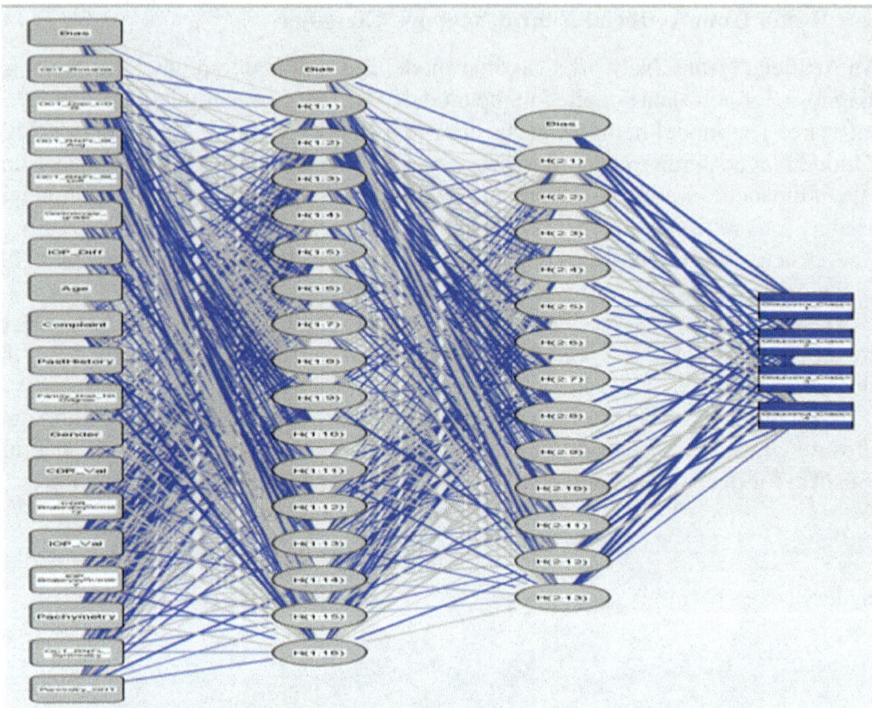

Fig. 3.14 Structure of ANN (multilayer perceptron) with 4 classes

The model accuracy was 89.9%. The Area Under the Curve (AUC) of ROC Curve are as shown in Table 3.7. The AUC were > 0.9, which shows that the classifier is excellent classifier (Fig. 3.15).

A structure of ANN (Multilayer Perceptron) with 7 classes—'No Glaucoma', 'Primary Open Angle Glaucoma', 'Primary Normal Tension Glaucoma', 'Primary Ocular Hypertension', 'Primary Angle closure', 'Angle closure Glaucoma', 'Secondary Glaucoma' is shown in Fig. 3.16.

The model accuracy was 92.6%. The Area Under the Curve (AUC) of ROC Curve are as shown in Table 3.8. The AUC were > 0.9, which shows that the classifier is excellent classifier (Fig. 3.17).

Table 3.7 Area under the curve for 4-classes

Glaucoma class	AUC
1. No glaucoma	1.000
2. Angle closure glaucoma	0.987
3. Open angle glaucoma	0.985
4. Glaucoma	0.934

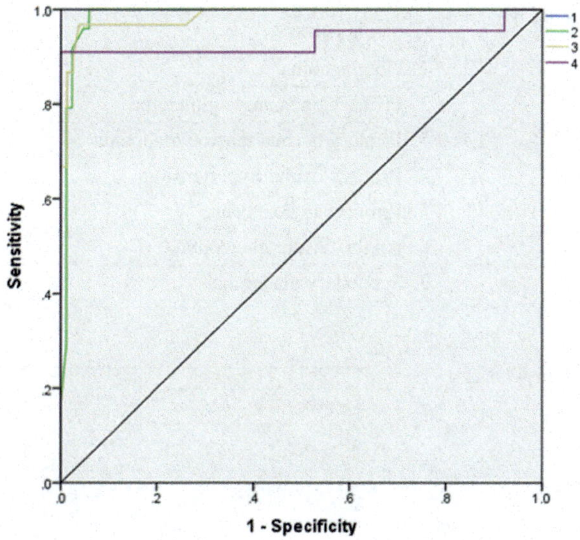

Fig. 3.15 ROC curve for 4-classes

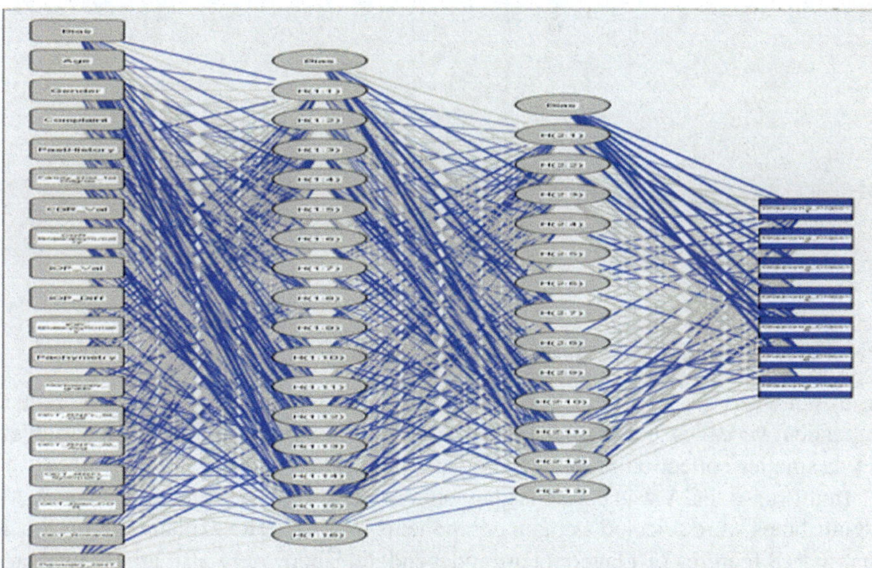

Fig. 3.16 Structure of ANN (multilayer perceptron) with 7 classes

Table 3.8 Area under the curve for 7-classes

Glaucoma class	AUC
1. No glaucoma	1.000
2. Primary open angle glaucoma	0.954
3. Primary normal tension glaucoma	0.970
4. Primary ocular hypertension	0.987
5. Primary angle closure	0.996
6. Angle closure glaucoma	0.968
7. Secondary glaucoma	0.964

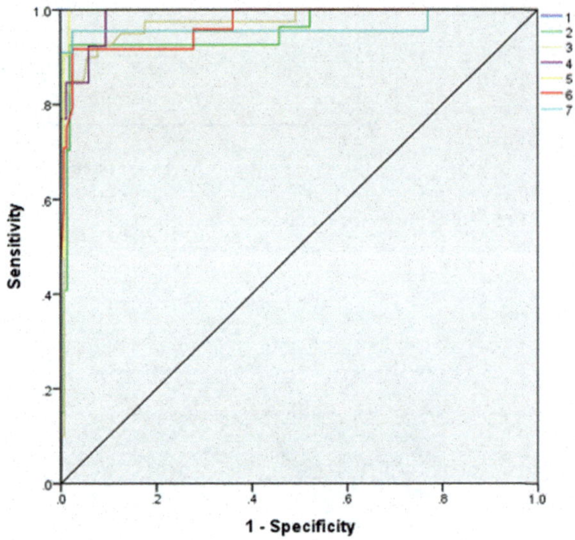

Fig. 3.17 ROC curve for 7-classes

Principal Component Analysis for Feature Selection

In a multi-various statistical testing technique PCA is commonly used to reduce dependent variables, based on the patterns of association among the initial variables, into a smaller collection of underlying variables (called components).

In this case, PCA was used for glaucoma to search for attribute dependence. All 18 attributes were selected as main components by PCA. The explanation for this is that the 18 features of glaucoma are independent. There were also all 18 attributes used for classification and prediction of glaucoma diagnosis for input to various classification algorithms.

Result from Artificial Neural Network Classifier using WEKA

The implementation of ANN using WEKA for 7 classes was done with the help of MLP (Multilayer Perceptron) function available in WEKA. The MLP configuration had learning rate 0.4 and momentum 0.6. The accuracy of MLP in WEKA was 90%.

Result from J48-Decision Tree Classifier(C 4.5) using WEKA

The implementation of J48, which is a decision tree classifier that implements C 4.5 classifier in WEKA was done in WEKA. The J48 configuration had confidence factor 0.25 and minimum number of objects was 2. The accuracy of J48 in WEKA was 86.50%.

Result from Naïve Bayes Classifier using WEKA

Naïve Bayes is a probability based classifier. It takes a strong assumption of attribute independence. The implementation of Naïve Bayes classifier was done in WEKA using configuration with kernel estimator and with supervised discretization. The accuracy of Naïve Bayes with kernel estimator was 94.32%, while with supervised discretization the accuracy improved to 98.58%.

Result from SMO Classifier using WEKA

The implementation of Sequential Minimal Optimization (SMO), which is a Support Vector Machine (SVM) classifier implementation in WEKA was done using poly-kernel by normalizing the data. The accuracy of SMO in WEKA was 92.63%.

Result from IBk-KNN Classifier using WEKA

The implementation of Instance Based Learner (IBk), which is a K-Nearest Neighbor (KNN) classifier implementation in WEKA. The implementation of IBk was done using Euclidian distance to calculate nearest neighbors. The accuracy of IBk in WEKA was 87.73%.

Result from RF Classifier using WEKA

The classifier Random Forest (RF) is a combination of different decision trees used as classifier. The implementation of Random Forest was done using 100 as batch size in single execution slot. The accuracy of Random Forest in WEKA was 95.70%.

Table 3.9 represent the comparison of accuracies from single classifiers from different group of classifiers. It can be seen from the table that single classifiers such as, J48 and IBk has less accuracy compared to probability based Naïve Bayes

Table 3.9 Comparison of different group of classifiers

Classifier group	Classifier	Accuracy (%)
Decision Trees	J48	86.50
	Random Forest	95.70
Bayes	Naïve Bayes	94.32, 98.58
Functions	Multilayer perceptron	90
	SMO	92.63
Lazy	IBk	87.73

classifier and Multilayer Perceptron and SMO function based classifier. Random Forest being a multi classifier works well with good accuracy.

c. **Result from Vote-Ensemble Classifier using WEKA**

An ensemble method technique was used to render correct diagnoses of glaucoma to improve the precision of the classification. The powers of each classifier in this classification category are used to fix the group participants' weaknesses.

In this research the Vote ensemble consisted of 3 classifiers or simple learners. A single ensemble was used in conjunction with J48, MLP and Naïve Bayes. RF, MLP and Naïve Bayes were also used by the other ensemble. This meant combining the MLP with separate Tree Decision Classification and the Naïve Bayes Classification using data set re-sampling.

Tuning Learning Rate and momentum parameters optimizes the MLP. At the same time, the kernel calculator optimizes the Naive Bayes algorithm for numerical estimates. The combiner-vote set Multi-classifier is optimized in order for each classifier to use an average of probabilities. Average probability helps to minimize the transition of the simple learners. The Vote ensemble is used as a heterogeneous whole that allows classifiers from various categories to be combined to produce better results.

Table 3.10 shows current comparison of various assembly classifiers. The MLP, Naïve Bayes and J48 ensemble provides outcomes with maximum precision. This ensemble has been used with glaucoma data from seven different groups in the classification and prediction of various conditions. Of which 1 class is regular control class and 6 other classes are affected by glaucoma. Since glaucoma is an eye disorder, it is important to detect it at an early stage of appearance and not to lose vision. This disease is not reversible, however, if identified in the early stages its development may be avoided. This prevents constant vision loss.

Figure 3.18 shows the result as diagnosis of a condition of glaucoma. Here, the diagnosis of the values provided is 'Angle Closure Glaucoma'.

Table 3.11, present the AUC Curve for ensemble of MLP and RF. The accuracy of this ensemble is 98.77%. The Area Under the Curve shows 1.000, which represents most accurate classifier.

Table 3.12, present the AUC Curve for ensemble FGLAUC-99 of MLP, Naïve Bayes and J48.

Table 3.10 Comparison of different ensembles

Classifier group	Vote ensemble—base classifiers	Accuracy (%)
Decision Trees, Bayes, Probabilistic Functions	RF, Naïve Bayes, SMO	89.57
Bayes, Probabilistic Functions, Decision Trees	Naïve Bayes, MLP, RF	93.86
Probabilistic Functions, Decision Trees	MLP, RF	98.77
Bayes, Probabilistic Functions, Decision Trees,	Naïve Bayes, MLP, J48	99.18

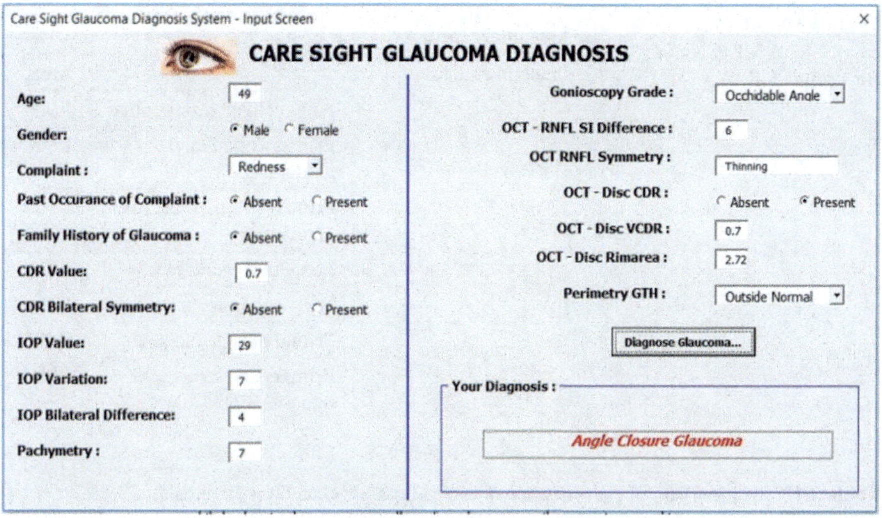

Fig. 3.18 Diagnosis of condition of glaucoma

Table 3.11 AUC for ensemble of MLP and RF

Area under the curve (ROC)		
Glaucoma class		Area
	Angle closure glaucoma	1.000
	Primary open angle galucoma	1.000
	Primary normal tension glaucoma	1.000
	Secondary galucoma	1.000
	No glaucoma	1.000
	Primary angle closure	1.000
	Primary ocular hypertension	1.000

The accuracy of this ensemble is 99.18%. The Area Under the Curve shows 1.000, which represents most accurate classifier.

Table 3.13, present comparison of performance achieved by proposed Vote ensemble classifier that uses Naïve Bayes, MLP and J48 algorithms with the performance of other algorithms or techniques available in literature. It is not possible carry out one to one comparison with a specific classifier, as the data set is customized dataset. The neural network classifier is designed according to the parameters of the customized dataset. Although the table shows accuracy obtained from proposed algorithm with accuracy obtained from various classifiers for glaucoma diagnosis

Table 3.12 AUC for ensemble of MLP, Naïve Bayes and J48

Area under the curve (ROC)		
Glaucoma class		Area
	Angle closure glaucoma	1.000
	Primary open angle galucoma	1.000
	Primary normal tension glaucoma	1.000
	Secondary galucoma	1.000
	No glaucoma	1.000
	Primary angle closure	1.000
	Primary ocular hypertension	1.000

Table 3.13 Comparison of performance of algorithms for glaucoma diagnosis

Sr. No.	Author	Parameters	Classifier	Measuring parameter
1	Nagarajan et al. [12]	Visual disc	ANN	Sensitivity–95% Specificity–94% Accuracy–94%
2	Bizioz et al. [13]	Optic nerve head	ANN	Sensitivity–93% Specificity–94%
3	Nayak et al. [14]	Optic disc, blood vessel	ANN	Sensitivity–100% Specificity–80%
4	Huang et al. [15]	RNFL Thickness	ANN	Area ROC–0.97
5	Chauhan et al. [16]	CDR, Perimetry, OCT	SVM	Sensitivity–84.1% Specificity-96.3% Accuracy–92.6%
6	Eddine Benzebouchi et al. [17]	Fundus image	Convolutional Neural Network	96.9%
7	Kim et al. [18]	RNFL, OCT, Visual Field	Random Forest	98%
8	Fu et al. [19]	CDR, Optic Disc from Fundus Image	Ensemble Network	77%
9	Barella et al. [20]	RNFL, OCT	CTREE	87.77%

available so far. These algorithms provide diagnosis based on some parameters used in proposed classifier, while, few classifier use fundus images for classification.

From Table 3.13, it can be derived that the performance accuracy obtained by the proposed ensemble classifier FGLAUC-99 is better than the classifiers used in literature. The results so far seem promising and shows improvement in accuracy of classification, in turn, improvement in automated diagnosis of various conditions of glaucoma.

The ensemble classifier FGLAUC-99, developed in this research, use machine learning classifiers and intelligently provide accurate diagnosis of various conditions of glaucoma.

3.7 Conclusion

The research encompasses study of different classification algorithm, finding out the most suitable and accurate algorithm by applying these algorithms to the glaucoma dataset, developing an ensemble algorithm to improve the accuracy of classification for glaucoma diagnosis.

The data obtained from practitioner shows that patients with age group 45–64 formed the major part of the data, which was 56.44%. From the total data, 63.05% patients were male, representing the eye aliment. The ensemble classifier FGLAUC-99 shows accuracy 99.18%, which is higher than accuracy of other techniques found in literature review.

Amongst the list of algorithms used for classification, C5.0, RF, SVM, KNN, Naïve Bayes and ANN, probability based algorithm Naïve Bayes and decision tree based algorithm Random Forest gave better accuracy than the other algorithms. The accuracy obtained for J48 decision tree classifier was 86.50%, while neural network based Multilayer Perceptron classifier accuracy was 90%. The accuracy obtained from Support Vector Machine based classifier SMO was 92.63%. Accuracy of IBk-KNN based classifier gave 87.73%. Accuracy for Naïve Bayes algorithm was 94.32% and accuracy for decision tree based Random Forest classifier was 95.70%, which is better than the accuracy of other algorithms considered for the study.

Ensemble of classifiers shown improved accuracy compared to single individual classifiers. Different ensembles of classification algorithm were developed. The ensembles were using Vote method for classification. Ensemble of Random Forest, Naïve Bayes and SMO gave 89.57% accuracy, while ensemble of Random Forest, Naïve Bayes and Multi layer perceptron shown accuracy of 93.86%. Ensemble of decision tree based and neural network based classifiers shown accuracy of 98.77%.

The classifiers were selected from group of classifiers. First classifier was included in ensemble from Probability based classifier such as, Naïve Bayes, As the probabilistic classifiers gives more accuracy. Second classifier was selected from Decision Tree based classifier, such as, Random Forest, J48. Decision Tree based classifier provides good interpretability. Third classifier was selected from neural Network based classifier, such as, MLP. Neural network based classifiers provide better prediction.

Ensemble of RF, Naïve Bayes and SMO shown accuracy of 89.57%. The second ensemble of Naïve Bayes, MLP and RF shown 93.86% accuracy, while ensemble of MLP and RF gave accuracy of 98.77%

The proposed ensemble FGLAUC-99 with J48, Naïve Bayes and MLP classifiers gave the best accuracy of 99.18%.

References

1. Pal, S.K., Dillon, T.S., Yeung, D.S.: Soft Computing in Case Based Reasoning. Springer, U.K (2000)
2. Breiman, L.: Classification and Regression Trees. Wadsworth International group, Belmont (1984)
3. Mangasarian, O.L., Street, W.N., Wolberg, W.H.: Breast cancer diagnosis and prognosis via linear programming. Oper. Res. **43**(4), 570–577
4. Yao, X., Liu, Y.: Neural networks for breast cancer diagnosis. In: Proceedings of the 1999 Congress on Evolutionary Computation, vol. 3, pp. 1760–1767. IEEE Press (1996)
5. Abbass H.A.: An evolutionary artificial neural networks approach for breast cancer diagnosis. Artif. Intell. Med. 25(3):265–281 (2002)
6. Thylefors, B., Negrel, A.D., Pararajasegaram, R., Dadzie, K.Y.: Global data on blindness. Bull World Health Org 73(1):115–121 (1995)
7. Resnikoff, S., Pascolini, D., Etyaale, D., Kocur, I., Pararajasegaram, R., Pokharel, G.P., Mariotti, S.P.: Global data on visual impairment in the year 2002. Bull. World Health Org. **82**(11), 844–851 (2004)
8. Quigley, H.A.: The number of people with glaucoma worldwide in 2010 and 2020. British J. Ophthal 90(3):262–267 (2006)
9. Correspondence to: Silvio P. Mariottio, World Health Organization, 20 Avenue Appia, 1211 Geneva 27, Switzerland. Retrieved from: http://www.who.int/blindness/GLOBALDATAFINALforweb.pdf
10. Ranadive, F., Sharma, P: OpthoABM—an intelligent agent based model for diagnosis of ophthalmic diseases. Int. J. Eng. Comput. Sci. **3**(12), 9667–9670 (2014). ISSN: 2319-7242
11. Caruana, R.: An empirical comparison of supervised learning algorithms. In: Proceedings of the 23rd International Conference on Machine Learning, 25–29 June 2006, pp. 161–168. ACM: Pittsburgh USA (2006)
12. Nagarajan, R., et al.: Neural network model for early detection of glaucoma using multi-focal visual evoked potential (M-Vep). Invest. Ophthalmol. Vis. Sci. **42** (2002)
13. Bizios, D., Heijl, A., Bengtsson, B.: Integration and fusion of standard automated perimetry and optical coherence tomography data for improved automated glaucoma diagnostics. BMC Ophthalmol. **11**(1), 20 (2011)
14. Nayak, J., et al.: Automated diagnosis of glaucoma using digital fundus images. J. Med. Syst. **33**(5), 337–346 (2009)
15. Huang, M.-L., Chen, H.-Y., Huang, J.-J.: Glaucoma detection using adaptive neuro-fuzzy inference system. Exp. Syst. Appl. **32458–468** (2007)
16. Chauhan, K.P., et al.: Data mining techniques for diagnostic support of glaucoma using stratus OCT and perimetric data. Int. J. Comput. Appl. (0975–8887) **151**(8) (2016)
17. Eddine Benzebouchi, N., Azizi, N., Bouziane, S.E.: Glaucoma diagnosis using cooperative convolutional neural networks. Int. J. Adv. Electron. Comput. Sci. (IJAECS) (2018)
18. Kim, S.J., Cho, K.J., Oh, S.: Development of machine learning models for diagnosis of glaucoma. PloS One **12**(5), e0177726 (2017). https://doi.org/10.1371/journal.pone.0177726
19. Fu, H., et al.: Disc-aware ensemble network for glaucoma screening from fundus image. arXiv: 1805.07549v1 [cs.CV]
20. Barella, K.A., et al.: Glaucoma diagnostic accuracy of machine learning classifiers using retinal nerve fiber layer and optic nerve data from SD-OCT. J. Ophthalmol. 2013, Article ID-789129, 7 p. https://doi.org/10.1155/2013/789129

Chapter 4
Diagnosis and Analysis of Tuberculosis Disease Using Simple Neural Network and Deep Learning Approach for Chest X-Ray Images

Ketki C. Pathak, Swathi S. Kundaram, Jignesh N. Sarvaiya, and A. D. Darji

Abstract Artificial Intelligence (AI) based diagnosis of Tuberculosis (TB) disease has experienced large developments with the application of Machine Learning (ML) and Deep Learning (DL) methods to classify the disease as well as to detect it. TB is a contagious infection that usually presents on lungs and is identified through the initial level symptoms by conducting tests using Chest Radiographs (CXRs) and microscopic images. Powerful Diagnosis of Tuberculosis depends on rigorous analysis of radiological patterns realized in CXR. However, due to high number of patients burden and lack of resources in underdeveloped country is high chance of human error in analyzing the CXRs and hence, the diagnosis of TB becomes difficult. Our aim is to develop a computer aided diagnosis (CAD) system for TB disease classification, which can help in early diagnosis of the disease. Nowadays, deep learning based automatic feature extractors are used, when large dataset is concerned for accurate classification instead of using hand crafted features. Our work deals with above mentioned methods to have a justified explanation when working with small- and large-scale data learning problems. This work proposes two approaches for classification of TB disease using X-Ray dataset. In first approach, we have utilized handcrafted features in simple neural network with Support Vector Machine (SVM) to classify the disease as TB images and normal images. We have designed ANN system for 13 input neurons, 10 hidden neurons and 2 output neurons to train features, which are fed as input to SVM for classification purpose. Experiments are conducted on MATLAB 2014b. ANN-SVM based classification gives accuracy of 94.6% when all features are fed to it. This method is thereby awarding increase in efficiency and

K. C. Pathak (✉) · S. S. Kundaram
S.C.E.T., Athwalines, Surat, Gujarat, India
e-mail: ketki.joshi@scet.ac.in

J. N. Sarvaiya · A. D. Darji
S.V.N.I.T., Ichhanath, Surat, Gujarat, India
e-mail: jns@eced.svnit.ac.in

A. D. Darji
e-mail: add@eced.svnit.ac.in

M. Mehta et al. (eds.), *Tracking and Preventing Diseases with Artificial Intelligence*,
Intelligent Systems Reference Library 206,
https://doi.org/10.1007/978-3-030-76732-7_4

reduced diagnosis time. In second, we implemented Deep learning technique, which is capable of training high level features from dataset compared to handcrafted feature method to classify the TB disease. In which we have developed binary classification using Deep Convolutional Neural Network (DCNN). Google collab notebooks are used to model DCNN with GPU based Keras library and Tensor flow as back end. Experiments are conducted on Tuberculosis Chest X-ray dataset obtained from Kaggle community and showed output classification accuracy of 99.24%.

4.1 Introduction

Tuberculosis (TB) is one of the major infections for people suffering from Human Immuno-deficiency Virus (HIV) worldwide. From about 9 million incidents identified in the medical report of 2016 [31], approximately 1.8 million people deaths were caused by the disease worldwide respectively. According to the World Health Organization, a report released showed that around ten million people are suffering from tuberculosis globally, while only India reports around 27% of these cases. TB is curable and avoidable but due lack of resources in poor and marginalized communities having weak hospital infrastructure as well as unspecialized equipment, it becomes difficult to provide better diagnosis and conduct follow up treatments. One of the major causes of TB disease is smoking, which is increasing throughout the world, for example; Individuals living together in the same household with smoking parents are naturally at higher risk of catching disease and Coughing in heavy smokers can lead to the infectious stage and longer exposure can severely promote bacillary transmission [6].

Usually TB is identified through the initial level symptoms and the tests are being done for accurate results, the testes are conducted using Chest Radiographs (CXRs) and microscopic test. But the accuracy of microscopic test is nearly half than the CXRs. So, we need to relay on CXRs for accurate identification of the disease. Once the disease is recognized, the vaccine for such becomes easy to develop. With the early symptoms of the disease and analysis through the radiological technology the therapy for tuberculosis can be developed. However, due to high number of patients burden and lack of resources in underdeveloped countries, there is high chance of human error in analyzing the CXRs and hence, the diagnosis of TB becomes difficult. As the solution, we require a strong computer-aided diagnosis (CAD) system that gives significant result with reduced error during the testing and efficiency could be increased even in the underdeveloped region where they cannot afford high budget radiological observation tools. Figure 4.1 shows the World Health Organization (WHO) report on number of tuberculosis patients worldwide.

The CAD is a step in the detection of the disease through computer examination and obtained results were used for efficient detection of prostate cancer, breast cancer, lungs cancer, colon cancer, congenital heart defects, coronary artery disease, Tuberculosis etc., [7, 8, 13, 22]. The CAD method is also helpful with the medical representatives to examine and make their final decision as the accuracy is increased

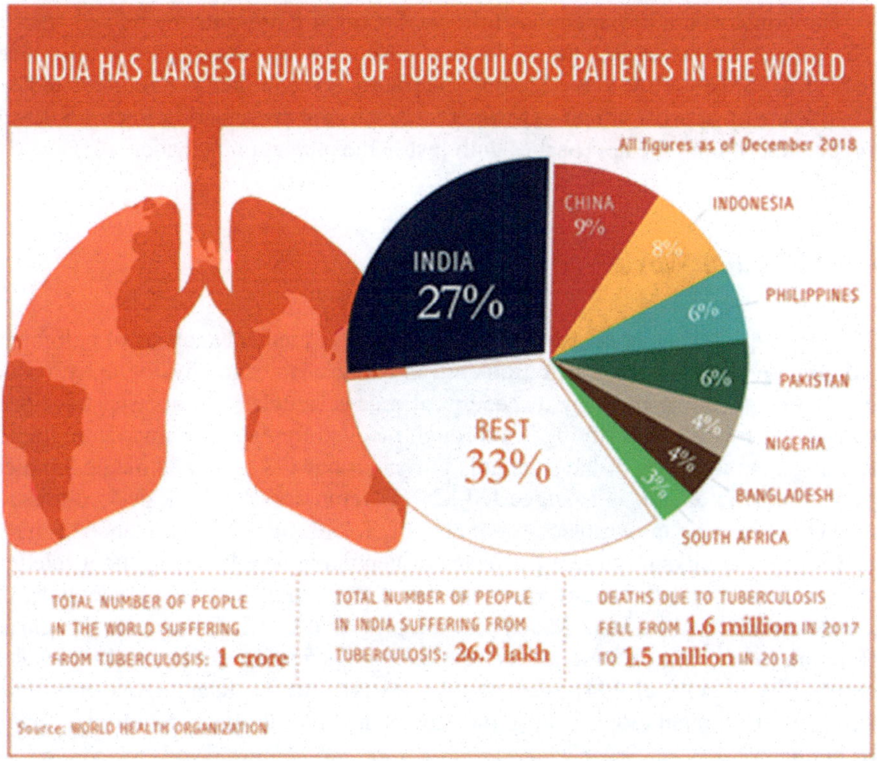

Fig. 4.1 WHO report 2018 on tuberculosis patients worldwide [3]

and it becomes more easy to analyze and interpret. The diagnosis is completely dependent on the radiologist ability; this is just the tool that helps the radiologist with significant results and easy way to access the data and to inspect them throughout. Commonly, there are four stages in CAD systems which are pre-processing, segmentation, feature extraction and classification. In these systems, first step is to segment the region of interest (ROI) image and a feature vector is formed by manually extracting features. With the introduction of deep learning in 2012, researchers shifted away from the above strategy, as its end-to-end architecture gives out best features naturally therefore the requirement of segmentation part is eliminated in these methods, which is an advantage of deep learning methods.

In this paper, we have proposed two approaches for classification of TB disease as normal image or TB image. Our first approach is purely based on simple neural network (feature training) cum support vector machine (SVM) for classification with all pre-processing task mentioned in CAD system. Other approach consists of deep learning-based method, which uses deep convolutional neural network (DCNN) which is one of the most widely used deep learning technique for image classification.

The organization of paper is as follows. Section 4.2 presents the related work of TB disease. Pre-processing task of Proposed Neural network approach is introduced in Sect. 4.3. ANN-SVM classification description is mentioned in Sects. 4.4 and 4.5. Deep learning approach with experimental results are presented in Sect. 4.6. Lastly conclusion of both the approaches with justified explanation is made.

4.2 Related Work

The techniques that can be used for the non-invasive medical analysis of pulmonary TB are conducted over medical traits or indications, rib cage physiography, tissue-based airing, culture and tiny inspection of mucus sample. Various tries have been carried out towards identifying TB through medical implications and demographic trainings. On the basis of the standards of numerous vital traits and medical indications, probable diagnosis is suggested [25]. Elveren and Yumusak desired towards existing a learning taking place for analyzing T.B. through the assistance of multi-layer neuronal links proficient by means of genomic procedure on patient role epicrisis information [9]. However around stay many symptom-based credentials, the inspection directed by Ruiz-Manzano et al. proposed that TB cannot be recognized from the-se oncological units on the foundation of scientific ciphers and indications alone [25]. Patil et. al. [24] estimated the texture features using active form classical based segmentation on lung pictures using gray level co-occurrence matrix technique.

Osman et al. proposed the geometric topographies and final minute invariants which was extracted from the tissue pictures. These topographies are supplied to fusion multi-layered perceptron system to distinguish between non-TB bacilli and TB bacilli [23]. Labor-intensive and time consuming are the glitches testified through labor-intensive airing procedure specifically intended for broadcast of the negative photographs. Thus, this leads for a mechanized TB judgment where large number of cases can be handled quickly with the same accuracy. At present, computer aided diagnosis in microscopic images play a vital role as they able to provide diagnostic information close to 100% [18] yet there involves uncertainty and impreciseness while making decision on disease.

To see the trials tangled in recognizing the TB objects (bacilli and outliers) from the mucus slur pictures, many computerized mechanized techniques are developed. In order to classify normal X ray image and TB kind of images, numerous image processing and design recognition algorithms have also been established. To find bacilli in images of auramine-stained saliva, Santiago-Mozos et al. utilized pixel classification and represented each pixel by a square patch of neighboring pixels. To lessen the number of pixels given as input to a support vector machine classifier, they used prime component analysis [26]. The literature also states that in order to classify TB objects into bacilli and outliers, other additional morphological operations are performed by employing edited bacillus data set. Active contour method has been implemented for image segmentation in this work.

Sub-pixel correctness of the article borders can be achieved by active contour models and preceding information approximately form and for robust image segmentation, intensity distribution can be incorporated. The subsequent object outlines are moderately consistent which are suitable for additional requests, such as shape investigation, detection, and cataloging [25]. Shape descriptors in ascending order signify individually region or edge in the segment-ed picture which are a set of parameters. These signifiers can be sorted into two classes, region and contour based [2]. For the classification feature selection plays a vital role by removing insignificant features from the data set to provide better diagnosis which is considered as an important requirement in medical applications. To select the most prominent features, a feature selection technique based on fuzzy entropy measure is used. The fuzziness of a fuzzy set is observed in this technique which is a measure of the information contained in it [21]. To differentiate bacilli from outliers, significant FDs are used as input to the classifier. Techniques based on machine learning provide support for the decision-making process in many areas of health care including prognosis, diagnosis and screening [21]. In TB detection, the use of image processing and artificial neural networks has received considerable attention over the past several years and still is an active research area.

To categorize amongst bacilli and non-bacilli, quite a few classifiers ought to be situated such as k-nearest neighbor, circular basis function grid, Multi-Layer Perceptron (MLP) network and kernel regression [5]. To improve its generalization capability, a Support Vector Machine(SVM) based training method for MLP classifier is working in this work. The unique pipe-line effort flow concerning the identification and cataloging at both object and image level of the TB kind of X-Ray images is considered as the novelty of this work.

Dissection and withdrawal of Grey-Level Co-occurrence Mmatrix (GLCM) textural features were the basic methods that were included during earlier studies for CAD [5]. To differentiate CXRs as TB or non-TB, Ginnekenet al. used the GLCM textural features on two datasets, but the performance of both varied significantly [10]. Where, respectively 90% and 50% accuracy were obtained. To develop a TB classification model, recognition of clavicle and abnormalities in texture and shape in the CXRs were combined by Hogeweget et.al. [15]. For the method, 0.86 was the area under the receiver Automatic operating Characteristic Curve (AUC) [20]. A prediction accuracy of 92.9% with sensitivity of 0.91 was provided by Semi-automated segmentation-based classification model using textural features [5].

Now-a-days, the CAD methods are used usually detect the region or the object [5]. Segmentation and proficiency against noise are the major requirements for such method [5]. So there seems to have less possible errors in the outcome of the image most importantly while working with medical images. Due to complexity of medical images, most of the time even most sufficient systems shows some errors [5]. Normally, segmentation is done manually or semi-automatically which still shows the dependency on human efforts. Fully automated segmentation methods can be developed for further improvement. But these methods fail to show accuracy, according

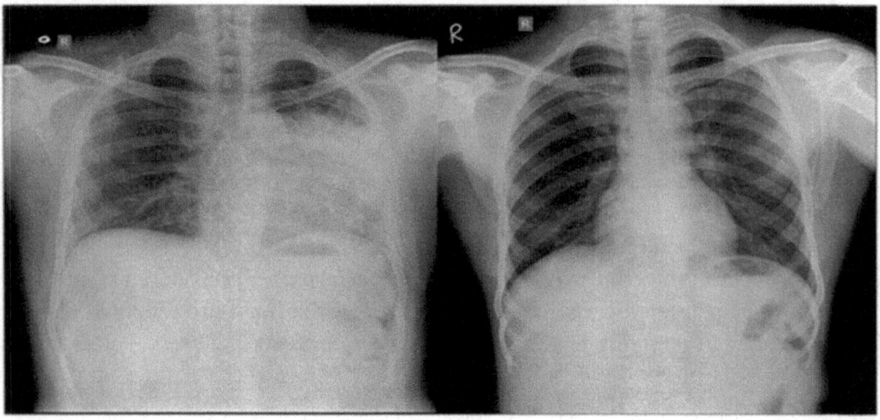

Fig. 4.2 TB and non-TB CXR image [22]

to some studies and sometimes have different outcomes. To identify the difference between the TB and non-TB CXRs we require having some texture features and these are some specific features that shows the difference between the two indicated in Fig. 4.2.

We need to check the ability of them correlating with the disease, to make the effective use of the textual features. Since these features are with accordance to the whole image as it is expected to have more information than the generally used images. Therefore, for the TB CXRs study the contextual based Gist features are used. Geologically reasonable topographies keen on a guarantor low-dimensional trajectory are captured in this method [5]. We are able to identify the pixels contain different values than the neighborhood pixels with these features and also the difference of the other pixel values in the entire image. Gabor, histogram of oriented gradients (HOG) [5], and pyramid histogram of oriented gradients (PHOG) [5] structures were also used in this study. For X-ray apparatuses for fully automated detection of TB this method can be used. The parameters defined are correlation, entropy and standard deviation. Many attempts were done to capture image and study it by microscope as human eye is sensitive to low frequencies and are unable to observe with naked eye [27]. Such techniques are being developed to enhance the observation and detection of such small things to have automatic detection of tuberculosis (TB) [27]. Following paragraph describes the difference between two approaches with advantages and limitations: Deep learning-based approach has certain advantageous features like:

1. It provides uniform feature extraction for classification purpose instead of complex and time-consuming handcrafted features.
2. Deep learning-based approach is appropriate for large scale database and complex learning algorithms

There are certain limitations of Deep learning approach compared to convention NN based approach which are as follows:

1. Automatic extracted features are sometime failing to specific task of classification or detection, they can be able to scarcely usage heuristics to guide attribute removal owing to the computerized feature learning.
2. For small dataset, this approach suffers from over fitting problem.

Medicinal image cataloging glitches habitually have a comparatively slight training dataset due to the laborious type of medicinal image annotation. When deep models are applied to small-sample learning problems, they ought to be standardized by means of the pictorial signifiers haul out by means of the supervision of heuristics. Several researchers have analyzed the neuronal network-based image depiction and its linking to handcrafted features. Most recently, a deep learning technique employing CNN with an artificial bee colony algorithm is successfully proposed for TB disease classification [31]. In [32], author has shown the inherent advantage of ensemble learning by constructing non-linear decision-making functions to gain visual recognition of TB using CNN model. Lately, with the rising popularity of deep-learning models coupled with the superb CNN performance, pre-trained networks such as GoogLeNet and AlexNet are used to classify the images as having manifestations of pulmonary TB or as healthy [19].

4.3 Proposed Methodology of Neural Network (NN)-Based Approach of TB Disease Classification

In this approach, we developed ANN based SVM classification method on X-ray images to detect the tuberculosis disease. To reduce the speckle noise on these images, pre-processing task is done so that visual quality can be enhanced for correct diagnosis. Filtering is done for accurate feature extraction to avoid mis leaded verdict towards remedy of the ailments and it is one of the imperative pre-processing steps. K-means gathering method is applied to have proper segmentation, thereby can obtain region of interest (ROI) image. The gathering practices epitomize non-managed form sorting into sets otherwise modules, which is the body of a set of patterns (trajectory of dimensions or a point in a d-dimensional area) into bunches based on similarity [12]. In the framework of image subdivision, the set of outlines can be embodied by an image in a d-dimensional space be subject to on the numeral of topographies used to signify the picture element, where to each point in this d-dimensional space will be so-called a picture element design. In the identical framework, the bunches correspond to some semantic sense in the image, which is devoted to as an object. Therefore, the chief goal line of the gathering procedure is to attain assemblies or classes from an unlabeled data set based on their resemblances to ease additional data abstraction. The comparison is assessed bestowing to a distance measure amongst the patterns and the architecture types or middles of the assemblies, and individually pattern is allocated to the adjacent or maximum alike architecture type. Grim level conjunction matrices (GLCM) are used to extracted texture features and these features are applied to train ANN model. Lastly, SVM classifier is used to have a decision of normal image or TB image. To have higher performance in terms of accuracy, most of the studies showed that SVMs are more capable compared to other classification algorithms such as Naïve Bayes classifier, Nearest Neighborhood classifier, Decision Tree based classifiers.

4.3.1 Image Preprocessing

Images usually contain noise and which results in the degradation of the image quality. For which certain pre-processing steps are done to remove the noise so that the constructed the image will be very similar to original image without having any change in its information and this will be easy to analyze for further processing. Figure 4.3 shows the overall flow of NN based algorithm and in the following sub section each block is explained thoroughly. Pre-processing task normally done using Gamma Adjustment or Image Enhancement or by Gray-scale Image. In our work, we depicted the gamma adjusted for pre-processing to have brightness and contrast adjustment in the medical image. Medicinal imageries are developed via many sensory systems alike CT scan, X-Ray, ultra-sound imagery etc. Several demonstration strategies/ screens can show the imageries with diverse strength as per their peculiar gamma modified standards. This gamma alteration is used to illustrate dissimilarity in concentration values of imageries. Distinction of γ disturbing pictorial enrichment of the images. Visual enhancement can be done with the used of Gamma transformation on images which is stated in subsequent Eq. (4.1)

$$S = cr^{\gamma} \tag{4.1}$$

where s represents output luminance value, c represents input luminance and r stands for gamma. Gamma standards larger than 14 effect to data cost in a few portions, the things (stones) incline to vanish later gamma modification, while values fewer than 14 is unsuccessful to lessen undesirable things even later gamma adjustment.

4.3.2 Image Segmentation

Segmentation as its name suggests, divides the image into the segments or region. It is the procedure to assign specific labels to the image in terms of identifying the objects (by detecting the edge, curves or lines).

4.3.2.1 Otsu Method

To differentiate the background and foreground of the image based on pixel values, Otsu method is one of the promising segmentation methods [24, 25] by calculating

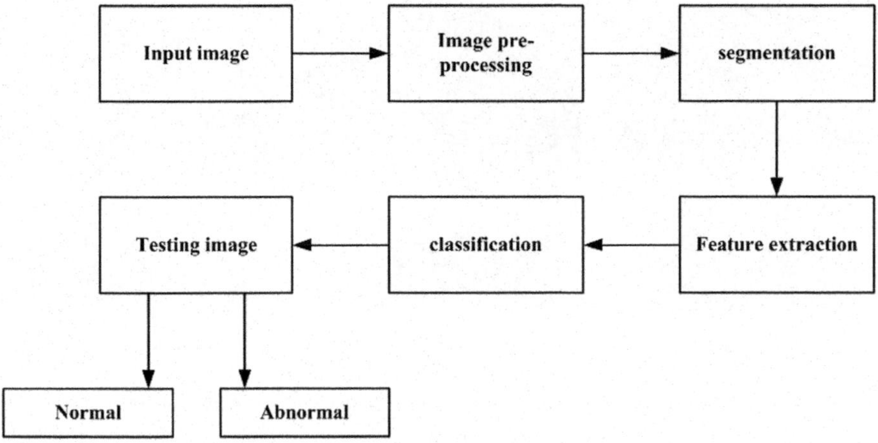

Fig. 4.3 Overall flow of processing the algorithms

the threshold values automatically based on the input image [27]. Otsu model is one of the statistical models which performs its function based on probability of the image pixel. The technique described in Otsu's method is to determine a variable which can segregate two or more regions which arise in the image.

4.3.2.2 K-Means Clustering

K-mean clustering is the segmentation method. Generally, this method is used for kidney stone detection. K-mean clustering arranges the large data into small clusters with same or nearly same values. In this method a single data could belong to one or more clusters [4]. The output image will be obtained in equal values of gray level. It works with large data that can be arranged in the small clusters with same values and that makes it more compatible with the application of detection of kidney stone.

4.3.2.3 Hough Transform

Any image contains texture, edge and smooth regions. Medical images are having abundant curvature and edge regions. Curve detection can be done very precisely by Hough Transform (HT). HT is one of the global features which is used to differentiate the pixel belonging to region of interest of background. It is also very helpful to identify the feature of region of interest or object which is to be detected. This is parametric approach in which it is making checking point on pixel whether that pixels set available on curve of region of interest or not [30]. Preselected threshold value is required for proper detection of object using HT method. HT is very efficient to detect curves as well as lines in presence of noise also.

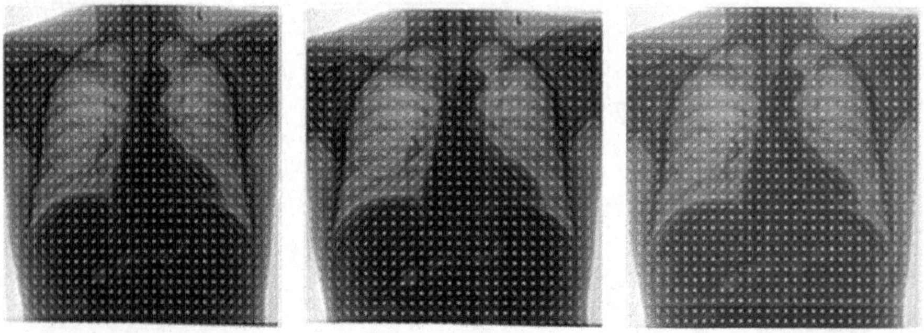

(a) Hough Featured (b) 4 x 4 Cell size (c)16 x 16 cell size

Fig. 4.4 Different operation performed on CXR lung image using Hough transform [30]

In this transform image space (p, q) is transformed into a (ρ, θ) parameter space. While applying HT on any image, the point (p, q) can be represented in polar coordinates as (r, α). That: $p = r\cos\alpha$ and $q = r\sin\alpha$, which can be represented by the following Equation (4.2).

$$\rho = r\cos\alpha = p\cos\theta + q\sin\theta \tag{4.2}$$

With the use of such parametric approach, the problem of finding lines within the image space to have the peak information can be solved. For which, initially HT requires binary images to produce best image enhancement. Performance if HT depends on the preprocessed input image. Figure 4.4 shows the different operation performed using Hough transform on CXR image.

4.4 Feature Extraction

The images used for medical analysis are considered unique as the images contain with more edge and texture information compared to smooth region. It is difficult to analyze texture represented features as they are in more number which is a tedious process. The objective of feature extraction is to differentiate one input arrangement from another by evaluating particular properties or features on original image dataset to have a reduced image for further processing steps. These are the features that illustrate the significant properties of the image, which are used to indicate and segregate input sample ones from large dataset [3]. Gray Level Co-occurrence Matrix (GLCM) is one of the texture features which are based on statistical distribution of pixel intensities whose positions are relative to each other. GLCM is second order statistical texture feature which is used to differentiate pixel intensity as per its specific location [5].

Table 4.1 Feature names and their equations

Feature names	Equations
Entropy	$\sum_{x=0}^{N-1}\sum_{y=0}^{N-1} -S(x,y)\log(S(x,y))$
Contrast	$\sum_{x=0}^{N}\sum_{y=0}^{N} \lvert x-y\rvert^2 S(x,y)$
Correlation	$\frac{\sum_{x,y=0}^{N-1}(x*y)s(x,y)-\{\mu_x*\mu_y\}}{\sigma_x*\sigma_y}$
Energy	$\sum_{x=0}^{N}\sum_{y=0}^{N} S(x,y)$
Homegeneity	$\frac{\sum_{x,y=1}^{N} S(x,y)}{1-(x-y)^2}$
Kurtosis	$\frac{\sum_{i=1}^{N}\frac{(X_i-\bar{X})}{N}}{s^4}$
GLCM mean	$\mu_x = \sum_{x,y=0}^{N-1} x(S_{x,y})$
Standard deviation	$S_D = \sigma_b = [\sum_{b=0}^{L-1}(b-\bar{b})^2 p(b)]^{1/2}$
Variance	$\sum_{x=0}^{N-1}\sum_{y=0}^{N-1}(x-\mu)^2 S(x,y)$
Inverse different moment	$\sum_{x,y=0}^{N-1} S(x,y)\frac{1}{1+(x-y)^2}$
Skewness	$\frac{n}{(n-1)(n-2)}\sum\left(\frac{x_j-\bar{x}}{s}\right)^4$
Smoothness	$R = 1 - \frac{1}{(1+\sigma^2)}$

Statistical texture feature can be extracted with the help of GLCM method [4]. Most of the texture features can be obtained with GLCM method which shows intensity values on the pixel pairs in spatial form, which extracts features such as standard deviation, mean, correlation, variance, energy, homogeneity, contrast, smoothness, kurtosis, entropy, inverse differentiate method (IDM) and skewness. All these features are extracted in our NN based approach on TB CXR images using the above described method. The information of these features in form of equations are shown in Table 4.1. Where x and y represents the particular row and column in the image and S is the pixel of the image.

4.4.1 Classification

Artificial Neural Network (ANN) is one part of Artificial Intelligence (AI). ANN is well-defined as a statistics processing scheme that has features similar to human neuronic networks and represents the human brain which always tries to simulate the learning process in the human brain. It is the most flexible system which does not have any direct relationship between internal and external parameters [11, 14]. There are three layers in ANN namely input, hidden and output layers. Input layer yields input information and maintain interface with external root; hidden layer does complex computation to build a fixed link to have desired output and output layer provides final decision on fed data. Many neural network models with definite algorithm and characteristics can be used for TB disease diagnosis [25].

In our approach, we have used Feed forward neural network and back propagation as ANN model. extracted features are fed for building a trained model. Back propagation is applied to train the ANN model. And overall performance is co-related with accuracy considering size of training data and time consumed to build model. Following points describes the functionality of feed forward and back propagation algorithm:

1. The solo deposit feed frontward grid comprises of a solitary layer of hefts, where the inputs are right associated to the outputs. The synaptic links booming weights fix every input to every output, but no other way. This way it is considered a network of feed forward type.
2. The use of back-propagation is to minimize the errors at the output caused in the neural network. The network is trained by back propagation to ensure balance of ability between recognizing the patterns used during training and respond to the same correctly to the inputted patterns which are similar to the patterns used during the process, where v_{ji} is the line weight of the input unit x_i to the hidden unit layer z_j (v_{j0} is the line weight that connects the bias in the input unit to the hidden unit layer z_j). w_{kj} is the weight of the hidden layer z_j to the output unit y_k (w_{k0} is the weight of the bias in the hidden layer to the output unit z_k). The backpropagation algorithm uses an output error to change the value of its weights in the backward direction. To get this error the forward propagation stage must be done.
3. SVM uses optimization procedures on the way to trace the optimum borders amongst classes. The finest borders ought to be comprehensive to unobserved tasters with tiniest faults midst all probable restrictions unraveling the classes, therefore minimizing the mis perception between classes [29]. The applicability of SVM for image classification is explored in this study. This concept is shown in Fig. 4.5.

Development of SVM: The support vector machine (SVM) is a mechanism studying algorithm built on arithmetical knowledge scheme. There are a number of journals specifying the accurate construction and algorithm development of the SVM [1, 32]. The inductive belief late SVM is organizational risk minimization (SRM). The risk of a learning machine (R) is constrained by the summation of the experiential risk estimated from training samples (Remp) and a confidence interval (Ψ) : $R \leq Remp + \Psi$ [32]. The approach of SRM is to keep the experiential risk (Remp) fixed and to minimize the confidence interval (Ψ), or to maximize the margin between a separating hyper plane and closest data points (Fig. 4.8). A separating hyperplane refers to a plane in a multi-dimensional space that separates the data samples of two classes. The optimal separating hyperplane is the separating hyperplane that maximizes the margin from closest data points to the plane. Currently, one SVM classifier is able to separate only two classes. Integration strategies are needed to extend this method to classifying multiple classes.

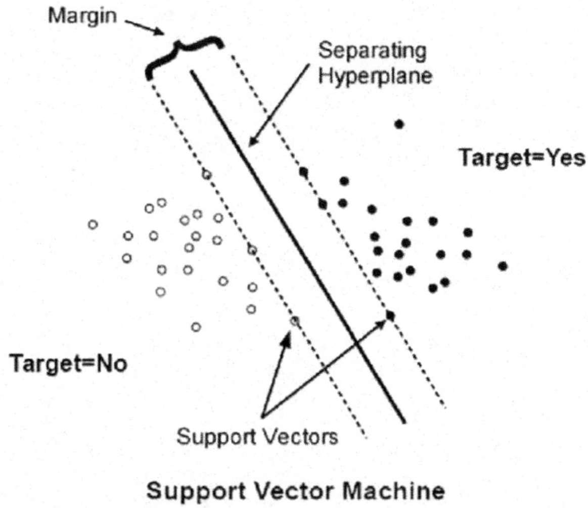

Fig. 4.5 Optimal hyper plane using the SVM algorithm

4.5 Result Analysis of Proposed NN Based TB Disease Classification

Analysis is performed on MATLAB R2014a with image processing platform for our proposed work. The performance is evaluated for 25 patient images. To have precise segmentation and evident analysis on image, pre-processing work is done at first. Following paragraph gives the detail description of result analysis.

Dataset: The dataset used for this work is taken from the National Institutes of Health (NIH)[24] .there are two different datasets available the first one is Shenzhen Hospital Data, in that total of 662 x-ray images available in different sizes and formats Among the 662 images, 336 images are that of patients suffering from tuberculosis and the other 326 are that of normal patients. The other one is Montgomery dataset, in that there are 138 x-rays present in different sizes, among that 80 images are of normal patients and the other 58 are of patients suffering from tuberculosis. both the datasets are openly available on the web. All the collected data was pre-processed and segmented using the above-mentioned techniques. 70% of the collected data was used for training and other 30% was used for testing We have implemented our proposed neural network architecture in MATLAB 2014. As shown in above Figure 4.6, the proposed neural network consist of 13 input parameter as hand crafted features, total 10 hidden layers and 2 output nodes as binary classifiers (Feed Forward Neural Network with 13 input node, 10 hidden nodes and 2 output nodes). Figure 4.7 shows best validation performance graph for proposed neural network based TB classification. It indicates error on y axis and number of epochs on x axis. Figure 4.8 shows training state graph. This graph demonstrates the variation in the gradient point corresponding to the number of epoch consumed by neural network based proposed model during training process, which varies on the gradient value with every epoch.

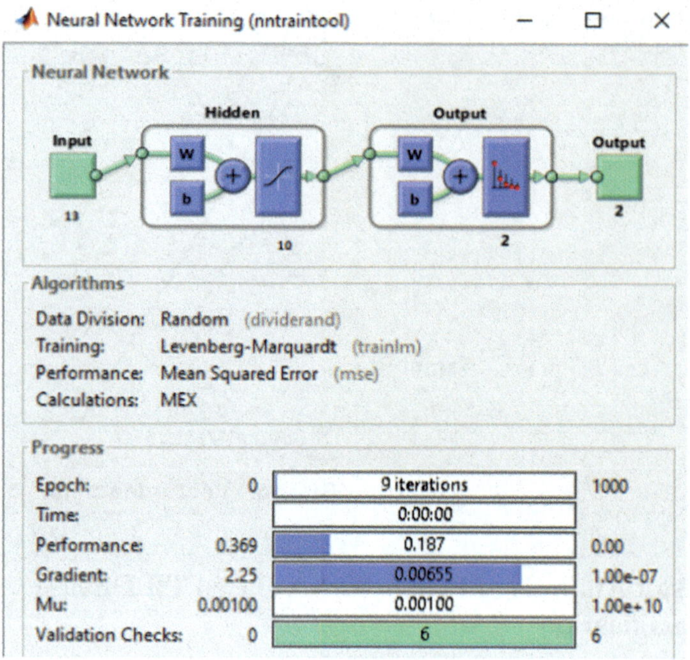

Fig. 4.6 Neural network block

This graph is also helpful to identify the minimum value of gradient factor occurrence time and failure condition of validation per epoch. Figure 4.9 shows the Confusion matrix. Notation used in the Confusion matrix is as follows like, green color square represents sensitivity parameter, red color square demonstrates specificity parameter and blue color square highlights overall accuracy. As this matrix combines all three kind of data representation (Training data, testing data and validation data) in single matrix, that's why the name of the matrix is given as confusion matrix. We have demonstrated three kinds of Confusion matrices such as training, testing and validation matrix. Our proposed method produces very accurate outputs. High numbers of correct response are available in green squares and incorrect responses which are low in number are represented in red squares in the confusion matrix. Overall accuracy of the proposed model is demonstrated as blue square. Table 4.2 gives the specification of confusion matrix.

Where, TP-Truly positive value, FP-False Positive Value, FN-False Negative Value and TN-Truly Negative Value. From Fig. 4.9, It is observed in the all confusion matrix that 726 images are truly predicted as having TB disease and 33 images are falsely predicted. Also 6 images are correctly predicted as normal image out of 15 images, hence overall accuracy of NN based model leads to 94.6%. .

The optimal operating point of the classifier is one of the key points on ROC curve, which provided perfect ratio of sensitivity and specificity. It defines a cost function

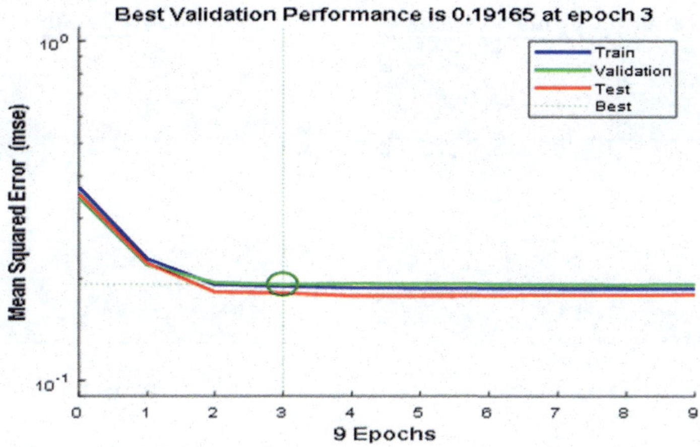

Fig. 4.7 Best validation performance of proposed NN based TB classification

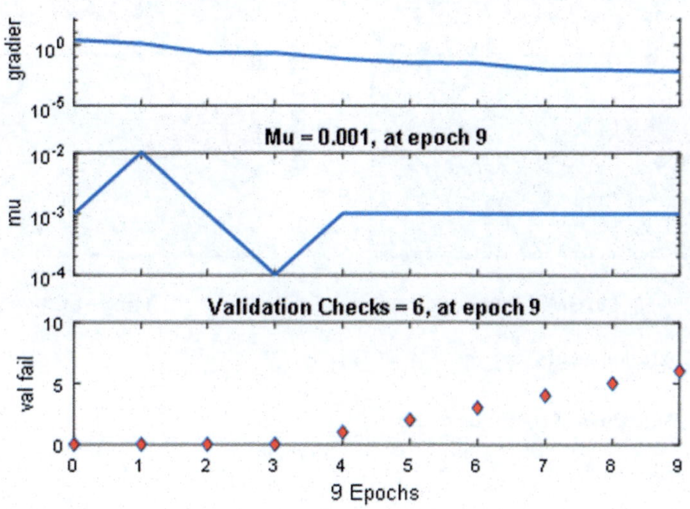

Fig. 4.8 Training state graph

which describes the missing TB affected cases. Figure 4.10 shows ROC graph of training data, testing data, validation data and overall data. ROC discriminated normal and abnormal classes. The ROC graph is the curve of the false positive rate (specificity) versus the true positive rate (sensitivity) with the variation of threshold value. An ideal graph gives the values near upper-left corner, with 100% specificity and 100% sensitivity. For this problem, the network outperforms. Like Confusion matrix, ROC curve is also very useful graph for classification of normal and abnormal images. Table 4.3 gives the value of sensitivity, specificity and accuracy values of NN based model. We have conducted another experiment on our NN based model

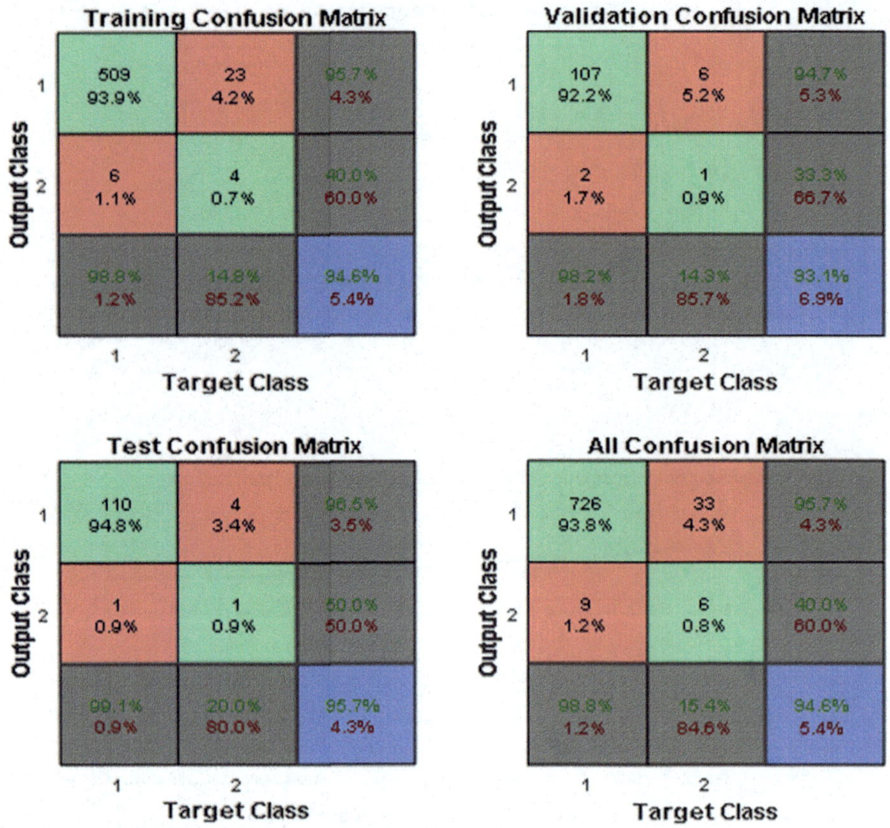

Fig. 4.9 Confusion matrix

Table 4.2 Parameters of confusion matrix

Measure	Formulas
Specificity	TN/TN+FP
Recall	TP/TP+TN
Precision	TP/TP+FP
Accuracy	TP+TN/TP+TN+FP+FN
Sensitivity	TP/TP+FN

to see its performance on real world data, which are collected from Horizon imaging centre, Surat. This dataset is in DICOM format, which is converted into JPEG to fed the NN model. Different number of datasets are considered and it is observed that as number of images are increased, accuracy and other parameters are giving better value of performance (approximately 90%). Hence, we can conclude that our NN based model is fit to predict the performance on new data.

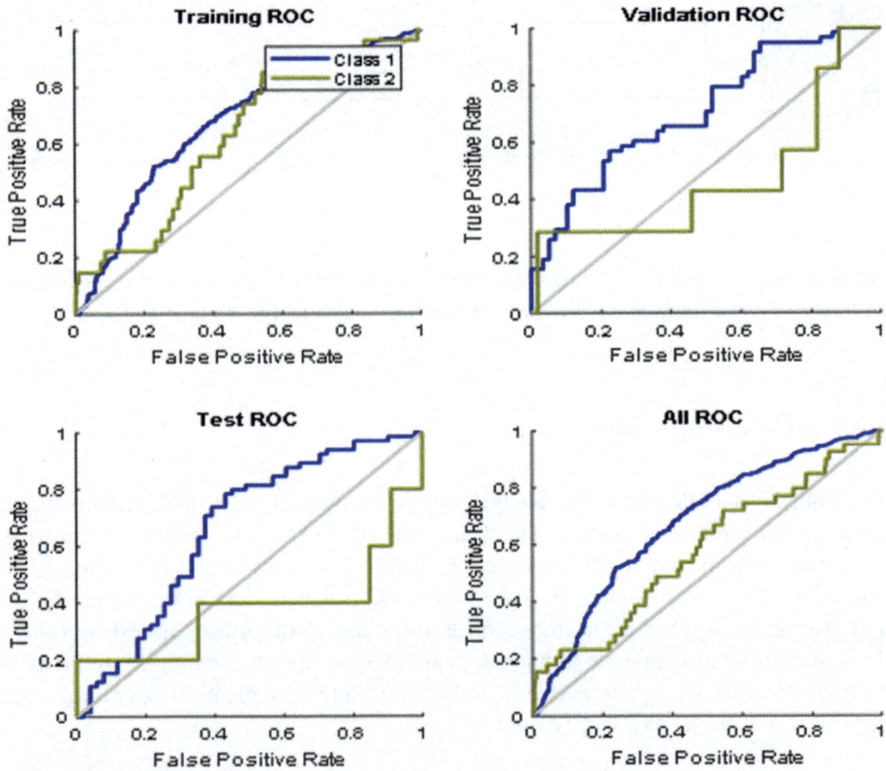

Fig. 4.10 Graph of training set, validation set, testing set and overall data

Table 4.3 Performance of NN based approch

No. of data sets	Sensitivity (%)	Specificity (%)	Accuracy (%)
661	90.2	90.9	90.90
450	62	78	67
313	60	52	54
151	51	53	63
60	53	51	63

4.6 Deep Learning Approach of TB Disease Classification

In this approach, we present a deep convolutional neural network architecture trimmed to tuberculosis diagnosis. Deep learning models are powerful enough for demonstrating high representational features. With this approach there will be significant reduction in the requirement of computation complexity and memory, without sacrificing the classification performance. Figure 4.11 shows the flow of Deep learn-

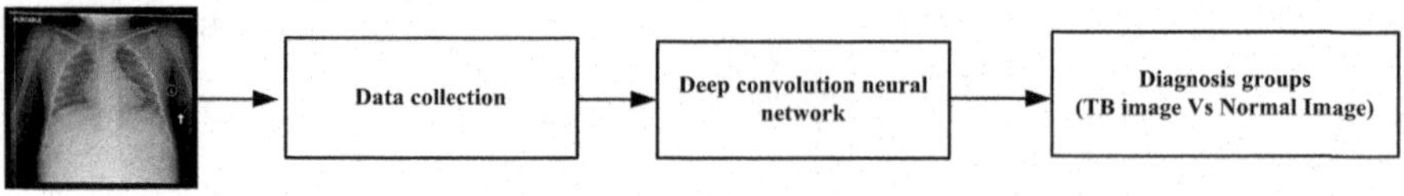

Fig. 4.11 The flow of deep learning for classification of tuberculosis disease as normal image and TB image

ing approach of TB disease classification, which consists of two steps; pre-processing and network training. The following subsection explains the detailing.

4.6.1 Data Collection

The dataset is collected from Kaggle online community for data scientists and machine learning practitioners. There are total of 7000 CXR images contained in the dataset. From which 3500 images are labelled as normal and remaining 3500 images as TB. These images are fed to network in two folders namely training set and testing set; total of 4900 images in training and 2100 images in testing folder. Re-scaling operation is done to have data augmentation of these images. In the next section, we will show our network architecture and optimization operations that improved our model to a large extent.

4.6.2 Network Architecture

The human visual system has small number of neuron cells which are sensitive to some particular field for example only few neurons in the brain are fired in the existence of edged in specific direction. Such operation is described in Convolutional neural networks (CNN). Wherefore, we can say that CNNs systems are developed in distinction to the fundamental working of human visual system. CNN consist of three layers namely convolutional layer, pooling layer and fully connected layer. Element wise multiplication is used in convolutional layer to extract feature maps on input images with application filter alongside on entire image. To have network being generalized instead of memorizing the provided data called as over fitting problem, to avoild this pooling layer is used. Which neuron will get fired is depended on the Rectified linear unit (ReLU) activation and that determines the output of neural network. With the combination of three layers (convolutional-ReLU-pooling); many feature maps are extracted and after that these feature maps are fed to dense layer to have final decision of model. The proposed model architecture is shown in Fig. 4.12. Our proposed Network architecture used for training consist of three layers, out of which three are convolutional layers followed by max pooling and two are fully

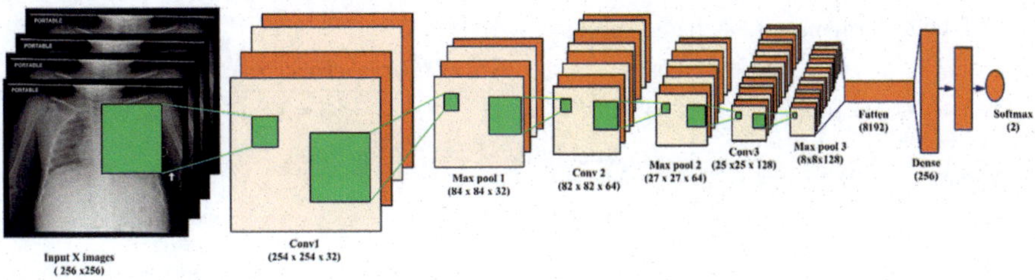

Fig. 4.12 Architecture of deep convolutional neural network (DCNN)

connected layers, shown in Fig. 4.12. Convolutional layer and max pooling layers are represented as Convx and Max pool x, where x represents the layer number e.g., first convolutional layer is represented by Conv1 and similarly max pooling is represented by Max pool 1 whereas fully connected layers are described separately as Dense. Output of fully connected layer is fed to SoftMax function to have a probability distribution of two classes labelled as normal image and TB image. Hence, probabilities of vector size 1 x 2 is obtained, where each vector element resemble to a class of dataset. All the images in the dataset are first pre-processed to reduce complexity which arise in handling the image dimension, pre-processing involves scaling down the size of input images to 256 x 256. The input for architecture is a grayscale image having dimension of 256 x 256, which is fed to conv1, where 32 features with kernel size having size 3 x 3 along stride equal to one are filtered out. Stride is a component of CNN, which is used to modify the dimension of the output image volume. The output of Conv1 is passed to non-linearity and then applied to spatial max pool 1 to have summarizing of neighboring neurons with the obtained dimension of image as 84 x 84 x 32. Rectified linear unit (ReLU) [14] nonlinearity is applied to all convolutional and fully connected layers. ReLU has the ability to train the network much faster compared to its equivalent tanh units [11] and it also allows to go deeper with vanishing gradient problems. The obtained dimension of image is from 256 x 25 6 x 1 to 254 x 254 x32, referring to the dimension formula given Eq. (4.3)

$$n^{[l]} = \frac{(n^{[l-1]} + 2 * p^{[l-1]} - f^{[l]})}{s^{[l]}} + 1 \tag{4.3}$$

where n represents the size of input layer, p is the size of padding, s is for size of stride, f is kernel size and l represent to current layer. Conv2 layer contains 64 filters with kernel size 3 x 3 and it gets input from Max pool 1. so, image dimensions become 82 x 82 x 64, which is applied to Max pool 2 with kernel size 3 x 3, dimension reduces to 27 x 27 x 64. Next, Conv3 layer contains total 128 feature maps having kernel size 3 x 3 and stride equal to one; layer dimension obtained is 25 x 25 x 128 and max pool 3 gives feature reduction dimension as 8 x 8 x 128. Hence, total parameters obtained are 8192, which are formed to one vector function. The first dense layer consist of 256 feature maps along with the kernel size equal to one. Each of these 256 neurons are connected to all the 8192 (8 x 8 x 128) neurons in the Max pool 3. s dense layer

Table 4.4 Performance of NN based approch

Layer	Input size	Output size	No. of filters	Filter size	Parameters
Conv 1 + ReLU	256 x 256 x 1	254 x 254 x 32	32	3 x 3 x 1	896
Max Pool 1	254 x 254 x 32	84 x 84 x 32	32	3 x 3 x 1	0
Conv 2 + ReLU	84 x 84 x 32	82 x 82 x 64	64	3 x 3 x 1	18496
Max Pool 2	82 x 82 x 64	27 x 27 x 64	64	3 x 3 x 1	0
Conv 3 + ReLU	27 x 27 x 64	25 x 25 x 128	128	3 x 3 x 1	73856
Max Pool 3	25 x 25 x 128	8 x 8 x 128	128	3 x 3 x 1	0
Flatten	8 x 8 x 128	1 x 8192	–	–	0
Dense + ReLU	1 x 8192	1 x 256	–	–	2097408
Dropout	Probability = 0.2				
Dense	1 x 256	1 x 2	–	–	514
Total params					2191170

is fully connected with 256 units. It becomes difficult for the network to operate on overlapping pooling operation as network model starts to overfit during training. To avoid further overfitting of neurons, a drop out regularization layer [28] is used after first and second dense layer with neuron drop out probability of $1 - p$, where p is the probability of neurons kept during network training. The neurons which are dropped out does not engage in further as well as in backward pass, i.e., all the neurons going into and coming out of drop out layer are removed in training phase. During testing phase, all the neurons are considered without any of them being dropped out. Finally, output of second Dense layer is fed to input of softmax function which scales the value of the vector in the range of (0,1) and summation of its vector value gives a value of one, which is assigned to probability distribution of each class. Specification of the network architecture shown in Table 4.4.

4.6.3 Experiments and Results Discussion

Intense understanding styles be in favor of cracking the challenge end to end slightly crashing the difficulty into unalike fragments. Owing to which it flops to understand the perceptive overdue the outcome gotten; because of the indefinite labor ended by the communal nerve cell in arrears the compact structure of system. Henceforth, to recover the execution of the paradigm succeeding considerations are accessible for regulation

- Loss function
- Layers
- Epochs
- Optimizers
- Augmentation
- Drop out

The loss function is the director to the topography, telling the optimizer as soon as it is affecting in the correct otherwise incorrect track. Optimizers outline and mold the model into its maximum precise likely form by futzing with the weights. Dropout is used to avert the system on or after overfitting by deactivating the nerve cell on persistence; which is finished by means of about possibility; usually 0.2 probability drop is favored. Data expansion theatrically rise the scope of drill set to evade regulation of the system. By fine-tuning the number of layers and epochs, precision of the model can be amplified. We experimented several cases of classifying our model and evaluate their performance. We used different optimizers to train the model and comparison table is demonstrated to have a best performance of the network architecture.

4.6.4 Experimental Setup and Evaluation

Network architecture is realized with the keras collection with Tensorflow back end. The experiments are performed on Dell Intel corei7 laptop with 8GB RAM. Model is trained on Google collab notebooks having NVIDIA K80/T4 GPU with 12GB/16GB memory and 0.82GHz/1.59GHz clock speed. Relu initiation is applicable over each neuron of CNN. Yield categorized as Normal imageries and TB imageries. There are overall of 4900 imageries are used for drilling the system and 2100 images for challenging. Batch size used is 32 and loss function considered is binary cross entropy. Different optimizers are taken for improving the model namely Adagrad, Adam, SGD, Nadam, Rmsprop, Adadelta. The whole network is trained for 10 epochs with drop out probability taken is 0.2. lasty, softmax activation function is used at fully connected dense layer. Table 4.5 demonstrates the consequences of the planned DCNN model. implementation is assessed in conditions of exactness and loss for aiming as well as authentication set. Loss outlines the finest information of how fit is the model. Out of all the optimizers, Adam proved to give finest accurateness with a smaller amount loss since it does not want physical modification of learning rate

Table 4.5 Performance of proposed network architecture

Optimizer	Training accuracy	Validation accuracy	Training loss	Validation loss
Adam	99.24	96.92	2.02	10.64

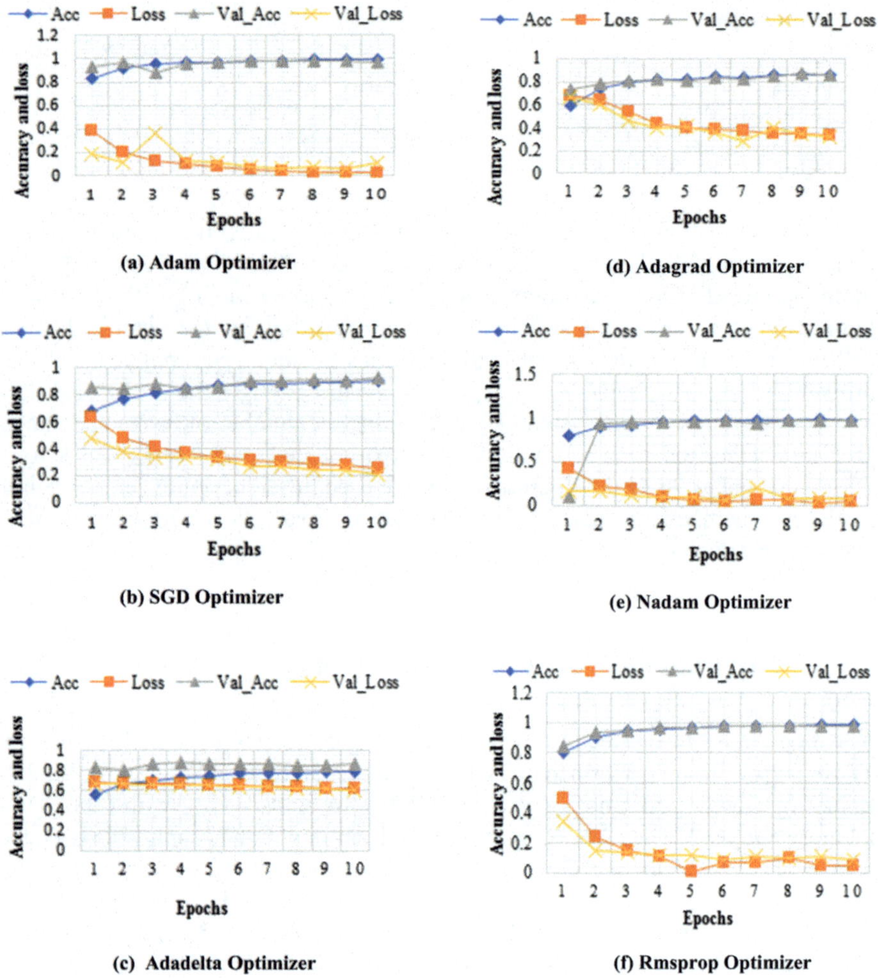

Fig. 4.13 Accuracy versus loss (training and validation set) of six different optimizers for 10 epochs

as it makes slight apprises for frequent constraints and big updates for uncommon parameters. The accuracy versus epoch and loss versus epoch graph for both training and validation set is shown in Fig. 4.13. It can be realized that training set has achieved an accuracy of 99.24% with loss nearly equal to zero. This gives an idea of accurateness of the model being trained. Meanwhile the validation set provides the understanding of the measure of quality of the model that how fit is the model to predict on new data. We have achieved 96.92% validation accuracy which represents that with 96.92% accuracy, model can figure out the detection on new data. We have tried our dataset over diverse optimizers with identical format deliberated at the commencement of section. Yet also the uppermost exactness is accomplished

Table 4.6 Performance of proposed network architecture

Performance parameters (%)	Adam	Nadam	Rmsprop	SGD	Adadelta	Adagrad
Overall accuracy	99.24	98.25	98.69	90.53	79.42	86.26
Validation accuracy	96.92	97.31	98.22	92.45	86.87	86.39
Over loss	2.02	5.62	4.56	25.39	61.82	34.15
Validation loss	10.64	8.07	9.26	21.40	60.54	32.11

Table 4.7 Comparison with other paper result in terms of accuracy

Authors	No. of images	Dataset	Accuracy (%)
Jaeger et al. [17]	138	Montgomery set (MC)	75.00
Hooda et al. [16]	800	MC & Shenzhen set	82.09
Chang et al. [20]	4701	Health care Peru [10]	85.68
Guo et al. [12]	662	NIH CXR	95.49
Proposed	7000	Kaggle	99.24

by nadam and rms prop which can be portrayed from Table 4.6 but the corroboration loss is repetitively cumulative i.e. the model has started to over fit with being prone to noise. This will sooner or later reduce the capability of the model on predicting the new data. Figures 4.13 shows Accuracy versus Loss (training and validation set) of all the optimizers for 10 epochs Table 4.7 gives the accomplishment assessment of planned prototypical with supplementary tactics forth with practices and data modalities. Amongst all the tactics, our proposed model has achieved accurateness as high as 99.24% deprived of any pre-learnt structures. We have conducted another small experiment on ChexPert data collected from Kaggle community. Total of 170 images are collected, out of which 140 images are used for training the deep learning based NN model and 30 images are used for testing the model. Figure 4.14 shows the confusion matrix and test data with label. From confusion matrix, we can observe that 13 images are truly classified as TB out of 15 images. Similarly, 8 images are correctly classified as normal image. The configuration of the model is kept same without changing any parameter, experiment conducted on Adam optimizer which is giving superior performance compared to other optimizers. The accuracy on this dataset is 76% and loss is 50%, this is due to very a smaller number of images in the dataset that means model is fitting towards noise. Hence, in deep learning models, large dataset can provide the desired accuracy.

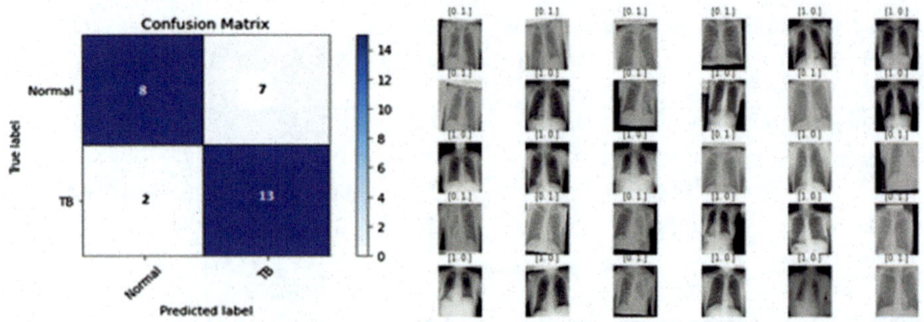

(a) Confusion Matrix (b) Test data with label

Fig. 4.14 Confusion matrix and test data of ChexPert dataset experimented on deep learning model

4.7 Conclusion

In this work, we have implemented two approaches for TB disease classification. One approach is based on ANN-SVM (NN based) algorithm, in which to have enhanced image quality and efficient reduction of speckle noise, different pre-processing techniques namely gray scale, median filtering, and gamma correction are used. Gray Level Co-occurrence Method is used to extract the texture features and region of interest is obtained with the help of K-means clustering. ANN is trained using these features. The whole system is able to classify the images into normal or TB with an accuracy of 94.6% on TB CXR image data-base. The performance of the SVM-ANN classifier surpass other existing classifiers. Other approach is based on DCNN, where we have conferred a scheme based on deep learning for Tuberculosis disease classification in terms of accuracy. We have trained the network on Kaggle dataset and have achieved 99.24 %accuracy without any handcrafted features. Validation accuracy comes out to be 96.92%. Both approaches showed their best results, when small data set is concerned NN based approach work very well but due to the inspection of larger data set available in hospitals, first approach would not suffice as handcrafted features are sometimes not accurate enough to distinguish the disease on image. That's when our second approach works pretty well without any handcrafted features. Future work will concentrate on improving the DCNN model by achieving performance parameters such as Sensitivity, Specificity, F1-score and Recall. And other experiment can be realized to validate the prognosis of disease on different staged.

References

1. Ahmed, S., Kabir, M., Arif, M., Ali, Z., Ali, F., Swati, Z.N.K.: Improving secretory proteins prediction in mycobacterium tuberculosis using the unbiased dipeptide composition with support vector machine. Int. J. Data Min. Bioinform. **21**(3), 212–229 (2018)

2. Airouche, M., Bentabet, L., Zelmat, M.: Image segmentation using active contour model and level set method applied to detect oil spills. In: Proceedings of the World Congress on Engineering, Lecture Notes in Engineering and Computer Science, vol. 1, pp. 1–3 (2009)
3. Chai, B., Cao, W., Gou, Q.: A method of medical ultrasound image enhancement based on wavelet adaptive transform. Acta Microscopica **28**(6) (2019)
4. Chak, P., Navadiya, P., Parikh, B., Pathak, K.C.: Neural network and svm based kidney stone based medical image classification. In: International Conference on Computer Vision and Image Processing, Springer, pp. 158–173 (2019)
5. Chauhan, A., Chauhan, D., Rout, C.: Role of gist and phog features in computer-aided diagnosis of tuberculosis without segmentation. PLoS One **9**(11), e112980 (2014)
6. Costa, M.G., Costa Filho, C.F., Sena, J.F., Salem, J., de Lima, M.O.: Automatic identification of mycobacterium tuberculosis with conventional light microscopy. In: 2008 30th Annual International Conference of the IEEE Engineering in Medicine and Biology Society, IEEE, pp. 382–385 (2008)
7. Dean, J.C., Ilvento, C.C.: Improved cancer detection using computer-aided detection with diagnostic and screening mammography: prospective study of 104 cancers. Am. J. Roentgenol. **187**(1), 20–28 (2006)
8. Doi, K.: Computer-aided diagnosis in medical imaging: historical review, current status and future potential. Comput. Med. Imaging Graph. **31**(4-5), 198–211 (2007)
9. Elveren, E., Yumuşak, N.: Tuberculosis disease diagnosis using artificial neural network trained with genetic algorithm. J. Med. Syst. **35**(3), 329–332 (2011)
10. Ginneken, B., Hogeweg, L., Maduskar, P., Peters-Bax, L., Dawson, R., et al.: Performance of inexperienced and experienced observers in detection of active tuberculosis on digital chest radiographs with and without the use of computer-aided diagnosis. In: Annual Meeting of the Radiological Society of North America (2012)
11. Glorot, X., Bordes, A., Bengio, Y.: Deep sparse rectifier neural networks. In: Proceedings of the Fourteenth International Conference on Artificial Intelligence And Statistics, JMLR Workshop and Conference Proceedings, pp. 315–323 (2011)
12. Guo, R., Passi, K., Jain, C.K.: Tuberculosis diagnostics and localization in chest x-rays via deep learning models. Front. Artif. Intell. **3**, 74 (2020)
13. Gur, D., Sumkin, J.H., Rockette, H.E., Ganott, M., Hakim, C., Hardesty, L., Poller, W.R., Shah, R., Wallace, L.: Changes in breast cancer detection and mammography recall rates after the introduction of a computer-aided detection system. J. Natl Cancer Inst. **96**(3), 185–190 (2004)
14. Hahnloser, R.H., Sarpeshkar, R., Mahowald, M.A., Douglas, R.J., Seung, H.S.: Digital selection and analogue amplification coexist in a cortex-inspired silicon circuit. Nature **405**(6789), 947–951 (2000)
15. Hogeweg, L., Mol, C., de Jong, P.A., Dawson, R., Ayles, H., van Ginneken, B.: Fusion of local and global detection systems to detect tuberculosis in chest radiographs. In: International Conference On Medical Image Computing and Computer-Assisted Intervention, Springer, pp. 650–657 (2010)
16. Hooda, R., Sofat, S., Kaur, S., Mittal, A., Meriaudeau, F.: Deep-learning: a potential method for tuberculosis detection using chest radiography. In: 2017 IEEE International Conference on Signal and Image Processing Applications (ICSIPA), IEEE, pp. 497–502 (2017)
17. Jaeger, S., Karargyris, A., Antani, S., Thoma, G.: Detecting tuberculosis in radiographs using combined lung masks. In: 2012 Annual International Conference of the IEEE Engineering in Medicine and Biology Society, IEEE, pp. 4978–4981 (2012)
18. Khutlang, R., Krishnan, S., Whitelaw, A., Douglas, T.S.: Automated detection of tuberculosis in ziehl-neelsen-stained sputum smears using two one-class classifiers. J. Microsc. **237**(1), 96–102 (2010)
19. Lakhani, P., Sundaram, B.: Deep learning at chest radiography: automated classification of pulmonary tuberculosis by using convolutional neural networks. Radiology **284**(2), 574–582 (2017)
20. Liu, C., Cao, Y., Alcantara, M., Liu, B., Brunette, M., Peinado, J., Curioso, W.: TX-CNN: Detecting tuberculosis in chest x-ray images using convolutional neural network. In: 2017 IEEE International Conference on Image Processing (ICIP), IEEE, pp. 2314–2318 (2017)

21. Luukka, P.: Feature selection using fuzzy entropy measures with similarity classifier. Expert Syst. Appl. **38**(4), 4600–4607 (2011)
22. Noble, M., Bruening, W., Uhl, S., Schoelles, K.: Computer-aided detection mammography for breast cancer screening: systematic review and meta-analysis. Arch. Gynecol. Obstet. **279**(6), 881–890 (2009)
23. Osman, M.K., Mashor, M.Y., Jaafar, H.: Performance comparison of extreme learning machine algorithms for mycobacterium tuberculosis detection in tissue sections. J. Med. Imag. Health Inform. **2**(3), 307–312 (2012)
24. Patil, S., Udupi, V.: Geometrical and texture features estimation of lung cancer and tb images using chest x-ray database. Int. J. Biomed. Eng. Technol. **6**(1), 58–75 (2011)
25. Priya, E., Srinivasan, S.: Automated object and image level classification of tb images using support vector neural network classifier. Biocybern. Biomed. Eng. **36**(4), 670–678 (2016)
26. Santiago-Mozos, R., Fernández-Lorenzana, R., Perez-Cruz, F., Artes-Rodriguez, A.: On the uncertainty in sequential hypothesis testing. In: 2008 5th IEEE International Symposium on Biomedical Imaging: From Nano to Macro, IEEE, pp. 1223–1226 (2008)
27. Shen, R., Cheng, I., Basu, A.: A hybrid knowledge-guided detection technique for screening of infectious pulmonary tuberculosis from chest radiographs. IEEE Trans. Biomed. Eng. **57**(11), 2646–2656 (2010)
28. Shin, H.C., Roth, H.R., Gao, M., Lu, L., Xu, Z., Nogues, I., Yao, J., Mollura, D., Summers, R.M.: Deep convolutional neural networks for computer-aided detection: CNN architectures, dataset characteristics and transfer learning. IEEE Trans. Med. Imaging **35**(5), 1285–1298 (2016)
29. Vijayarani, S., Dhayanand, S., Phil, M.: Kidney disease prediction using svm and ann algorithms. Int. J. Comput. Bus. Res. (IJCBR) **6**(2), 1–12 (2015)
30. Vishwakarma, H., Katiyar, S.K.: An approach for line feature extraction using hough transform for remote sensing images
31. World Health Organization et al.: Global tuberculosis report 2016. 2016. Google Scholar 214 (2016)
32. Zulvia, F.E., Kuo, R., Roflin, E.: An initial screening method for tuberculosis diseases using a multi-objective gradient evolution-based support vector machine and c5. 0 decision tree. In: 2017 IEEE 41st Annual Computer Software and Applications Conference (COMPSAC), vol. 2, IEEE, pp 204–209 (2017)

Chapter 5
Adaptive Machine Learning Algorithm and Analytics of Big Genomic Data for Gene Prediction

Oluwafemi A. Sarumi and Carson K. Leung

Abstract Artificial intelligence helps in tracking and preventing diseases. For instance, machine learning algorithms can analyze big genomic data and predict genes, which helps researchers and scientists to gain deep insights about protein-coding genes in viruses that cause certain diseases. To elaborate, prediction of protein-coding genes from the genome of organisms is important to the synthesis of protein and the understating of the regulatory function of the non-coding region. Over the past few years, researchers have developed methods for finding protein-coding genes. Notwithstanding, the recent data explosion in genomics accentuates the need for efficient gene prediction algorithms. This book chapter presents an adaptive naive Bayes-based machine learning (NBML) algorithm to deploy over a cluster of the Apache Spark framework for efficient prediction of genes in the genome of eukaryotic organisms. To evaluate the NBML algorithm on its discovery of the protein-coding genes from the human genome chromosome GRCh37, a confusion matrix was constructed and its results show that NBML led to high specificity, precision and accuracy of 94.01%, 95.04% and 96.02%, respectively. Moreover, the algorithm can be effective for transfer knowledge in new genomic datasets.

Keywords Genomics · Machine learning · Data science · Gene prediction · Protein synthesis · Big data · Apache Spark · Data analytics

5.1 Introduction

In many centuries and around the world, the spread of infectious diseases—such as, Black Death in 1331–1353, Spanish flu in 1918–1920, severe acute respiratory

---

O. A. Sarumi · C. K. Leung (✉)
University of Manitoba, Winnipeg, MB, Canada
e-mail: kleung@cs.umanitoba.ca

O. A. Sarumi
e-mail: oasarumi@futa.edu.ng

O. A. Sarumi
The Federal University of Technology—Akure (FUTA), Akure, Nigeria

M. Mehta et al. (eds.), *Tracking and Preventing Diseases with Artificial Intelligence*,
Intelligent Systems Reference Library 206,
https://doi.org/10.1007/978-3-030-76732-7_5

syndrome (SARS) in 2002–2004, Zika virus disease in 2015–2016, and the recent coronavirus disease 2019 (COVID-19)—has been a major concern. These infectious diseases directly affect people health. In recent decades, artificial intelligence (AI) [2, 96]—especially machine learning [52]—has been adapted, in conjunction with good use of data mining [16, 32, 47, 48, 51] and big data science [40, 44, 54, 89], to assist physicians in diagnosis, disease tracking, prevention and control. Examples include computer-aided diagnosis of thyroid dysfunction [78], hiding of sensitive electronic health information [102], development of brain computer interface [79]. Moreover, in terms of data types, these works range from analytics on epidemiological data (e.g., COVID-19 data [17, 53], Zika data [94]) to bioinformatics on genomics data.

Over the past decade, bioinformatics have been revolutionized partially due to technological advances and scientific discoveries. For instance, thanks to the emergence of high-throughput next-generation sequencing (NGS) technologies (e.g., Illumina Genome Analyzer, Illumina HiSeq X), the sequencing time has been reduced tremendously, while the production and collection of huge volumes of omics data (e.g., genomics, metagenomics, proteomics, transcriptomics) has been increased. In genomics, petabytes of complete sequenced genomes for many organisms (e.g., eukaryotes and prokaryotes) are available currently in public repositories. This avalanche of genomic datasets has led to new challenges for researchers, and has demanded for improved computational approaches for bioinformatics tasks like gene analytics [59, 86], gene classification [26], gene prediction [60, 70], motifs discovery [3, 41, 87, 95], omic analytics [80], palindrome identification in gene sequence [85], sequence alignment [22, 43], and sequence assembly [31].

Gene prediction involves the process of locating the regions to encode a protein-coding genes and other functional elements in genomic datasets [6, 30]. For eukaryotic organisms [35], gene prediction can be a rigorous task partially because of (a) the inconsistency problem between genes and (b) a very limited amount of genes found in most genomes. As a concrete example, protein-coding gene (cf. non-coding elements [29, 90]) accounts for smaller than 5% of the entire human genome. Moreover, the long distance between these coding regions (i.e., *exons*), the limited knowledge of promoters, as well as an alternative splicing site after the transcription [65], all further complicate the problem. Due to their wide distance and undefined length, non-coding regions (i.e., *introns*) that separate the exons and the splicing sites (i.e., a region that divides the exons from the introns) are also difficult to be identified.

Conversely, for prokaryotic organisms [64], their exons can be identified less rigorously. More specifically, they can be identified in a contiguous sequence [106]—known as an *open reading frame* (OFR)—without interrupting the introns.

As prediction of gene is more rigorous in eukaryotic organisms, researchers has used two conventional methods to identify protein-coding [7, 21] from their gene sequence in eukaryotic organisms. The first method is the *ab initio gene prediction* [82], which identifies new genes using gene structures as a prototype through a keenly investigated process that splits the signal sensors [66] from other striking biological patterns—content sensors [37]. Besides, this method is capable to differentiate gene area within a single input sequence. Several ab initio gene prediction methods [14, 69, 103, 107] have been proposed in the literature by researchers. However, there are limitations on these methods [66, 99] due to the availability of limited knowledge of gene structures, particularly for newly sequenced genomes. In addition, identification of periodicity and additional familiar content traits of protein-coding genes can be arduous too.

The second method for identifying protein-coding from their gene sequence is hinged on *sequence similarity searches* [10, 71], which aims to find resemblance from gene sequences between expressed sequence tags (ESTs), proteins, or other genomes to the input sequence. This method assumes that the coding regions are more evolutionarily conserved [68] than non-coding regions. Different sequence similarity search-based gene prediction methods [1, 34, 42, 73] have been proposed. Given that ESTs correspond to only little fragments of the gene sequence, estimation of the wholesome gene structure of a given region can be laborious. This explains why EST-based sequence similarity usually leads to bottlenecks [36, 99].

Recently, gene has been predicted by applying machine learning (ML) algorithms and data mining techniques [18, 19, 76]. Key advantages of applying ML algorithms include the potentials to instinctively recognize patterns in data [105]. This is highly significant when the expert knowledge is inadequate or inaccurate–this implies that the ML algorithms can discover new or hidden patterns in the dataset that were previously not known or explicit to the experts . Also, when the volume of data available is enormous to be processed manually, and when there are peculiar situations that does not follow the normal trends. Besides, ML algorithms are effective for transfer learning process [9, 15, 55, 72]. Transfer learning paradigm provides an opportunity for knowledge reusability—such that knowledge acquired in solving a problem can be employed in a related domain. Also, transfer learning can employed for big data applications [100]. Several ML algorithms [45, 84, 88, 93] have been proposed for predicting gene in genomes of organisms. Key limitations of these algorithms include (a) their unscalability for handling huge volumes of genomic datasets, and (b) their high restriction to one type of organism or dataset. Consequently, more efficient and adaptive ML algorithms are needed. Their ability to improve inherently with experience in gene prediction from huge volumes of a genomic dataset is desirable. Our ***key contribution*** of this book chapter is our adaptive naive Bayes-based machine learning algorithm for identification of protein-coding regions in huge volumes of genome for eukaryotic organisms.

We organize the remaining sections of this book chapter as follows. The next section describes machine learning algorithms and public referenced genome databases for bioinformatics tasks. Then, we present the data preprocessing actions and our adaptive NBML algorithm. Afterwards, we show and discuss experimental results, and then draws our conclusions.

5.2 Background: Common Machine Learning Algorithms and Public Referenced Genome Databases for Bioinformatics Tasks

There is a major paradigm shift in bioinformatics research after completing the Human Genome Project (HGP) [98]. The success of the HGP project and the introduction of several high technological next-generation sequencing machines accentuate the production of big data in the bioinformatics domain at high speed and very low cost. Researchers can now leverage the availability of these big omics data in conjunction with machine learning algorithms to improve on several bioinformatics tasks. In general, machine learning algorithms [13, 57, 58, 74] help develop models to learn from historical experience and to discover novel patterns in future data. They are broadly categorized into unsupervised learning [8, 12, 28, 46, 92] and supervised learning [4, 20, 49, 50, 56]. Methods in the latter category include the following that are commonly used for bioinformatics tasks:

1. **Logistic regression (LR)**, which is a statistical model that uses a set of independent variables to predict a binary outcome. The two possible binary outcomes—0 or 1 are employed in training the LR model to predict the possibility of a second event. There are other complex forms of LR models—such as polychotomous or multinomial logistic regression [11]—for handling events in which the predicted variable with more than two categories. LR models have widely been used for models that involve the identification of a disease state in a suspected carrier, which could either be positive or negative. In bioinformatics LR have been used for the modeling of relationship between mortality rate and iatrogenic illness in patients [27], single nucleotide polymorphisms (SNP)-SNP interactions [38], and the usage of logistic regression in examination of relationship between congenital tract infections and maternal risk factors [91].
2. **Support vector machine (SVM)**, which is a non-probabilistic classification model for providing efficient solutions to both classification and regression problems. SVM is capable for modeling multidimensional borderlines that are unsequential, non-straightforward, or difficultly susceptible to disproportionately complex situations (e.g., having numerous parameters when compared to the number of observations). In developing an SVM model [62], the input vector is mapped onto a high-dimensional space for the construction of a maximal separating hyperplane between two parallel hyperplanes. Better performance of the algorithm is achieved when there is a larger margin of separation between the parallel hyperplanes. Some of the bioinformatics task performed with SVM include the integration of clinical and microarray data [24], the selection of features and the classification of mass spectrometry and microarray samples [108], classification of gene expression [97], detection of orthologs in yeast species [33], genomic feature learning [5] and, evolutionary feature selection [81].
3. **Decision tree (DT) algorithm**, which generates a directed tree consisting of the root node, internal nodes, and leaf nodes. Its goal is to produce a DT model, and to train the produced model to forecast values of goal variables based on the input

variables. Each leaf node indicates the value of a goal variable computed based on the values of input variables along the path leading from the root node to the leaf node. Each internal node communicates with some input variables [39], and acts as a decision node indicating a test on attribute or subset of attributes. Given a set of training input, the DT algorithm applies a measurement function to all the attributes for finding the best splitting attribute [23]. Upon identifying the splitting attribute, the DT algorithm divides the instance space into several parts. If all the training instances belong to the same class, the DT algorithm terminates. Or else, the DT algorithm recursively performs the splitting process until it assigns the whole partition to a class label [75]. Once the DT is generated, classification rules can be deduced to classify the new instances of data having unknown class labels. Several DT algorithms are used for classification including ID3, logic model tree, random forest, alternating decision (AD) tree, ID3 C4.5, and C5.0. Each algorithm differs in the method of selecting the most significant attribute for constructing a DT. Although DTs are one of the simplest and popular methods of data classification, it has several drawbacks. The algorithm fails to handle missing values and has issues due to overfitting of data [83]. Common bioinformatics tasks performed with DT include the integration of clinical data and gene expression to identify disease endotypes [101], and predictive models for analyzing racial disparities in breast cancer [77].

4. **Naive Bayes (NB) algorithms** [25, 61], which are developed using probabilistic classifiers from the Bayes theorem. The NB algorithms are established on the premise that the addition or removal of a given attribute in the model is not dependent on the addition or removal of any other attribute and their quota towards the probability is independent of each other. A key benefit of NB classifiers is that they only require a limited amount of training data to estimate the classification parameters. The goal of NB algorithms is to develop an intelligent model that can automatically assign new features to a known class.

The availability of several forms of genomic datasets in many public databases has enhanced the process of using these machine learning algorithms for building new knowledge and intelligence from the genomic dataset. Some of the available public genome databases include:

- DNA Data Bank of Japan (DDBJ),[1]
- European Molecular Biology Laboratory—European Bioinformatics Institute (EMBL-EBI),[2]
- FlyBase,[3]
- microRNA database (miRBase),[4]
- retired Stanford Microarray Database, and

[1]https://www.ddbj.nig.ac.jp/index-e.html.

[2]https://www.ebi.ac.uk/.

[3]https://flybase.org/.

[4]http://mirbase.org/.

- US National Centre for Biotechnology Information (NCBI).[5]

These genomic database contains:

- deoxyribonucleic acid (DNA), ribonucleic acid (RNA), and protein sequence data;
- gene expression data;
- gene ontology (GO) key data types that are copious in size and are frequently used in bioinformatics research;
- pathway data; and
- protein-protein interaction (PPI) data.

Other forms of bioinformatics data include networks on human diseases, as well as on disease gene associations, which are used also for disease diagnosis and other related research. These avalanche volumes of genomic data are helpful for more accurate analytics and big data computation in the field of bioinformatics. Besides, the advent of big data frameworks like Apache Kafka, Apache Storm, and Apache Spark has reduced the complexities and speed of processing these big data. In processing big data, the Apache Spark framework [87] is very useful due to its nice properties listed below:

1. It makes use of a dedicated resource dispenser and a result accumulator called the driver program.
2. It partitions that are lost can be reconfigured effortlessly without depleting the information.
3. It stores intermediate results in the memory in place of the disk.
4. It provides an ample support for system workloads (e.g., batch processing, graph processing, interactive processes, iterative procedures, machine learning).
5. It applies computing units (i.e., worker nodes) to handle sub-tasks.

5.3 Our Naive Bayes Algorithm

5.3.1 Apache Spark Framework

In this research, a Naive Bayes algorithm is adapted for modeling an intelligent model for gene prediction and is scalable for big data analytics application via its deployment on the Apache Spark framework. Such a framework has triggered a lot of cognizance among many researches over the past decade due to its capacity for handling copious amounts of data over a distributed and parallel systems. Spark application runs as maverick sets of individual processes on each worker nodes on a cluster, supervised by the SparkContext object on the master node through the driver program (which is responsible for describing the specifications for different transformations and invoking corresponding actions on the worker nodes, and thus establishes a connection to the worker nodes via the cluster manager application).

[5]https://www.ncbi.nlm.nih.gov/.

A set of phases, which are parted by distributed operations, are used to perform the Spark actions.

5.3.2 *Data Preprocessing and Munging*

Here, we preprocess and mung the two human genome (*Homo sapiens*) assemblies datasets obtained from the Ensembl data repository (www.ensembl.org):

1. **Genome Reference Consortium Human Build 37 (GRCh37) patch 13 (GRCh37.p13)**, which was created by UC Santa Cruz (UCSC) as its human genome version 19 (hg19) database. The 104,763 protein-coding sequences (PCS) and 24,513 non-coding sequences (NCS) in this database occupy 3.2 GB.
2. **GRCh38 patch 10 (GRCh38.p10)**, which was also created by UCSC but as its human genome version 38 (hg38) database. The 102,915 PCS and 28,321 NCS in it occupy 3.4 GB.

Short sequences were filtered from the human genome GRCh37 and GRCh38 as shown in Fig. 5.1. Afterwards, 94,830 PCS and 24,266 NCS were obtained from the GRCh37 genome and 92,716 PCS and 28,024 NCS from the GRCh38 genome. A total of 40,000 sequences were selected from GRCh37 consisting of 20,000 PCS and 20,000 NCS. Also, 44,000 sequences containing of 22,000 PCS and 22,000 NCS were selected from the GRCh38 for the purpose of developing the gene predictive model. Our dataset was then converted into set of codons—which are groups of

```
AGAAGTTGTTAGTCTACGTGGACCGACAAGAACAGTTTCGAATCGGA
AGCTTGCTTAACGTAGTTCTAACAGTTTTTATTAGAGAGCAGATCTC
TGATGAACAACCAACGGAAAAAGACGGGTCGACCGTCTTTCAATATG
CTGAAACGCGCGAGAAACCGCGTGTCAACTGTTTCACAGTTGGCGAA
GAGATTCTCAAAAGGATTGCTTTCAGGCCAAGGACCCATGAAATTGG
TGATGGCTTTTATAGCATTCCTAAGATTTCTAGCCATACCTCCAACAG
CAGGAATTTTGGCTAGATGGGGCTCATTCAAGAAGAATGGAGCGATC
AAAGTGTTACGGGGTTTCAAGAAAGAAATCTCAAACATGTTGAACAT
AATGAACAGGAGGAAAAGATCTGTGACCATGCTCCTCATGCTGCTGC
CCACAGCCCTGGCGTTCCATCTGACCACCCGAGGGGGAGAGCCGCAC
ATGATAGTTAGCAAGCAGGAAAGAGGAAAATCACTTTTGTTTAAGAC
CTCTGCAGGTGTCAACATGTGCACCCTTATTGCAATGGATTTGGGAG
AGTTATGTGAGGACACAATGACCTACAAATGCCCCCGGATCACTGAG
ACGGAACCAGATGACGTTGACTGTTGGTGCAATGCCACGGAGACATG
GGTGACCTATGGAACATGTTCTCAAACTGGTGAACACCGACGAGACA
AACGTTCCGTCGCACTGGCACCACACGTAGGGCTTGGTCTAGAAACA
```

Fig. 5.1 Sample of the genome sequence

AGA	AGT	TGT	TAG	TCT	ACG	TGG	ACC	GAC	AAG	AAC	AGT	TTC	GAA
TCG	GAA	GCT	TGC	TTA	ACG	TAG	TTC	TAA	CAG	TTT	TTT	ATT	AGA
GAG	CAG	ATC	TCT	GAT	GAA	CAA	CCA	ACG	GAA	AAA	GAC	GGG	TCG
ACC	GTC	TTT	CAA	TAT	GCT	GAA	ACG	CGC	GAG	AAA	CCG	CGT	GTC
AAC	TGT	TTC	ACA	GTT	GGC	GAA	GAG	ATT	CTC	AAA	AGG	ATT	GCT
TTC	AGG	CCA	AGG	ACC	CAT	GAA	ATT	GGT	GAT	GGC	TTT	TAT	AGC
ATT	CCT	AAG	ATT	TCT	AGC	CAT	ACC	TCC	AAC	AGC	AGG	AAT	TTT
GGC	TAG	ATG	GGG	CTC	ATT	CAA	GAA	GAA	TGG	AGC	GAT	CAA	AGT
GTT	ACG	GGG	TTT	CAA	GAA	AGA	AAT	CTC	AAA	CAT	GTT	GAA	CAT
AAT	GAA	CAG	GAG	GAA	AAG	ATC	TGT	GAC	CAT	GCT	CCT	CAT	GCT
GCT	GCC	CAC	AGC	CCT	GGC	GTT	CCA	TCT	GAC	CAC	CCG	AGG	GGG
AGA	GCC	GCA	CAT	GAT	AGT	TAG	CAA	GCA	GGA	AAG	AGG	AAA	ATC
ACT	TTT	GTT	TAA	GAC	CTC	TGC	AGG	TGT	CAA	CAT	GTG	CAC	CCT
TAT	TGC	AAT	GGA	TTT	GGG	AGA	GTT	ATG	TGA	GGA	CAC	AAT	GAC
CTA	CAA	ATG	CCC	CCG	GAT	CAC	TGA	GAC	GGA	ACC	AGA	TGA	CGT
TGA	CTG	TTG	GTG	CAA	TGC	CAC	GGA	GAC	ATG	GGT	GAC	CTA	TGG
AAC	ATG	TTC	TCA	AAC	TGG	TGA	ACA	CCG	ACG	AGA	CAA	ACG	TTC
CGT	CGC	ACT	GGC	ACC	ACA	CGT	AGG	GCT	TGG	TCT	AGA	AAC	AAG
AAC	CGA	AAC	GTG	GAT	GTC	CTC	TGA	AGG	CGC	TTG	GAA	ACA	AAT

Fig. 5.2 Sample codon from the genome sequence

nucleotides that specifies one amino acid—as shown in Fig. 5.2. Furthermore, our dataset is then discretized and labeled into an acceptable format for the NB algorithm training as shown in Table 5.1. All the PCS were labeled as 1 and the NCS labeled as 0.

5.3.3 *Our Adaptive NBML Algorithm*

Given that there are k possible classes of sequence $C = \{c_1, c_2, c_3, \dots, c_k\}$ in a genome of sequence $S = \{s_1, s_2, s_3, \dots, s_n\}$, if $T = \{t_1, t_2, \dots, t_m\}$ be a set of unique codons appear once or more times in the genome sequence S, then the probability of a genome sequence s to be in class c can be determined by using the Bayes rules as shown in Eq. (5.1):

$$P(c|s) = \frac{P(c)P(s|c)}{P(s)} \tag{5.1}$$

where $P(s)$ is a constant for the known genome sequence size, and not often calculated for a maximal a-posteriori estimation problem in Bayesian statistics.

Therefore, with NB, each codon t_j in the sequence independently occurs given the class c. Consequently, Eq. (5.1) can be written as:

Table 5.1 Sample of the discretized codon dataset

Non-coding sequence	Coding sequence
0:4040606213	1:5712918261
0:0246318452	1:6262833345
0:3360314114	1:5284441412
0:1530381463	1:1820556143
0:6164240363	1:2477614054
0:4427103617	1:0922504348
0:8422019134	1:5048316256
0:1516138151	1:4043244542
0:1084729423	1:0423443294
0:2753581161	1:2271510274
0:5622572926	1:3102481040
0:2662476073	1:2148195345
0:4481546140	1:5022218083
0:1262163614	1:3173850566
0:4551927723	1:8341544502
0:2777332224	1:2543930264
0:2633324234	1:9503915364
0:2425431606	1:4221634270
0:3403755631	1:1529422812
0:3414449373	1:4137910942

$$P(c|s) \propto \left\{ P(c) \prod_{j=1}^{n_s} \left[P(t_j|c) \right]^{f_j} \right\} \tag{5.2}$$

where

- n_s is represent the figure of unique codons in the genome sequence s, and
- f_j is the frequency of each codon t_j.

We adopt an analogous of Eq. (5.2) as shown in Eq. (5.3) to circumvent a floating point underflow error that can occur using Eq. (5.2):

$$\log P(c|s) \propto \left\{ \log P(c) + \sum_{j=1}^{n_s} \left[f_j \log P(t_j|c) \right] \right\} \tag{5.3}$$

If the class of genome sequence c^* is the class maximizing $\log P(c|s)$ in Eq. (5.3), then c^* can be described by Eq. (5.4):

$$c^* = \text{argmax}_{c \in \mathbb{C}} \left\{ \log P(c) + \sum_{j=1}^{n_s} [f_j \log P(t_j|c)] \right\} \tag{5.4}$$

Therefore, using NB classifiers, we can estimate the probabilities of $P(c)$ and $P(t_j|c)$ by Eqs. (5.5) and (5.6), respectively;

$$\widehat{P}(c) = \frac{W_c}{W} \tag{5.5}$$

such that

- W_c represents the amount of the sequence in class c,
- W gives the sum of genome sequence, and

Note that $\widehat{P}(c)$ is a constant because the numbers of protein-coding sequence and of non-coding sequence are the same.

$$\widehat{P}(t_j|c) = \frac{W_{t_j}}{\sum_{t_i} \in \mathbb{T} W_{t_i}} \tag{5.6}$$

such that W_{t_i} is the frequency of a codon t_i in a class. The genome sequence can be classified into two classes:

- class 1 for the protein-coding genes, and
- class 0 for the non-coding genes.

Here, $P(t_j|c)$ returns the frequency of codon t_j in all sequence in c.

Algorithm 5.1 shows the pseudocode for our Spark-based adaptive NBML algorithm, which identifies protein-coding genes in genomes. Besides, an overview of our algorithm is as follows:

1. submission of the discretized and labelled training data through the master node
2. slit and parallelization of the submitted and discretized data to all worker nodes
3. training of slitted and parallelized data on the worker nodes in coordination with the master node
4. aggregation of the training results from worker nodes to the master node to obtain a predictive model
5. usage of the predictive model on test data.

5.4 Evaluation Results and Discussion

Our algorithm for predicting protein-coding genes from eukaryotic organism genome was evaluated using 80% of the sequence from the GRCh37 and GRCh38 dataset

Algorithm 5.1: Adaptive Spark-based Naive Bayes Model

Input: S[Codon(T_i), Label(C_i)]
Output: (Predictive Model)
start Spark Cluster;
initialization;
`Function` (*Parallelize(S) on all* **worker_nodes**);
while *Sequence S is on each worker_node*$[W_n]$ **do**
  for all *set* T_i, C_i *in the sequence S* **do**
    map(t_i, c_i);
  if *map == True on each worker_node*$[W_n]$ **then**
    call Procedure NBayes.train(*data S*);
    compare $[t, c]$;
    for *t, c in enum(label)* **do**
      check *T label*$[\{t\}$ == predict $\{k\}]$;
      append *T*;
  if *predict == PositiveCodon* **then**
    reduce {Accuracy ← $k\times$ compare.count};
  else
    Codon ← false;
collect output {**for all** Wn, cluster};
save model(*Master Node*);
call function NB.test (*model, test data*);
end Program;
stop Spark Cluster;

respectively as the training set and 20% as the testing set. Our algorithm was implemented in Python on a standalone cluster of Apache Spark version 2.4.0. The cluster consists of:

- one master node, which contains a 3.33×4 GHz processor and a two dual Intel core i5 with 16 GB of RAM; and
- four worker nodes, each of which contains a processor of 4.2 GHz, 56 cores with 125 GB of RAM;

for a total of five computing nodes. Our Spark framework was configured on Ubuntu-18.10 and 64-bit operating system.

Genes that were correctly predicted as a member of a known gene are referred to as *true positives (TP)*, and those that were newly and accurately predicted are called *false positives (FP)*. Genes that were wrongly predicted as non-gene sequences are referred to *false negatives (FN)*, and those that were correctly predicted as non-gene sequences are called *true negatives (TN)*. With these basic performance metrics (TP, FP, FN, and TN), it is encouraging to observe from Fig. 5.3a, d that the numbers of both TP and TN increase when the number of sequences in the dataset increases. It is also encouraging to observe from Fig. 5.3b, c that the numbers of both FP and FN decrease when the number of sequences in the dataset increases.

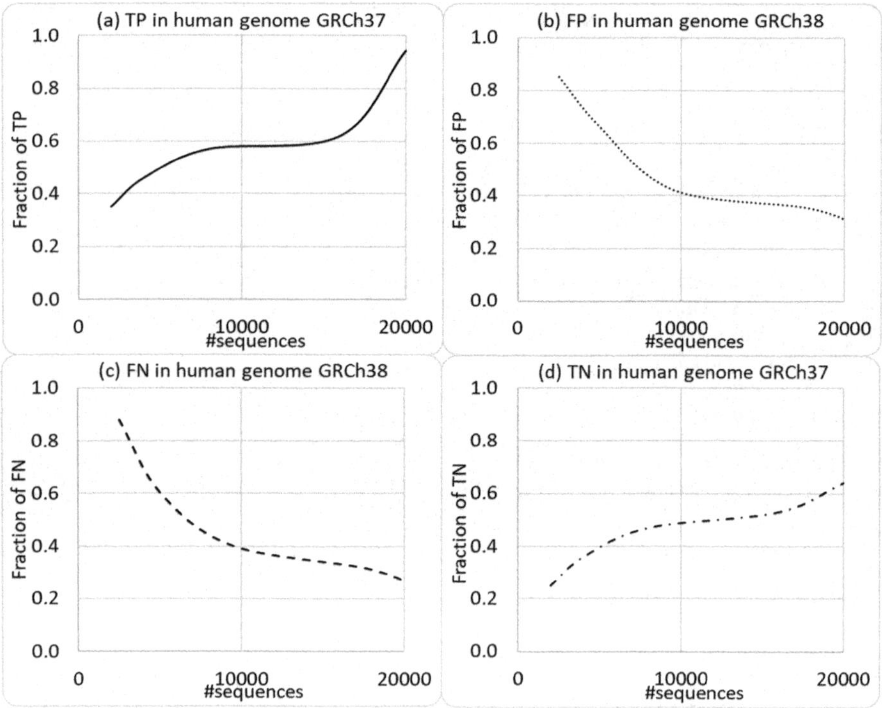

Fig. 5.3 Evaluation results: **a** TP, **b** FP, **c** FN, and **d** TN

With the aforementioned four basic performance metrics (TP, FP, FN, and TN), standard performance metrics (e.g., sensitivity, specificity, precision, accuracy) can be derived to quantify the performance of our algorithm. To elaborate, sensitivity measures the percentage of the coding sequence in the genome that is accurately predicted as the coding sequence:

$$\text{Sensitivity} = \frac{\text{TP}}{\text{TP} + \text{FN}} \tag{5.7}$$

Specificity measures the percentage of the non-coding sequence in the genome that is correctly predicted as a non-coding sequence:

$$\text{Specificity} = \frac{\text{TN}}{\text{TN} + \text{FP}} \tag{5.8}$$

Precision describes the number of the correctly predicted coding and non-coding sequences that actually turned out to be correct:

$$\text{Precision} = \frac{\text{TP}}{\text{TP} + \text{FP}} \tag{5.9}$$

Table 5.2 Performance evaluation of our predictive model

Evaluation metrics	GRCh37 (%)	GRCh38 (%)
Specificity	94.01	92.05
Sensitivity	81.52	78.09
Precision	95.04	96.04
Accuracy	96.02	97.08

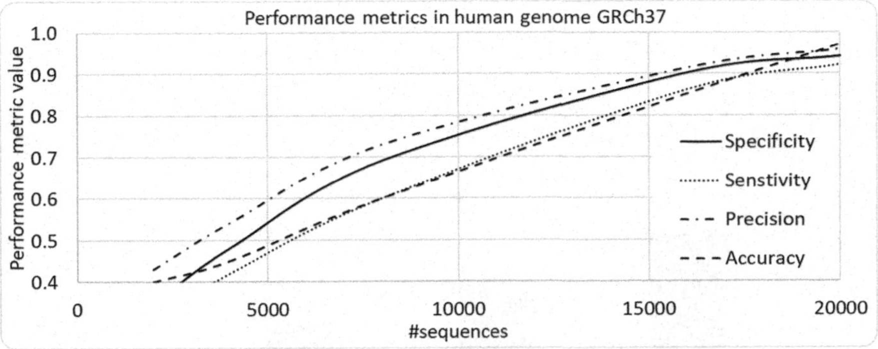

Fig. 5.4 Evaluation results: specificity, sensitivity, precision, accuracy on GRCh37

Finally, accuracy measures the overall correct predictions in the genome:

$$\text{Accuracy} = \frac{\text{TP} + \text{TN}}{\text{TP} + \text{TN} + \text{FP} + \text{FN}} \tag{5.10}$$

Table 5.2 shows the performance evaluation result of our adaptive naive Bayes model in predicting coding and non-coding genes from GRCh 37 and GRCh 38 genome sequences. Recall from Fig. 5.3 that, when the number of sequences in the dataset increases, both TP and TN increase, but both FP and FN decrease. Hence, it is logical for Fig. 5.4 to show that all the four aforementioned standard performance evaluation metrics—namely, specificity, sensitivity, precision, and accuracy—of our NBML algorithm increase when the number of GRCh 37 genome sequences in the dataset increases. Each point along the four curves on these figures is a computed average of six runs of the algorithm. Evaluation results demonstrate that our algorithm performs better with larger datasets. For instance, when there are more than 15,000 sequences in the datasets, all these performance metrics achieve values over 0.8 (i.e., over 80%). Similarly, Fig. 5.5 shows that the accuracy of our NBML algorithm also increases when the number of GRCh 38 genome sequences in the dataset increases.

In addition, we also demonstrated the scalability of our NBML algorithm in Fig. 5.6. The figure shows that, when the number of worker nodes in the cluster increases, our algorithm requires shorter runtime.

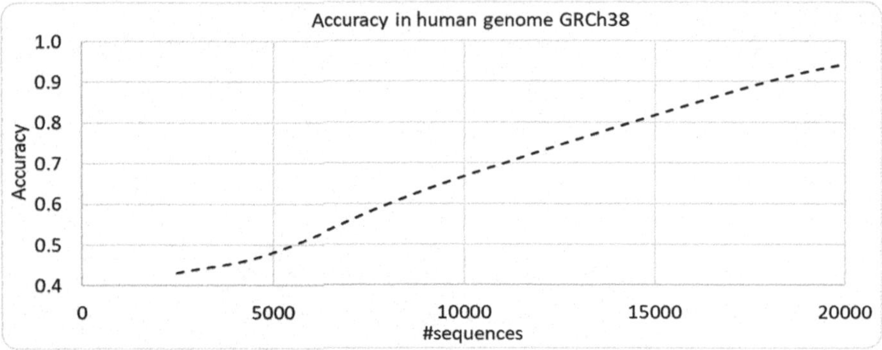

Fig. 5.5 Evaluation results: accuracy on GRCh38

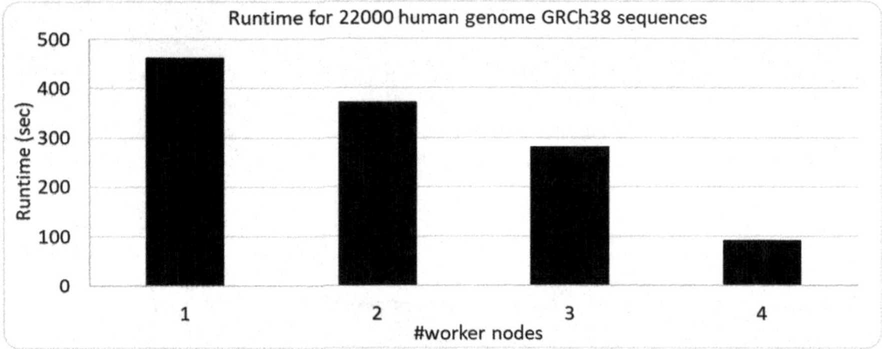

Fig. 5.6 Evaluation results: scalability on GRCh38

To evaluate our NBML algorithm with related works, Fig. 5.7 shows a comparative analysis in terms of the sensitivity, specificity and accuracy of the result of our adaptive NBML algorithm with the results obtained using FGENEH [63] and GRAIL II [67] algorithms. Note that the FGENES is a species-specific gene prediction or estimation program developed using the Viterbi algorithm, and the GRAIL II algorithm was developed for gene prediction using the neural network model. Results show that our adaptive NBML algorithm performed better than FGENEH and GRAIL II algorithms in terms of specificity, sensitivity, and accuracy.

5.5 Conclusions

Massive amounts of a vast variety of very valuable data can be easily generated and collected from a wide range of rich data sources of different levels of veracity at a high speed. In genomics, the use of next-generation sequencings (NGS) technolo-

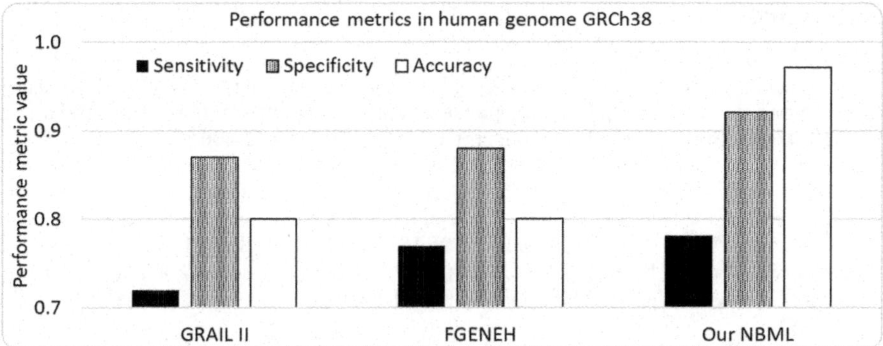

Fig. 5.7 Evaluation results: comparative analysis of related works (e.g., GRAIL II [67], FGENEH [63]) with our NBML on GRCh38

gies has reduced the sequencing time tremendously and has led to availability of petabytes of complete sequenced genomes for eukaryotic and prokaryotic organisms in several public repositories. These copious genomic datasets have introduced new challenges for researchers and demands for improved computational approaches and methods for some bioinformatics tasks. In this book chapter, we presented a Spark-based adaptive naive Bayes machine learning (NBML) algorithm to predict efficiently protein-coding genes in the genome of eukaryotic organisms. Moreover, evaluation of our algorithm on datasets—specifically, human genome GRCh 37 and GRCh 38—-shows that our algorithm is accurate, sensitive, specific, and scalable. Moreover, we leverage the Spark fault tolerance [104] capacity is highly efficient for big data analytics computations, we further enhance our adaptive algorithm by taking advantages of the inherent Spark fault tolerance ability, which allows Spark actions to degrade gracefully without depleting information in the case of nodes failure during a computing process. As ongoing and future work, we enhance our adaptive NBML algorithm by using the knowledge of a protein-coding genes in a previous genomic dataset to identify a set of protein-coding in a new genomic data, i.e., via transfer learning. Moreover, we also adapt ensemble machine learning approach—which combines the strength of at least two machine learning algorithms—to develop a more sturdy gene prediction algorithm with further enhanced performance.

Acknowledgements This project is partially supported by (a) Association of Commonwealth Universities (ACU), (b) Natural Sciences and Engineering Research Council of Canada (NSERC), and (c) University of Manitoba.

References

1. Abbasi, O., Rostami, A., Karimian, G.: Identification of exonic regions in DNA sequences using cross-correlation and noise suppression by discrete wavelet transform. BMC Bioinform. **12,** 430:1–430:14 (2011). https://doi.org/10.1186/1471-2105-12-430
2. Ahn, S., Couture, S.V., Cuzzocrea, A., Dam, K., Grasso, G.M., Leung, C.K., Kaleigh L. McCormick, Bryan H. Wodi: A fuzzy logic based machine learning tool for supporting big data business analytics in complex artificial intelligence environments. FUZZ-IEEE 2019, 1259–1264 (2019). https://doi.org/10.1109/FUZZ-IEEE.2019.8858791
3. Alaee, S., Kamgar, K., Keogh, E.J.: Matrix profile XXII: exact discovery of time series motifs under DTW. IEEE ICDM **2020**, 900–905 (2020). https://doi.org/10.1109/ICDM50108.2020.00099
4. Alam, M.T., Ahmed, C.F., Samiullah, M., Leung, C.K.: Discriminating frequent pattern based supervised graph embedding for classification. PAKDD 2021 Part II, 16–28 (2021). https://doi.org/10.1007/978-3-030-75765-6_2
5. Anaissi, A., Goyal, M., Catchpoole, D.R., Braytee, A., Kennedy, P.J.: Ensemble feature learning of genomic data using support vector machine, PLOS ONE **11**(6), e0157330:1–e0157330:17 (2016). https://doi.org/10.1371/journal.pone.0157330
6. Awe, O.I., Makolo, A., Fatumo, S.: Computational prediction of protein-coding regions in human transcriptomes: an application to the elderly. IREHI **2017**, 29–32 (2017). https://doi.org/10.1109/IREEHI.2017.8350465
7. Bandyopadhyay, S., Maulik, U., Roy, D.: Gene identification: classical and computational intelligence approaches. IEEE TSMCC **38**(1), 55–68 (2008). https://doi.org/10.1109/TSMCC.2007.906066
8. Bauckhage, C., Drachen, A., Sifa, R.: Clustering game behavior data. IEEE TCIAIG **7**(3), 266–278 (2015). https://doi.org/10.1109/TCIAIG.2014.2376982
9. Benchaira, K., Bitam, S., Mellouk, A., Tahri, A., Okbi, R.: AfibPred: a novel atrial fibrillation prediction approach based on short single-lead ECG using deep transfer knowledge. BDIoT 2019, 26:1–26:6 (2019). https://doi.org/10.1145/3372938.3372964
10. Birney, E., Durbin, R.: Using GeneWise in the Drosophila annotation experiment. Gen. Res. **10**(4), 547–548 (2000). https://doi.org/10.1101/gr.10.4.547
11. Boateng, E.Y., Oduro, F.T.: Predicting microfinance credit default: a study of Nsoatreman Rural Bank Ghana. J. Adv. Math. Comput. Sci. **26**(1), 33569:1–33569:9 (2018). https://doi.org/10.9734/JAMCS/2018/33569
12. Braun, P., Cuzzocrea, A., Keding, T.D., Leung, C.K., Pazdor, A.G.M., Sayson, D.: Game data mining: clustering and visualization of online game data in cyber-physical worlds. Proc. Comput. Sci. **112**, 2259–2268 (2017). https://doi.org/10.1016/j.procs.2017.08.141
13. Brown, J.A., Cuzzocrea, A., Kresta, M., Kristjanson, K.D.L., Leung, C.K., Tebinka, T.W.: A machine learning system for supporting advanced knowledge discovery from chess game data. IEEE ICMLA **2017**, 649–654 (2017). https://doi.org/10.1109/ICMLA.2017.00-87
14. Burge, C., Karlin, S.: Prediction of complete gene structures in human genomic DNA. J. Mole. Biol. **268**(1), 78–94 (1997). https://doi.org/10.1006/jmbi.1997.0951
15. Chalmers, E., Contreras, E.B., Robertson, B., Luczak, A., Gruber, A.: Learning to predict consequences as a method of knowledge transfer in reinforcement learning. IEEE TNNLS **29**(6), 2259–2270 (2018). https://doi.org/10.1109/TNNLS.2017.2690910
16. Chanda, A.K., Ahmed, C.F., Samiullah, M., Leung, C.K.: A new framework for mining weighted periodic patterns in time series databases. ESWA **79**, 207–224 (2017). https://doi.org/10.1016/j.eswa.2017.02.028
17. Chen, Y., Leung, C.K., Shang, S., Wen, Q.: Temporal data analytics on COVID-19 data with ubiquitous computing. IEEE ISPA-BDCloud-SocialCom-SustainCom 2020, 958–965 (2020). https://doi.org/10.1109/ISPA-BDCloud-SocialCom-SustainCom51426.2020.00146
18. Cheng, J.: Machine Learning Algorithms for Protein Structure Prediction. University of California, Irvine, USA (2007). PhD thesis

19. Cheng, J., Tegge, A.N., Baldi, P.: Machine learning methods for protein structure prediction. IEEE RBME **1**, 41–49 (2008). https://doi.org/10.1109/RBME.2008.2008239
20. Choudhary, R., Gianey, H.K.: Comprehensive review on supervised machine learning algorithms. MLDS **2017**, 37–43 (2017). https://doi.org/10.1109/MLDS.2017.11
21. Claverie, J.: Computational methods for the identification of genes in vertebrate, genomic sequences. Human Mole. Gen. **6**(10), 1735–1744 (1997). https://doi.org/10.1093/hmg/6.10.1735
22. Cuong, P., Binh, K., Tran, N.T.: A high-performance FPGA-based BWA-MEM DNA sequence alignment. CCPE **33**(2) (2021). https://doi.org/10.1002/cpe.5328
23. Dada, E.G., Bassi, J.S, Chiroma, H., Abdulhamid, S.M, Adetunmbi, A.O, Ajibuwa, O.E.: Machine learning for email spam filtering: review, approaches and open research problems. Heliyon **5**(6), e01802:1–e01802:23 (2019). https://doi.org/10.1016/j.heliyon.2019.e01802
24. Daemen, A., Gevaert, O., De Moor, B.: Integration of clinical and microarray data with kernel methods. IEEE EMBS **2007**, 5411–5415 (2007). https://doi.org/10.1109/IEMBS.2007.4353566
25. Dai, W., Xue, G., Yang, Q., Yu, Y.: Transferring naive Bayes classifiers for text classification. AAAI **2007**, 540–545 (2007)
26. De Guia, J., Devaraj, M., Leung, C.K.: DeepGx: deep learning using gene expression for cancer classification. IEEE/ACM ASONAM **2019**, 913–920 (2019). https://doi.org/10.1145/3341161.3343516
27. De Vries, et al.: Effect of a comprehensive surgical system on patient outcomes. New England J. Med. **363**(20), 1928–1937 (2010). https://doi.org/10.1056/nejmsa0911535
28. Dierckens, K.E., Harrison, A.B., Leung, C.K., Pind, A.V.: A data science and engineering solution for fast k-means clustering of big data. IEEE TrustCom-BigDataSE-ICESS **2017**, 925–932 (2017). https://doi.org/10.1109/Trustcom/BigDataSE/ICESS.2017.332
29. Do, J.H., Choi, D.K.: Computational approaches to gene prediction. J. Microbiol. **44**(2), 137–144 (2006)
30. Domeniconi, G., Masseroli, M., Moro, G., Pinoli, P.: Cross-organism learning method to discover new gene functionalities. Comput. Methods Progr. Biomed. **12**, 20–34 (2016). https://doi.org/10.1016/j.cmpb.2015.12.002
31. Ekblom, R., Wolf, J.B.: A field guide to whole-genome sequencing, assembly and annotation. Evol. Appl. **7**(9), 1026–1042 (2014). https://doi.org/10.1111/eva.12178
32. Fariha, A., Ahmed, C.F., Leung, C.K., Abdullah, S.M., Cao, L.: Mining frequent patterns from human interactions in meetings using directed acyclic graphs. PAKDD 2013, Part I, 38–49 (2013). https://doi.org/10.1007/978-3-642-37453-1_4
33. Galpert, D., del Río, S., Herrera, F., Ancede-Gallardo, E., Antunes, A., Agüero-Chapin, G.: An effective big data supervised imbalanced classification approach for ortholog detection in related yeast species. BioMed. Res. Int. **2015**, 748681:1–748681:12 (2015). https://doi.org/10.1155/2015/748681
34. Gelfand, M.S.: Gene recognition via spliced sequence alignment. PNAS **93**(17), 9061–9066 (1996). https://doi.org/10.1073/pnas.93.17.9061
35. Gross, T., Faull, J., Ketteridge, S., Springham, D.: Eukaryotic microorganisms. In: Introductory Microbiology, pp. 241–286 (1995). https://doi.org/10.1007/978-1-4899-7194-4_9
36. Guigo, R., Agarwal, P., Abril, J.F., Burset, M., Fickett, J.W.: An assessment of gene prediction accuracy in large DNA sequences. Genome Res. **10**(10), 1631–1642 (2000). https://doi.org/10.1101/gr.122800
37. Gunawan, T.S., Epps, J., Ambikairajah, E.: Boosting approach to exon detection in DNA sequences. Electron. Lett. **44**(4), 323–324 (2008). https://doi.org/10.1049/el:20082343
38. Heidema, A.G., Boer, J.M.A., Nagelkerke, N., Mariman, E.C.M., van der A, D.L., Feskens, E.J.M.: The challenge for genetic epidemiologists: how to analyze large number of SNPs in relation to complex diseases. BMC Gen. **7**, 23:1–23:15 (2006). https://doi.org/10.1186/1471-2156-7-23
39. Holmes, G., Pfahringer, G., Kirkby, B., Frank, R., Hall, E.M.: Multiclass alternating decision trees. ECML **2002**, 161–172 (2002). https://doi.org/10.1007/3-540-36755-1_14

40. Jiang, F., Leung, C.K.: A data analytic algorithm for managing, querying, and processing uncertain big data in cloud environments. Algorithms **8**(4), 1175–1194 (2015). https://doi.org/10.3390/a8041175
41. Jiang, F., Leung, C.K., Sarumi, O.A., Zhang, C.Y.: Mining sequential patterns from uncertain big DNA in the Spark framework. IEEE BIBM, 874–88 (2016). https://doi.org/10.1109/BIBM.2016.7822641
42. Kan, Z., Rouchka, E.C., Gish, W.R., States, D.J.: Gene structure prediction and alternative splicing analysis using genomically aligned ESTs. Genome Res. **11**(5), 889–900 (2001). https://doi.org/10.1101/gr.155001
43. Kaya, M., Sarhan, A., Alhajj, R.: Multiple sequence alignment with affine gap by using multi-objective genetic algorithm. Comput. Methods Programs Biomed. **114**(1), 38–49 (2014). https://doi.org/10.1016/j.cmpb.2014.01.013
44. Kobusinska, A., Leung, C.K., Hsu, C., Raghavendra, S., Chang, V.: Emerging trends, issues and challenges in Internet of Things, big data and cloud computing. FGCS **87**, 416–419 (2018). https://doi.org/10.1016/j.future.2018.05.021
45. Le, D.H., Xuan, H.N., Kwon, Y.K.: A comparative study of classification-based machine learning methods for novel disease gene prediction. KSE 2014, 577–588 (2015). https://doi.org/10.1007/978-3-319-11680-8_46
46. Lee, R.C., Cuzzocrea, A., Lee, W., Leung, C.K.: An innovative majority voting mechanism in interactive social network clustering. ACM WIMS 2017, 14:1–14:10 (2017). https://doi.org/10.1145/3102254.3102268
47. Leung, C.K.: Big data analysis and mining. In: Advanced Methodologies and Technologies in Network Architecture, Mobile Computing, and Data Analytics, pp. 15–27 (2019). https://doi.org/10.4018/978-1-5225-7598-6.ch002
48. Leung, C.K.: Uncertain frequent pattern mining. In: Frequent Pattern Mining, pp. 417–453 (2014). https://doi.org/10.1007/978-3-319-07821-2_14
49. Leung, C.K., Braun, P., Cuzzocrea, A.: AI-based sensor information fusion for supporting deep supervised learning. Sensors **19**(6), 1345:1–1345:12 (2019).https://doi.org/10.3390/s19061345
50. Leung, C.K., Braun, P., Pazdor, A.G.M.: Effective classification of ground transportation modes for urban data mining in smart cities. DaWaK **2018**, 83–97 (2018). https://doi.org/10.1007/978-3-319-98539-8_7
51. Leung, C.K., Carmichael, C.L.: FpVAT: a visual analytic tool for supporting frequent pattern mining. ACM SIGKDD Explorations **11**(2), 39–48 (2009). https://doi.org/10.1145/1809400.1809407
52. Leung, C.K., Chen, Y., Hoi, C.S.H., Shang, S., Cuzzocrea, A.: Machine learning and OLAP on big COVID-19 data. IEEE BigData **2020**, 5118–5127 (2020). https://doi.org/10.1109/BigData50022.2020.9378407
53. Leung, C.K., Chen, Y., Hoi, C.S.H., Shang, S., Wen, Y., Cuzzocrea, A.: Big data visualization and visual analytics of COVID-19 data. IV 2020, 415–420 (2020). https://doi.org/10.1109/IV51561.2020.00073
54. Leung, C.K., Chen, Y., Shang, S., Deng, D.: Big data science on COVID-19 data. IEEE BigDataSE **2020**, 14–21 (2020). https://doi.org/10.1109/BigDataSE50710.2020.00010
55. Leung, C.K., Cuzzocrea, A., Mai, J.J., Deng, D., Jiang, F.: Personalized DeepInf: enhanced social influence prediction with deep learning and transfer learning. IEEE BigData **2019**, 2871–2880 (2019). https://doi.org/10.1109/BigData47090.2019.9005969
56. Leung, C.K., Elias, J.D., Minuk, S.M., de Jesus, A.R.R., Cuzzocrea, A.: An innovative fuzzy logic-based machine learning algorithm for supporting predictive analytics on big transportation data. FUZZ-IEEE **2020**, 1905–1912 (2020). https://doi.org/10.1109/FUZZ48607.2020.9177823
57. Leung, C.K., Jiang, F., Zhang, Y.: Explainable machine learning and mining of influential patterns from sparse web. IEEE/WIC/ACM WI-IAT 2020 (2020)
58. Leung, C.K., MacKinnon, R.K., Wang, Y.: A machine learning approach for stock price prediction. IDEAS **2014**, 274–277 (2014). https://doi.org/10.1145/2628194.2628211

59. Leung, C.K., Sarumi, O.A., Zhang, C.Y.: Predictive analytics on genomic data with high-performance computing. IEEE BIBM **2020**, 2187–2194 (2020). https://doi.org/10.1109/BIBM49941.2020.9312982
60. Lim, H., Xie, L.: A new weighted imputed neighborhood-regularized tri-factorization one-class collaborative filtering algorithm: application to target gene prediction of transcription factors. IEEE/ACM TCBB **18**(1), 126–137 (2021). https://doi.org/10.1109/TCBB.2020.2968442
61. Liu, B., Blasch, E., Chen, Y., Shen, D., Chen, G.: Scalable sentiment classification for big data analysis using naive Bayes classifier. IEEE BigData **2013**, 99–104 (2013). https://doi.org/10.1109/BigData.2013.6691740
62. MacKinnon, R.K., Leung, C.K.: Stock price prediction in undirected graphs using a structural support vector machine. IEEE/WIC/ACM WI-IAT **2015**, 548–555 (2015). https://doi.org/10.1109/WI-IAT.2015.189
63. Maji, S., Garg, D.: Progress in gene prediction: principles and challenges. Curr. Bioinform. **8**(2), 226–243 (2013). https://doi.org/10.2174/1574893611308020011
64. Margulis, L.: The classification and evolution of prokaryotes and eukaryotes. In: Bacteria, Bacteriophages, and Fungi, pp. 1–41. (1974). https://doi.org/10.1007/978-1-4899-1710-2_1
65. Martins, P.V.L.: Gene Prediction Using Deep Learning. Master's dissertation, University of Porto, Portugal (2018). https://repositorio-aberto.up.pt/handle/10216/114372
66. Mathe, C., Sagot, M., Schiex, T., Rouze, P.: Current methods of gene prediction, their strengths and weaknesses. Nucleic Acids Res. **30**(19), 4103–4117 (2002). https://doi.org/10.1093/nar/gkf543
67. McElwain, M.: A Critical Review of Gene Prediction Software. BIOC 218 final paper, Stanford University, USA (2007)
68. Meisler, M.H.: Evolutionarily conserved noncoding DNA in the human genome: how much and what for? Genome Res. **11**(10), 1617–1618 (2000). https://doi.org/10.1101/gr.211401
69. Meyer, M., Durbin, R.: Comparative ab initio prediction of gene structures using pair HMMs. Bioinformatics **18**(10), 1309–1318 (2002). https://doi.org/10.1093/bioinformatics/18.10.1309
70. Miao, Y., Jiang, H., Liu, H., Yao, Y.: An Alzheimers disease related genes identification method based on multiple classifier integration. Comput. Methods Programs Biomed. **150**, 107–115 (2017). https://doi.org/10.1016/j.cmpb.2017.08.006
71. Mignone, F.: Computational identification of protein coding potential of conserved sequence tags through cross-species evolutionary analysis. Nucleic Acids Res. **31**(15), 4639–4645 (2003). https://doi.org/10.1093/nar/gkg483
72. Min, B., Oh, H., Ryu, G., Choi, S.H., Leung, C.K., Yoo, K.: Image classification for agricultural products using transfer learning. BigDAS **2020**, 48–52 (2020)
73. Min, X.J., Butler, G., Storms, R., Sang, A.T.: OrfPredictor: predicting protein-coding regions in EST-derived sequences. Nucleic Acids Res. **33**, W677–W680 (2005). https://doi.org/10.1093/nar/gki394
74. Morris, K.J., Egan, S.D., Linsangan, J.L., Leung, C.K., Cuzzocrea, A., Hoi, C.S.H.: Hoi: Token-based adaptive time-series prediction by ensembling linear and non-linear estimators: a machine learning approach for predictive analytics on big stock data. IEEE ICMLA **2018**, 1486–1491 (2018). https://doi.org/10.1109/ICMLA.2018.00242
75. Nagaraj, K., Sharvani, G.S., Sridhar, A.: Emerging trend of big data analytics in bioinformatics: a literature review. IJBRA **14**(1–2), 144–205 (2018). https://doi.org/10.1504/IJBRA.2018.089175
76. Olson, R.S., La Cava, W., Mustahsan, Z., Varik, A., Moore, J.H.: Data-driven advice for applying machine learning to bioinformatics problems. Biocomputing **2018**, 192–203 (2018). https://doi.org/10.1142/9789813235533_0018
77. Palit, I., Reddy, C.K., Schwartz, K.L.: Differential predictive modeling for racial disparities in breast cancer. IEEE BIBM **2009**, 239–245 (2009). https://doi.org/10.1109/BIBM.2009.89
78. Parmar, B.S., Mehta, M.A: Computer-aided diagnosis of thyroid dysfunction: a survey. BDA 2020, 164–189 (2020). https://doi.org/10.1007/978-3-030-66665-1_12

79. Patelia, V., Patel, M.S.: Brain computer interface: applications and P300 Speller overview. ICCCNT **2019**, 2129–2133 (2019). https://doi.org/10.1109/ICCCNT45670.2019.8944461
80. Pawliszak, T., Chua, M., Leung, C.K., Tremblay-Savard, O.: Operon-based approach for the inference of rRNA and tRNA evolutionary histories in bacteria. BMC Gen. **21**(Supplement 2), 252:1–252:14 (2020). https://doi.org/10.1186/s12864-020-6612-2
81. Peralta, D., del Río, S., Ramírez-Gallego, S., Triguero, I., Benitez, J.M., Herrera, F.: Evolutionary feature selection for big data classification: a MapReduce approach. Math. Probl. Eng. **2015**, 246139:1–246139:11 (2015). https://doi.org/10.1155/2015/246139
82. Picardi, E., Pesole, G.: Computational methods for ab initio and comparative gene finding. In: Data Mining Techniques for the Life Sciences, pp. 269–284 (2010). https://doi.org/10.1007/978-1-60327-241-4_16
83. Quinlan, J.R.: Decision trees and decision-making. IEEE TSMC **20**(2), 339–346 (1990). https://doi.org/10.1109/21.52545
84. Sacar, D., Allmer, J.: Machine learning methods for microRNA gene prediction. Methods Mol. Biol. **1107**, 177–187 (2014). https://doi.org/10.1007/978-1-62703-748-8_10
85. Sarumi, O.A., Leung, C.K.: Exploiting anti-monotonic constraints for mining palindromic motifs from big genomic data. IEEE BigData **2019**, 4864–4873 (2019). https://doi.org/10.1109/BigData47090.2019.9006397
86. Sarumi, O.A., Leung, C.K.: Scalable data science and machine learning algorithm for gene prediction. BigDAS **2019**, 118–126 (2019)
87. Sarumi, O.A., Leung, C.K., Adetunmbi, O.A.: Spark-based data analytics of sequence motifs in large omics data. Proc. Comput. Sci. **126**, 596–605 (2018). https://doi.org/10.1016/j.procs.2018.07.294
88. Schneider, H.W., Raiol, T., Brigido, M.M., Walter, M.E.M., Stadler, P.F.: A support vector machine based method to distinguish long non-coding RNAs from protein coding transcripts. BMC Gen. **18**(1), 804:1–804:14 (2017). https://doi.org/10.1186/s12864-017-4178-4
89. Shang, S., Chen, Y., Leung, C.K., Pazdor, A.G.M.: Spatial data science of COVID-19 data. IEEE HPCC-SmartCity-DSS 2020, 1370–1375 (2020). https://doi.org/10.1109/HPCC-SmartCity-DSS50907.2020.00177
90. She, R., Chu, J.S., Wang, K., Chen, N.: Fast and accurate gene prediction by decision tree classification. SIAM DM **2010**, 790–801 (2010). https://doi.org/10.1137/1.9781611972801.69
91. Shnorhavorian, M., Bittner, R., Wright, J.L., Schwartz, S.M.: Maternal risk factors for congenital urinary anomalies: results of a population-based case-control study. Urology **78**(5), 1156–1161 (2011). https://doi.org/10.1016/j.urology.2011.04.022
92. Singh, S.P., Leung, C.K., Hamilton, J.D.: Analytics of similar-sounding names from the web with phonetic based clustering. IEEE/WIC/ACM WI-IAT 2020 (2020)
93. Song, Y., Liu, C., Wang, Z.: A machine learning approach for accurate annotation of noncoding RNAs. IEEE/ACM TCBB **12**(3), 551–559 (2015). https://doi.org/10.1109/TCBB.2014.2366758
94. Souza, J., Leung, C.K., Cuzzocrea, A.: An innovative big data predictive analytics framework over hybrid big data sources with an application for disease analytics. AINA **2020**, 669–680 (2020). https://doi.org/10.1007/978-3-030-44041-1_59
95. Toivonen, J., Das, P.K., Taipale, J., Ukkonen, E.: MODER2: first-order Markov modeling and discovery of monomeric and dimeric binding motifs. Bioinformatics **36**(9), 2690–2696 (2020). https://doi.org/10.1093/bioinformatics/btaa045
96. van der Schaar, M., Alaa, A.M., Floto, R.A., Gimson, A., Scholtes, S., Wood, A.M., McKinney, E.F., Jarrett, D., Lió, P., Ercole, A.: How artificial intelligence and machine learning can help healthcare systems respond to COVID-19. Mach. Learn. **110**(1), 1–14 (2021). https://doi.org/10.1007/s10994-020-05928-x
97. Vanitha, C.D.A., Devaraj, D., Venkatesulu, M.: Gene expression data classification using support vector machine and mutual information-based gene selection. Proc. Comput. Sci. **47**, 13–21 (2015). https://doi.org/10.1016/j.procs.2015.03.178

98. Venter, J.C., et al.: The sequence of the human genome. Science **291**(5507), 1304–1351 (2001). https://doi.org/10.1126/science.1058040
99. Wang, Z., Chen, Y., Li, Y.: A brief review of computational gene prediction methods. Gen. Proteom. Bioinform. **2**(4), 216–221 (2004). https://doi.org/10.1016/s1672-0229(04)02028-5
100. Weiss, K., Khoshgoftaar, T.M., Wang, D.: A survey of transfer learning. J. Big Data **3**, 9:1–9:40 (2016). https://doi.org/10.1186/s40537-016-0043-6
101. Williams-DeVane, C.R., Reif, D.M., Cohen Hubal, E.C., Bushel, P.R., Hudgens, E.E., Gallagher, J.E., Edwards, S.W.: Decision tree-based method for integrating gene expression, demographic, and clinical data to determine disease endotypes. BMC Syst. Biol. **7**, 119:1–119:19 (2013). https://doi.org/10.1186/1752-0509-7-119
102. Wu, J.M., Srivastava, G., Jolfaei, A., Fournier-Viger, P., Lin, J.C.: Hiding sensitive information in eHealth datasets. FGCS **117**, 169–180 (2021). https://doi.org/10.1016/j.future.2020.11.026
103. Yassour, M., Kaplan, T., Fraser, H.B., Levin, J.Z., Pfiner, J., Adiconis, X., Schroth, G., Luo, S., Khrebtukova, I., Gnirke, A.: Ab initio construction of a eukaryotic transcriptome by massively parallel mRNA sequencing. PNAS **106**(9), 3264–3269 (2009). https://doi.org/10.1073/pnas.0812841106
104. Ying, C., Yu, J., He, J.: Towards fault tolerance optimization based on checkpoints of in-memory framework Spark. J. Ambient. Intell. Humaniz. Comput. (2018). https://doi.org/10.1007/s12652-018-1018-6
105. Yip, K.Y., Cheng C., Gerstein M.: Machine learning and genome annotation: a match meant to be? Gen. Biol. **14**(5), 205:1–205:10 (2013). https://doi.org/10.1186/gb-2013-14-5-205
106. Yu, N., Yu, Z., Li, B., Gu, F., Pan, Y.: A comprehensive review of emerging computational methods for gene identification. J. Inf. Process. Syst. **12**(1), 1–34 (2016). https://doi.org/10.3745/JIPS.04.0023
107. Zhang, C.T., Wang, J.: Recognition of protein coding genes in the yeast genome at better than 95% accuracy based on the Z curve. Nucleic Acids Res. **28**(14), 2804–2814 (2002). https://doi.org/10.1093/nar/28.14.2804
108. Zhang, X., Lu, X., Shi, Q., Xu, X-Q., Hon-chiu E.L., Harris, L.N., Iglehart, J.D., Miron, A., Liu, J.S., Wong, W.H.: Recursive SVM feature selection and sample classification for mass-spectrometry and microarray data. BMC Bioinform. **7**, 197:1–179:13 (2006). https://doi.org/10.1186/1471-2105-7-197

Chapter 6
Microscopic Analysis of Blood Cells for Disease Detection: A Review

Nilkanth Mukund Deshpande, Shilpa Shailesh Gite, and Rajanikanth Aluvalu

Abstract Any contamination in the human body can prompt changes in blood cell morphology and various parameters of cells. The minuscule images of blood cells are examined for recognizing the contamination inside the body with an expectation of maladies and variations from the norm. Appropriate segmentation of these cells makes the detection of a disease progressively exact and vigorous. Microscopic blood cell analysis is a critical activity in the pathological analysis. It highlights the investigation of appropriate malady after exact location followed by an order of abnormalities, which assumes an essential job in the analysis of various disorders, treatment arranging, and assessment of results of treatment. A survey on different areas where microscopic imaging of blood cells is used for disease detection is presented in this paper. A small note on Blood composition is presented, which is followed by a generalized methodology for microscopic blood image analysis for certain application of medical imaging. Comparison of existing methodologies proposed by researchers for disease detection using microscopic blood cell image analysis is discussed in this paper.

Keywords Blood cells · Microscopic images · Disease detection · Image processing · Red blood cells · White blood cells · Leukemia detection · Sickle cell

---

N. M. Deshpande
Department of Electronics and Telecommunication, Symbiosis Institute of Technology, Lavale, Pune 412115, India

Dr. Vithalrao Vikhe Patil College of Engineering, Ahmednagar, India

S. S. Gite (✉)
Department of Computer Science, Symbiosis Institute of Technology, Symbiosis Centre for Applied AI (SCAAI), Lavale, Pune, India
e-mail: shilpa.gite@sitpune.edu.in

N. M. Deshpande · S. S. Gite
Symbiosis International (Deemed University), Pune 412115, India

R. Aluvalu
Department of CSE, Vardhaman College of Engineering, Hyderabad, India

M. Mehta et al. (eds.), *Tracking and Preventing Diseases with Artificial Intelligence*, Intelligent Systems Reference Library 206,
https://doi.org/10.1007/978-3-030-76732-7_6

6.1 Introduction

6.1.1 Background

Normally health of any person is judged by the analysis of different features of blood cells and their counts. Previously manual methods of blood cells analysis were used by pathologists. This might cause error in disease prediction since manual methods are dependent on experience and skills of pathologists. Hence, it is proposed that an automated system of image processing be developed using different algorithms. Thus microscopic blood images could be analyzed for prediction and detection of particular diseases. A simplified, automated and cost effective method is required for detection of diseases. Thus the above components explained are analyzed for knowing health indication of human being and thereby detecting abnormalities related to health.

6.1.1.1 Blood and Its Composition

Blood, the most integral part of body is constituted of white blood cells (WBC), red blood cells (RBC), platelets and plasma. This can be further categorized as; cells and platelets are about 45% of human blood, whereas remaining 55% is filled by plasma (the yellow fluid in blood) [1, 2]. These components and their physical properties like size, shape, color and count in the whole blood, changes due to ingress of any foreign object or micro-organism that can lead to any sort of infections. There are different pathological procedures for detection of diseases [3]. In many cases, microscopic imaging plays a vital role in prediction and detection of abnormalities and occurrence of diseases within body.

Figure 6.1 shows the details of different blood components. Blood is made up of following elements- erythrocytes, known as red blood cells (RBC), leukocytes,

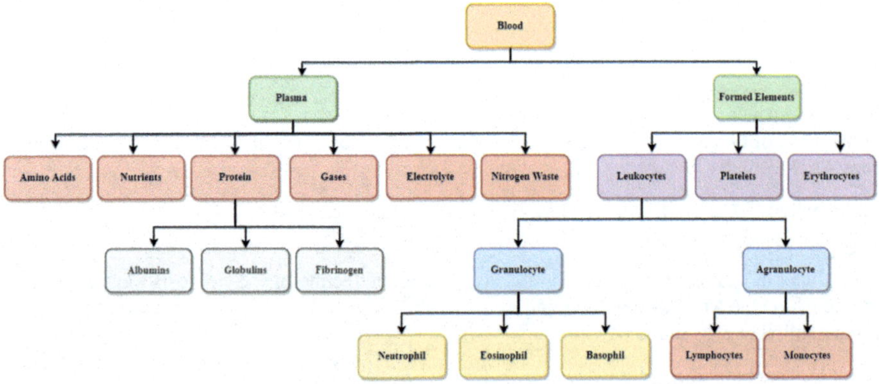

Fig. 6.1 Composition of blood. *Source* https://healthengine.com.au/info/blood-function-and-composition, assessed on 25th Sept. 2020 [4]

known as white blood cells (WBC) and platelets. These are combinedly present within the plasma membrane.

Leukocytes are further classified into two subcategories called granulocytes which consist of neutrophils, eosinophil and basophils and agranulocytes which consist of lymphocytes and monocytes. Blood plasma is a mixture of proteins, enzymes, nutrients, wastes, hormones and gases. Platelets are small fragments of bone marrow cells. The main function of red blood cells is to carry oxygen from lungs to different body organs. Carbon dioxide is carried back to the lungs, which will be exhaled afterwards. RBC count is the measure of different diseases in the human body. A low RBC count means anemia and high means polycythemia. White blood cells protect the body against infection. The different components of blood are identified to know about the health of a human being. Microscopic images of blood smear are analyzed for different disease detection.

6.1.1.2 Traditional Methods of Disease Detection

Disease detection is generally by two different ways traditionally. First is the detection through symptoms and second is through different tests. Routine symptoms of any disease include cough, fever, headache etc. Depending upon the prolonged symptoms, there is need to go for some tests those detect presence of some malady in the body. Different types of tests are shown below.

Imaging tests: Different imaging tests include X ray, computed tomography (CT) imaging, nuclear medicine, ultrasound, and microscopic imaging.

Chemical tests: Blood test and urine test.

In case of many diseases, microscopic analysis is preferred, that utilizes the blood cells.

6.1.1.3 Procedure of Microscopic Analysis of Blood

Trained pathologist collects the blood sample of a patient. The sample is to be collected carefully and the proper hygiene is to be taken in this process (Fig. 6.2).

For analysis of microscopic blood images, the blood film needs to be prepared. Glass slide is used for making of the blood film. For examination and analysis of this film under microscope, staining is required. Preparation of blood film requires a slide, a tube and a blood spreader. Generally weldge method is used for this purpose.

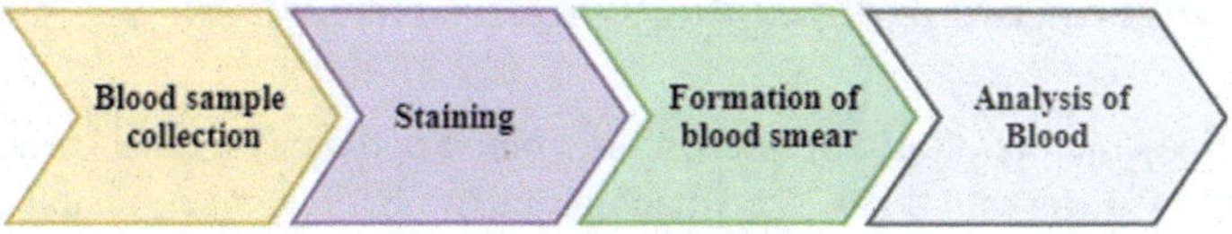

Fig. 6.2 Blood analysis procedure

On a base slide, a drop of blood is placed. A spreader slide is moved over this blood drop backwards to touch the blood to get the blood spread over the slide uniformly. To get perfection and accuracy in the smear, spreader slide should be inclined at an angle of 30°–45° to the blood base slide. Prepared blood smear is dried using air dryer and then staining is performed. Dried smear is fixed by absolute methanol or ethyl alcohol. Afterwards, it is stained using any of the staining methods—rewmanosky stain, leishmon stain, may-grawald giema or wright-giemsa stain, which differs with the liquid used for staining purpose. These stained slides are then used for analysis under microscope [1, 2, 5].

6.1.1.4 Open Source Datasets Available for Work

For the analysis of blood cells for detection and diagnosis of diseases there are different databases available. These databases include images of blood cells with blood cells of healthy subjects, infected cells, blast cells (in case of blood cancer), cells containing parasites and so on. Table 6.1 shows these different databases available for the work (Table 6.2).

6.2 Literature Review

6.2.1 Collection and Exclusion of Articles for Review

Google scholar platform is used for collection of different articles in the microscopic imaging area. Popular keywords such as white blood cell, red blood cell, machine learning, disease, deep learning, and image processing are used for the database searching. Large numbers of articles are obtained as a result of the search. Out of these, the articles those signify the unique contribution are shortlisted for writing the review. Generally articles utilizing the images processing and machine learning are considered. Articles from purely medical background are omitted from the review. Although for basic concepts related to blood and staining, the medical field articles are reviewed those added the correct conceptual interpretation of the basic terminologies related to the blood and different diseases.

6.2.2 Generalized Methodology of Disease Detection

A generalized methodology for microscopic blood cell analysis is shown in Fig. 6.3. It consists of different stages like image acquisition, image segmentation, feature extraction, and disease detection [56]. Blood sample is taken from the patient by a trained pathologist. After that, a slide is prepared to form a blood smear. The same

Table 6.1 Different open source databases of microscopic blood cells

Name		Image formats	Number of images	Color depth	Remark
BCCD database [6]		JPEG, xml, metadata	12,500	Not mentioned	Different sub-types of blood cells
ALL-IDB (acute lymphoblastic Leukemia Database [7–9])	ALL-IDB-1	JPEG	109 (510 lymphoblast)	24-bit, 2592 × 1944	Cancerous
	ALL-IDB-2	JPEG	260 (130 lymphoblast)	24-bit 257 × 257	Cancerous
Atlas of hematology by Mediros [10]		JPEG	300	Not mentioned	Visceral leishmaniasis, cellular similarity, morphologic similarities
ASH Image Bank [11]		JPEG	5084	Not mentioned	Cancerous and other different types of images
Leukocyte images for segmentation and classification (LISC)			400 (720 × 576)	Not mentioned	Healthy subjects with different sub-types of blood cells
C-NMC Dataset [12, 13]		BMP	15,135	Not mentioned	Normal and cancerous images of blood cells

slide is observed under the good quality microscope that will give an image. This image is taken either by camera directly or from an adaptor connected to a microscope. This image is considered for further analysis. Acquired images may have some unwanted regions and overlapping of different components of blood. This image is enhanced by applying a suitable image enhancement technique. So that, good quality image is now available for analysis. After pre-processing, separation of different components of blood is done which include separation of RBC, WBC, plasma, and platelets. Considering the generalized characteristics of blood components, segmentation is done. This will separate the region of interest for further classification. RBC, WBC and, other components are further classified into their respective subclasses. This will help to specify a particular sub-class image for extracting features of blood cells and depending upon the analysis in further stages such as classifier, detection of disease is done. After the segmentation, different features are extracted by considering different components of blood. Features include size, shape, color,

Table 6.2 Analysis of different applications of blood cells analysis

Application	WBC/RBC segmentation
References	[14–27]
Author opinion/potential of further research	In blood cell analysis, WBC and RBC segmentation is the major thrust. For diagnosis of different diseases, the morphology of these cells plays an important role. The segmentation is still progressing and there is a good potential of work in the segmentation of blood cells
Application	RBC/WBC counting
References	[14, 28–30]
Author opinion/potential of further research	Counting of RBC and WBC is the indication of different infections within the body. This process is generally less costly and is routinely done by equipment based analysis. Although for microscopic analysis computer aided framework is also been developed by many researchers. There is still potential in this area, as segmentation of different parts of the blood (RBC, WBC, and platelet) is still in the pipeline of improvement
Application	Anemia or Sickle Cell Detection
References	[31–35]
Author opinion/potential of further research	Anemia detection is primarily done with RBC counting. There are some shape changes in RBC also detects the anemia. In major cases.
Application	Malaria/dengue and other viral diseases detection
References	[36–39]
Author opinion/potential of further research	In viral diseases like malaria and dengue, the platelet count in blood comes into picture. Also there might be presence of the parasites due to these malady infections. Parasites detection is done by morphological analysis that is done with microscopic imaging. This work also has potential, as the amount of parasites and types of parasites can lead to severity of disease and will provide a distinct treatment guideline further
Application	Thalassemia detection
References	[40–42]
Author opinion/potential of further research	There are 12 different types of thalassemia depending upon the size and shapes of the RBC. In major cases, thalassemia is detected as infected cells and non-infected cell. There is still the potential in detection of this disease
Application	Leukemia detection
References	[7, 8, 10, 15, 26, 43–55]

(continued)

Table 6.2 (continued)

Application	WBC/RBC segmentation
Author opinion/potential of further research	In Leukemia, the white blood cells created by bone marrow are anomalous. It has two significant subtypes, acute Leukemia, and chronic Leukemia. Leukemia can further be classified into other following types namely, acute lymphocytic (ALL), acute myelogenous (AML), chronic lymphocytic (CLL), and chronic myelogenous (CML). Detection of Leukemia is done primarily by morphological analysis. This has different sub-types which led to different treatment guidelines. Many researchers worked on detection and diagnosis of Leukemia. Still many hybrid, optimized algorithms of machine learning and artificial intelligence are to be worked out for the improvement and trust of these existing frameworks

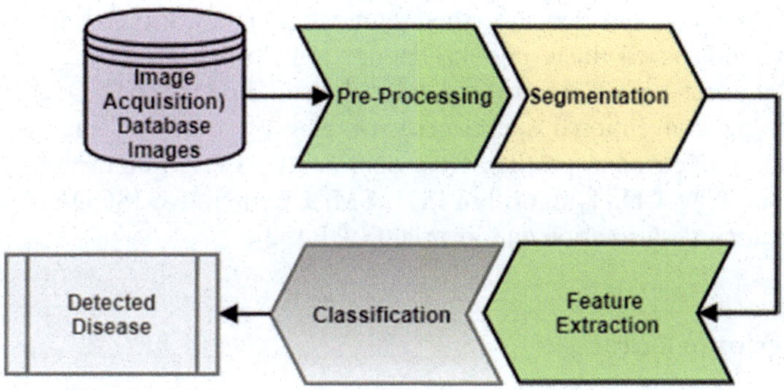

Fig. 6.3 Generalized methodology for blood cell analysis

count [21–25, 31, 32, 37, 50, 52, 54, 57–59] of different blood components like WBC, RBC. Analysis of these features will further detect the disease or count the cells. Depending upon different features extracted, the decision about the disease could be taken. To take decisions different classifiers could be designed.

6.2.3 State-of-the-Art Methods for Different Stages of Microscopic Analysis for Disease Detection

6.2.3.1 Image Pre-processing

The different methods used for pre-processing are, Self dual multi-scale morphological toggle (SMTT) block [60], Wiener filter [33], median filtering Gaussian filtering [6], gray scale transformation [32, 61–67] which has 3 types viz, linear, logarithmic and power-law, histogram stretching [32, 62–64], green color component from RGM image [44], morphological operations [32], edge detection [67].

6.2.3.2 Image Segmentation

The following are the different segmentation methods employed by the researchers. Watershed transform [31, 60] granulometric analysis and mathematical morphology (MM) operation, fuzzy logic approach [66], zack algorithm [43], k-means clustering [44], marker controlled watershed segmentation [45], stimulating discriminant measures (SDM) based clustering [45], Hough transform [57], iterative thresholding followed by watershed transform [67], edge thresholding [51, 64], Otsu's algorithm [18, 37, 65], a conventional neural network chen prepared laplacian of Gaussian (LOG) and coupled edge profile active contours(C- EPAC) algorithm [37], triangular thresholding DOST algorithm [51], SMACC, Iterative ISODATA clustering algorithm along with rotation and invariant LBP [52].

6.2.3.3 Feature Extraction

There are number of features that could be considered for feature extraction purpose. Some of them are given below.

Color Features: Color of the cell can be one of the features which can separate a cell from other types. For example: color of plasma is very different (yellow) than other blood components. In many cases, the color of the cell talks much about the abnormalities.

Geometric Features: These are the features based on the geometry or shape of the cell. These include following,

$$Elongation = 1 - \frac{Majoraxis}{Minoraxis} \tag{6.1}$$

$$Eccentricity = \frac{\sqrt{majoraxis^2 - monoraxis^2}}{minoraxis} \tag{6.2}$$

$$Rectangularity = \frac{area}{majoraxis \times minoraxis} \tag{6.3}$$

$$convexity = \frac{perimeterconvex}{perimeter} \tag{6.4}$$

$$Compactness = \frac{4 \times \pi \times area}{perimeter} \tag{6.5}$$

Statistical Features: Statistical moments such as, mean and standard deviation gives information about the appearance of distribution. Skewness and kurtosis shape the distribution along with the area and perimeter of the shape. The following are the different statistical features.

$$Mean, \overline{x} = \frac{1}{N}\sum_{i=1}^{N} x_i \tag{6.6}$$

Standard Deviation,

$$\sigma = \sqrt{\frac{1}{N-1}\sum_{i=1}^{n}(X_i - \overline{X})^2} \tag{6.7}$$

$$Skewness, SK = \frac{1}{N}\sum_{i=1}^{N}\frac{(x_i - \overline{x})^3}{\sigma^3} \tag{6.8}$$

$$Kurtosis, K = \frac{1}{N}\sum_{i=1}^{N}\frac{(x_i - \overline{x})^4}{\sigma^4} \tag{6.9}$$

Texture Features: There are different texture features that are defined such as entropy, correlation, energy, contrast, homogeneity, and so on.

Entropy generally defines randomness in the characterization of texture of an image. When co-occurrence elements are same, entropy leads to its maximum value. The equation of entropy as follows.

$$Entropy, Ent = \sum_{i=0}^{N-1}\sum_{j=0}^{N-1} M(i, j)(-\ln(M(i, j))) \tag{6.10}$$

Contrast is the intensity variations in the neighboring pixels in an image.

$$Contrast = \sum_{i=0}^{N-1}\sum_{j=0}^{N-1}(i - j)^2(M(i, j) \tag{6.11}$$

Energy (E) is the measure of the extent of repetitions of pixel pairs. It gives an uniformity of the image. It gives a larger value for similar pixels.

$$Energy, E = \sqrt{\sum_{i=0}^{N-1}\sum_{j=0}^{N-1} M^2(i, j)} \tag{6.12}$$

Correlation Features: The repetitive nature of the texture elements position in the image is an important. An auto-correlation function gives the coarseness in an image.

Auto-correlation,

$$P(x, y) = \frac{\sum_{u=0}^{N}\sum_{v=0}^{N} I(u, v)I(u + x, v + y)}{\sum_{u=0}^{N}\sum_{v=0}^{N} I^2(u, v)} \tag{6.13}$$

Inverse Difference Moment or Homogeneity gauges the local homogeneity of a picture. IDM features acquire the proportions of the closeness of the distribution of the GLCM components to the diagonal of GLCM. IDM has a scope of determining the image and classify it as textured or non-textured.

$$IDM = \sum_{i=0}^{N-1}\sum_{j=0}^{N-1} \frac{1}{1 + (i - j)^2} M(i, j) \tag{6.14}$$

Directional Moment: In this, the image alignment is considered with respect to the angle.

$$DM = \sum_{i=0}^{N-1}\sum_{j=0}^{N-1} M(i, j)|i = j| \tag{6.15}$$

6.2.3.4 Classifier for Disease Detection

There are different classifiers for the classification of images which are employed for microscopic imaging of blood cells. These include machine learning algorithms as below. Different classifiers include, Decision Tree Classifier, Random Forest, K-Nearest Neighbors (KNN) [57], Logistic Regression, Binary Logistic Regression model, Multinomial logistic regression, and Ordinal regression, Naïve Bayes Algorithms including Gaussian Naive Bayes, Multinomial Naive Bayes, and Bernoulli Naive Bayes, Support Vector Machine (SVM) [43, 45, 47, 68], Convolutional Neural Networks [54, 58].

These classifiers are utilized by different researchers for disease detection purpose during the microscopic analysis of blood. Depending upon the disease to be identified, the classifier is employed by researchers. Literature does not found any hard rule about using a particular classifier for a particular disease. Generally, SVM, decision tree, Naïve Bayes are the classifier those are well explained in their performances.

Also these classifiers need a comparatively less size of database for training purposes. The advanced machine learning classifiers such as NN and its other types, require larger database size for its training purpose. So this can increase the time required for taking the decisions. Moreover, NN is found to be efficient in major cases in terms of accuracy of classification and detection of diseases (Tables 6.3 and 6.4).

6.3 Research Gaps

After having reviewed the related literature, the following research gaps are obtained. Overlapping cells are not considered at the segmentation stage by many researchers. As many practical cases have the overlapping of cells, during the staining procedure. For segmentation, different bio-inspired algorithms could be employed, which may prove efficient. Different optimization techniques are yet to be applied for improvement in the classifier performance. Leukemia is the disease that proves very dangerous in its later stages. It has different types such as Acute Lymphoblastic Leukemia (ALL), Acute Mylogeneous Leukemia (AML), Chronic Lymphoblastic Leukemia (CLL), and Chronic Mylogeneous Leukemia (ACL) [69]. For detection of these types, is a big challenge for the pathologists, as only morphology speaks in this cases.

ALL is further sub-classified in to its sub-types such as L1, L2, and L3. This classification is based upon the morphological features of the blasts cells and WBC in the blood. These different sub-types of ALL are indicative of different infections and are suggestive of the different line of treatment in the patients. The identification of these sub-types is not considered by most of the researchers in this area. Similarly, AML has different subtypes such as M0 to M7. These sub-types also differ in regard to the treatment guidelines. Moreover, types T1, T2, and T3 are so similar that, the distinctness is still a challenge for the researchers. The diagnosis of these different sub-types is not considered in most of the cases. Performance measures are limited to accuracy in most of the cases. There is a scope of improvement in accuracy. Accuracy of different stages of blood cell analysis is tested on a limited database.

6.4 Conclusion

Blood cell analysis assumes a crucial job in location and expectation of various issue and maladies identified with person. There are distinctive neurotic strategies for the equivalent, which ends up being exorbitant and furthermore requires long understanding for location. Image processing and computer vision strategies are produced for investigation of blood cells and discovery of maladies. Microscopic blood cell analysis framework has various stages to be specific, pre-processing, segmentation, feature extraction, classifier and illness identification. Pre-processing comprises of improving the gained picture quality and commotion expulsion. This incorporates

Table 6.3 Comparison of different techniques for Leukemia detection

Author	Year	Methodology	Performance measure	Database	No. of images
Patel and Mishra [43]	2015	K-means clustering for detection of WBC. Histogram and Zack algorithm for grouping WBCs, SVM for classification	Efficiency: 93.57%	ALL-IDB	7
Neoh et al. [45]	2015	Multilayer perceptron, support vector machine (SVM) and Dempster Shafer	Accuracy: Dempster-Shafer method: 96.72%, SVM model: 96.67%	ALL-IDB2	180
Negm et al. [44]	2018	Panel selection for segmentation, K-means clustering for features extraction, and image refinement. Classification by morphological features of Leukemia cells detection	Accuracy: 99.517%, Sensitivity: 99.348%, Specificity: 99.529%	Private datasets	757
Shafique et al. [47]	2019	Histogram equalization, Zack algorithm, watershed segmentation, support vector machine (SVM) classification	Accuracy: 93.70%, Sensitivity: 92%, Specificity: 91%	ALL-IDB	108
Abbasi et al. [68]	2019	K-means and watershed algorithm, SVM, PCA	Accuracy, specificity, sensitivity, FNR, precision all are above 97%	Private	Not mentioned
Mishra et al. [51]	2019	Triangle thresholding, discrete orthogonal	Accuracy: 99.66%	ALL-IDB1	108
Kumar et al. [33]	2019	SMI based model, local directional pattern (LDP)chronological sine cosine algorithm (SCA)	Accuracy: 98.7%, TPR:987%, TNR:98%	AA-IDB2	Not mentioned

(continued)

Table 6.3 (continued)

Author	Year	Methodology	Performance measure	Database	No. of images
Iltaf et al. [53]	2019	Expectation maximization algorithm, PCA, sparse representation	Accuracy, specificity, sensitivity all more than 92%	ALL-IDB2	260
Ahmed et al. [58]	2019	CNN	Accuracy: 88% Leukemia cells and 81% for subtypes classification	ALL-IDB, ASH Image Bank	Not mentioned
Matek et al. [54]	2019	ResNeXt CNN	Accuracy, sensitivity and precision above 90%	Private	18,365
Sahlol et al. [59]	2020	VGGNet, statistically enhanced Salp Swarm Algorithm (SESSA)	Accuracy: 96% dataset 1 and 87.9% for dataset 2	ALL-IDB, C-NMC	Not mentioned

gray-scale conversion, thersholding, filtering, and histogram stretching, morphological operations. Pre-processed image is portioned to get the locale of interest for further processing. Here WBC, RBC and platelets are isolated. Distinctive computer vision techniques utilized for segmentation are edge detection, watershed transformation, mathematical morphology, zack algorithm, k-means clustering, SDM, HSV thresholding, otsu's algorithm. There are overlapping cells at the time of staining of blood smear. Expulsion of these overlapping cells at the time of segmentation is a difficult undertaking. Hough transform evacuates certain overlapping however it makes the framework slower. Segmented images are classified by algorithms like SVM, ANN classifier, ELM classifier, circular Hough transform. There are various databases accessible for experimentation and investigation of microscopic blood cell such as BCCD (Kaggle) Database, ALL-IDB1, ALL-IDB2, Atlas of Hematology by Nivaldo Meridos, Leukocyte pictures for division and characterization (LISC), Ash image bank, and C-NMC dataset. There are different application territories where microscopic blood cell examination assumes a crucial job. RBC, WBC count, blood group identification, leukemia detection, sickle cells detection, partition of various WBC sub-classes, malaria parasite detection could be performed utilizing complex image processing and computer vision techniques.

Table 6.4 Analysis of articles adding significant research contribution in the area

1	Application	Blood cancer diagnosis
	Background	Need of an automated technique for blood cancer detection
	Objective	Hybrid CNN model for ALL detection
	Data source	(2019) Journal of Digital Imaging, https://doi.org/10.1007/s10278-019-00288-y
	Interventions/participants	Anuj Sharma, Bala Buksh
	Methods	K-means clustering, Fuzzy based CNN with firefly optimization algorithm
	Results	97.01%
	Limitation	Accuracy needs to be increased
	Key findings	Hybridization of FOA as optimization technique is used here with the basis used is k means clustering and CNN classifier Histogram of Oriented Gradients (HOG) descriptor with FOA is used as feature extraction and selection mechanism from the Region of Blood Cell (ROBC)
	Authors opinion	Proposed method given good result in terms of accuracy. But the commonly used validation technique for classifiers such as Cross fold etc. is not mentioned
2	Application	Detection of Leukemia
	Background	Need of an automated technique for blood cancer detection
	Objective	Segmentation of WBCs, classification of normal and blasts cells
	Data source	(2019), International Journal of Engineering and Advanced Technology (IJEAT) ISSN: 2249-8958
	Interventions/participants	Roopa B. Hegde, Keerthana Prasad, Harishchandra Hebbar, Brij Mohan Kumar Singh, Sandhya
	Methods	Thresholding, morphology, SVM and NN classifier
	Results	98% for segmentation of WBC, 92.8% for Leukemia detection
	limitation	Blasts detection needs improvement
	Key findings	Segmentation is done by using traditional techniques such as thresholding, morphology and filtering. Classification is done with the combination of two classifiers SVM and NN
	Authors opinion	Validation of SVM classifier is done using hold-out validation but for NN classifier the validation technique is not mentioned. Also the combination of NN and SVM might need a proper exploration of a particular suitable validation technique
3	Application	Acute lymphoblastic Leukemia detection
	Background	Need to detect Leukemia with its subtypes L1, L2 and L3
	Objective	Detection of Leukemia, classification into its sub-types
	Data source	(2018), Technology in cancer research and treatment 17 1,533,033,818,802,789
	Interventions/participants	Shafique, Sarmad, and Samabia Tehsin

(continued)

Table 6.4 (continued)

	Methods	Deep CNN AlexNet architecture
	Results	Average accuracy of 98% for detection and 95–96% for sub-types classification
	limitation	Database images used are limited. Also pre-processing and feature extraction is not performed
	Key findings	Transfer learning approach is used with the Alexnet architecture that improve the performance of classification
	Authors opinion	Pre-processing and other enhancement techniques can improve the performance of system. Other DCNN architectures could also be explored and could be tested to find the best suitable architecture for this application
4	Application	ALL detection and diagnosis
	Background	Need for a computer assisted framework for ALL diagnosis
	objective	Segmentation of WBC, RBC, and platelets, feature extraction, classification into normal and blasts cells
	Data source	(2019), 2nd International Conference on Communication, Computing and Digital systems (C-CODE) (pp. 184–189). IEEE
	Interventions/participants	Shafique, S., Tehsin, S., Anas, S., & Masud, F.
	methods	Zack algorithm, SVM classifier
	Results	Accuracy of 93.7%
	Limitation	Dataset images are limited to 108 only
	Key findings	Color and shape features are used, compared with KNN SVM found improved performance slightly
	Authors opinion	Deep classifier can be applied further for improvement of framework but dataset is to be increased in that case. Different types of ALL such as L1, L2, and L3 could be detected in the future study
5	Application	WBC identification
	Background	Under-segmentation and over segmentation, complexity in feature extraction methods
	objective	Pre-processing, feature extraction and selection, classification, applying TLA approach, WBCs Net architecture
	Data source	(2019), White blood cells identification system based on convolutional deep neural learning networks. Computer methods and programs in biomedicine, 168, 69–80
	Interventions/participants	Shahin, A. I., Guo, Y., Amin, K. M., & Sharawi, A. A.
	methods	CNN, SVM, WBCsNet
	Results	Accuracy up to 96%
	Limitation	For higher dataset s accuracy is reduced to 92.6%
	Key findings	CNN and SVM are used in combination, WBCsNet improves accuracy

(continued)

Table 6.4 (continued)

	Authors opinion	As the deep learning is applied, data size to be increased to improve the performance of system
6	Application	ALL detection
	Background	Need of automated CAD framework for ALL detection
	Objective	Pre-processing, feature extraction, dimensionality reduction, classifier
	Data source	(2019), Texture feature based classification on microscopic blood smear for acute lymphoblastic Leukemia detection. Biomedical Signal Processing and Control, 47, 303–311
	Interventions/participants	Mishra, S., Majhi, B., & Sa, P. K.
	Methods	Triangle thresholding, discrete orthogonal S-transform, Adaboost algorithm with RF(ADBRF) classifier
	Results	Accuracy of about 99%
	Limitation	Only one dataset IDB1 is used with less number of cells of 799. ALL sub-types, L1, L2, and L3 are not detected
	Key findings	ADABOOST RF classifier found superior compared with SVM and NN classifier
	Authors opinion	The framework can be extended for acute myloid Leukemia detection with its sub-types. Also ALL subtypes can be detected with some improved and optimized version further
7	Application	ALL detection in single cell
	Background	Need to develop image processing framework for diagnosis with deep learning for improvement in accuracies of popular system
	Objective	Pre-processing, segmentation, feature extraction, classification
	Data source	(2019), Mutual Information based hybrid model and deep learning for Acute Lymphocytic Leukemia detection in single cell blood smear images. Computer Methods and Programs in Biomedicine, 179: 104987, 2019. ISSN 18727565. https://doi.org/10.1016/j.cmpb.2019.104987. URL https://doi.org/10.1016/j.cmpb.2019.104987
	Interventions/participants	Krishna Kumar Jha and Himadri Sekhar Dutta
	methods	MI based hybrid model, Deep CNN classifier with chronological Sine Cosine Algorithm (SCA), k fold validation
	Results	Accuracy of 98.7%
	Limitation	Only single cell is considered for analysis
	Conclusion	DCN N with SCA gives good results compared to current state of the art using NN with hybridized work
	Key findings	Hybrid segmentation with fuzzy means and active contour
	Authors opinion	Single cells identification is done in this work. The work can be extended to multiple cells. Also hybrid optimization algorithms can be employed for performance improvement further

(continued)

Table 6.4 (continued)

8	Application	ALL diagnosis
	Background	Leukocyte segmentation under uneven imaging conditions
	Objective	Leukocyte segmentation, feature extraction and classification
	Data source	(2018), An automatic and robust decision support system for accurate acute Leukemia diagnosis from blood microscopic images. Journal of digital imaging, 31(5), 702–717
	Interventions/participants	Moshavash, Z., Danyali, H., & Helfroush, M. S.
	Methods	Zack algorithm, SVM, KNN, Naïve Bayes, and decision tree
	Results	Accuracy of 97.6%
	Limitation	Overlapping cells can be detected by this framework. The system could not be assured for further classification of Leukemia in to its sub-classes
	Key findings	Two ensemble classifiers are combinedly used for classification purpose. Classifier 1 with 4 different classifier combinations and classifier 2 using SVM with different kernel functions Two types of features extractions are compared, GLCM and LBP. LBP achieved higher accuracy
	Authors opinion	The system is much complex as it utilizes two different ensemble classifiers with different combinations and also feature extraction with two methods
9	Application	ALL detection
	Background	Time consuming manual methods based on morphology of cells
	Objective	Segmentation of WBC, feature extraction, and classification as normal, and blast cell
	Data source	(2019), Computer-assisted Acute Lymphoblastic Leukemia detection and diagnosis." In 2019 2nd International Conference on Communication, Computing and Digital systems (C-CODE), pp. 184–189. IEEE
	Interventions/participants	Shafique, Sarmad, Samabia Tehsin, Syed Anas, and Farrukh Masud
	Methods	Histogram equalization, zack algorithm, watershed algorithm, SVM classifier
	Results	Accuracy of 93.7%
	Limitation	Database images used are lesser, 108 images
	Key findings	As shape features can be affected during segmentation and pre-processing stages, here color features are used along with shape features
	Authors opinion	There is scope to use different features, other than specified here for performance improvement. In order to apply deep learning algorithm further, the dataset images are to be increased. The classification done here is in terms of normal cell and blast cells only. The framework could not be justified for detection of other different subtypes of leukemia

(continued)

Table 6.4 (continued)

10	Application	ALL detection and classification
	Background	Manual detection of Leukemia is critical and challenging task
	Objective	Pre-processing, processing, post processing, data normalization, feature extraction, and classification
	Data source	(2018), A decision support system for Acute Leukaemia classification based on digital microscopic images. Alexandria engineering journal, 57(4), 2319–2332
	Interventions/participants	Negm, A. S., Hassan, O. A., & Kandil, A. H.
	Methods	Histogram equalization, k means algorithm, watershed algorithm, decision tree, and NN classifier
	Results	Overall accuracy of decision tree is 96.6%, and NN is 96.76%
	Limitation	Dataset size is limited to 115 images for public dataset
	Key findings	NN is performed better than decision tree, but decision tree proves to be faster than NN approach
	Authors opinion	There are different sub-types of both ALL and AML. Further expansion of this algorithm can explore these sub-types
11	Application	ALL detection
	Background	Need of improvement of diagnosis system of Leukemia
	Objective	Segmentation, feature extraction, classification
	Data source	(2019), Automatic detection of acute lymphoblastic leukaemia based on extending the multifractal features, IET Image Processing 14, no. 1 (2019): 132–137
	Interventions/participants	Abbasi, Mohamadreza, Saeed Kermani, Ardeshir Tajebib, Morteza Moradi Amin, and Manije Abbasi
	Methods	k-means algorithm, watershed transform, SVM classifier
	Results	Accuracy up to 99%
	Limitation	Dataset images are limited to 600. Popular datasets for this study such as ALL-IDB1, ALL-IDB2 etc. are not used in this work
	Key findings	Fractal features are used. Feature reduction is taken care by using PCA algorithm. Use of chaotic features at feature extraction stage increases the accuracy of classification
	Authors opinion	The similar framework could be extended for AML detection along with its sub-types. Exact selection of features is not explored in depth here. There can be the exploration related to complexity and processing time with regard to the use of PCA algorithm
12	Application	ALL detection
	Background	Need of an automatic and novel approach for ALL detection
	Objective	Pre-processing, feature extraction and selection, classification

(continued)

Table 6.4 (continued)

	Data source	"Automated acute lymphoblastic leukaemia detection system using microscopic images." IET Image Processing 13, no. 13 (2019): 2548–2553
	Interventions/participants	Sukhia, Komal Nain, Abdul Ghafoor, Muhammad Mohsin Riaz, and Naima Iltaf
	Methods	Diffused expectation maximization (DEM) algorithm, thresholding, sparse classifier
	Results	Accuracy of 94%
	Limitation	Database has limited dataset images, 260. Out of these, 160 used for training and 100 for validation purpose
	Key findings	Accuracy of local binary pattern (LBP), HD, and SFTA is more as compared with the color and shape features
	Authors opinion	The paper has given more stress upon the feature extraction and selection stage. Also sparse classifier is introduced for this work that proves to be good in terms of accuracy. Further the similar framework could be extended for diagnosis of other types of leukemia
13	Application	Anemia detection
	Background	Image processing techniques can prove more reliable and cost effective in analysis of anemia
	Objective	For Sickle Cell Detection: Ellipse detection, sickle cell detection for Thalassemia: image acquisition, pre-processing, segmentation, feature extraction, classification
	Data source	Detection of Sickle Cell Anemia and Thalassemia using Image Processing Techniques
	Interventions/participants	Lavanya, T. H., Tumkur Gubbi, and S. Sushritha
	Methods	Edge detection, midpoint ellipse algorithm(MEA), histogram thersholding, diffused expectation maximization (DEM), Otsu's thresholding, KNN classifier
	Results	Not evaluated in percentage of accuracy
	Limitation	Datasets are not mentioned
	Key findings	Ellipse detection and DE (diffused expectation) algorithm is used for anemia detection. Thalassemia different types are not considered here
	Authors opinion	Mentioned framework could not be justified as, accuracy and datasets are not clearly explored in system evaluation
14	Application	Detection of healthy and unhealthy RBC
	Background	Anemia detection by hemoglobin percentage and microscopic examination
	Objective	Detection of RBC, separation of overlapped cells, classification
	Data source	(2016), "Healthy and unhealthy red blood cell detection in human blood smears using neural networks." Micron 83: 32–41
	Interventions/participants	Elsalamony, Hany A

(continued)

Table 6.4 (continued)

	Methods	Morphological operations, circular hough transform, watershed transform, NN
	Results	Accuracy of detection of different cell types is greater than 97.8%
	Limitation	Dataset contains fewer images, limited to 160
	Key findings	Different anemia type including sickle cell, elliptocytosis, microsites and unknown shapes are detected. Green color is considered as a detection parameter for healthy cells
	Authors opinion	Different parameters of cells, area, convex area, perimeter, eccentricity, solidity, and ratio are considered for differentiating healthy and unhealthy cells
15	Application	Thalassemia detection
	Background	Requirement of accurate diagnosis system for the disease
	Objective	Image acquisition, enhancement, segmentation, and filtering
	Data source	(2015), "Unsupervised color image segmentation of red blood cell for thalassemia disease." In 2015 2nd International Conference on Biomedical Engineering (ICoBE), pp. 1–6. IEEE
	Interventions/participants	Rashid, Nurul Zhafikha Noor, Mohd Yusoff Mashor, and Rosline Hassan
	Methods	Global contrast technique, image color conversion, k-means clustering, median filtering, seed region growing area extraction (SRGAE) algorithm
	Results	Average segmentation accuracy is of 94.57%
	Limitation	Database has limited size, containing only 60 images
	Key findings	Alpha, beta, and thalassemia trait images are used and analyzed in the study. These two types are distinguished significantly. SRGAE algorithm is effectively used for getting region of interest
	Authors opinion	Contrast technique is used for image enhancement which is comparatively simple. Color conversion has given a good importance in the detection. Use of SRGAE added the novelty in this study. This algorithm could be further utilized for other applications of microscopic analysis
16	Application	Thalassemia identification
	Background	Need of an automated framework for detection
	Objective	Image acquisition, feature extraction, information processing, classification
	Data source	(2016), "Automated Thalassemia Identifier Using Image Processing."
	Interventions/participants	Sandanayake, T. C., A. T. P. M. N. Thalewela, H. P. Thilakesooriya, R. M. A. U. Rathnayake, and S. A. Y. A. Wimalasooriya

(continued)

Table 6.4 (continued)

	Methods	Gray scaling, thresholding, edge detection, ANN, and SVM classifier
	Results	Accuracy of 89% for female and 97% for male patients
	Limitation	Thalassemia has many types depending upon the shapes of RBC. All these types are not detected in this work
	Key findings	Six different parameters are considered for thalassemia analysis including RBC, hemoglobin (HB), mean corpuscular volume (MCV), mean corpuscular hemoglobin (MCH), and RBC distribution width (RDW)
	Authors opinion	Research can be extended to detect and diagnose all the types of thalassemia. Also severity of the disease can also be diagnosed by exploring the research in that direction
17	Application	Minor thalassemia detection
	Background	Need of an automated framework for thalassemia detection
	Objective	Preprocessing, segmentation, feature extraction, and classification
	Data source	(2017), "The classification of abnormal red blood cell on the minor thalassemia case using artificial neural network and convolutional neural network." In Proceedings of the International Conference on Video and Image Processing, pp. 228–233
	Interventions/participants	Tyas, Dyah Aruming, Tri Ratnaningsih, Agus Harjoko, and Sri Hartati
	Methods	Histogram equalization, morphological operations, back propagation NN
	Results	Accuracy reached to 92.55%
	Limitation	Number of images are limited in the database to 256 only
	Key findings	Texture features, color features, and shape features are used for feature extraction purpose
	Authors opinion	A total of 43 values of features are used for one cell. This shows the feature extraction is more precise. But this may increase the processing time
18	Application	Malaria detection and cell counting
	Background	Requirement of faster and trustable method for malaria diagnosis
	Objective	RBC detection, feature computation, cell classification
	Data source	(2018), Malaria parasite detection and cell counting for human and mouse using thin blood smear microscopy. Journal of Medical Imaging 5, no. 4: 044,506
	Interventions/participants	Poostchi, Mahdieh, Ilker Ersoy, Katie McMenamin, Emile Gordon, Nila Palaniappan, Susan Pierce, Richard J. Maude et al.

(continued)

Table 6.4 (continued)

	Methods	Microscopic imaging with thin blood smear, multiscale Laplacian of Gaussian (LOG) along with coupled edge profile active contours(C-EPAC), SVM, ANN
	Results	Accuracy of 98% and 99% for SVM and ANN respectively
	Limitation	Dataset has less number of images, 70 in this case
	Key findings	Multiscale LOG and C-EPAC are combined. ANN achieves higher accuracy compared to SVM
	Authors opinion	The work could be explored further for more types of parasites. The counts of parasites and there life stages could be detected to explore the severity of the disease
19	Application	WBC classification
	Background	WBC classification is important for different disease detections
	Objective	Pre-processing, classification
	Data source	(2019), "Classification of White Blood Cells by Deep Learning Methods for Diagnosing Disease." Revue d' Intelligence Artificielle 33, no. 5: 335–340
	Interventions/participants	Yildirim, Muhammed, and Ahmet Çinar
	methods	Median and Gaussian filtering, CNN classifier
	Results	Accuracy is in between 62 and 83% for different deep learning architectures
	Limitation	Accuracy is less as compared to other popular frameworks
	Key findings	Different architecutres of Deep learning are applied to the problem. Alexnet, RessNet, DenseNet201, GoogleNet are used for analysis. Original data as well as pre-processed data is analyzed for segmentation. Pre-processed data with Gaussian and median filtering gives an improvement in the accuracy
	Authors opinion	Pre-processing improves accuracy. Deep learning architecture could improve the performance by increasing the dataset size
20	Application	Detection of dengue
	Background	Requirement of a framework for viral disease detection
	Objective	Blood cell classification, dengue virus detection
	Data source	(2015), "Image processing for detection of dengue virus based on WBC classification and decision tree." In 2015 13th International Conference on ICT and Knowledge Engineering (ICT & Knowledge Engineering 2015), pp. 84–89. IEEE
	Interventions/participants	Tantikitti, Sarach, Sompong Tumswadi, and Wichian Premchaiswadi
	methods	Color transformation, multi-level thersholding, decision tree classification
	Results	Accuracy of 72% for dengue detection and 92% for WBC classification

(continued)

Table 6.4 (continued)

Limitation	Database images are limited. Sub-types of dengue virus are not detected
Key findings	Different types of parameters are considered such as number of Lymphocytes, number of Phagocyte, number of WBC, percentage of Lymphocytes, percentage of Phagocyte and percentage of Hct
Authors opinion	Decision tree algorithm is used for classification. It spends more time but is more trusted compared with NN

6.5 Future Scope

A powerful division of white and red cells in minuscule blood smear pictures to meet better precision could be actualized. To conquer overlapping cells issue at the hour of division will likewise end up being a significant extension. A viable feature extraction by utilizing distinctive image transforms will like- wise demonstrate to a significant degree. There are different optimization algorithms which could be utilized efficiently for classification of blood cells. Different deep learning algorithms, that may demonstrate productive and might give high accuracy to various phases of examination of blood cells. The designed algorithms must be tasted with various publicly accessible databases for precision. Precision of the calculation should be similar enough with all the databases. Another parameter like vigor can be presented for this reason. Relative accuracy of various databases can be determined. To gauge the exhibition of framework with various measures such as true positive, true negative, faults, sensitivity, specificity, precision, FI score, J-score in addition with accuracy.

Contribution is still needed for various ailments location, such as diabetes, viral diseases such as chickungunya and dengue, anemia diseases such as pancytopenia, thalassemia and leukemia.

References

1. Houwen, B.: Blood film preparation and staining procedures. Lab. Hematol. **6**, 1–7, 22 (2002), 1–14 (2000)
2. Adewoyin, A.S.: Peripheral blood film-a review. Ann. Ibadan Postgr. Med.**12**(2), 71–79 (2014)
3. Deshpande, N.M., Gite, S.S.: A brief bibliometric survey of explainable AI in medical field. Libr Philos Pract, 1–27 (2021)
4. https://healthengine.com.au/info/blood-function-and-composition. Assessed on 25th Sept 2020
5. Vives Corrons, J.L., Albarede, S., Flandrin, G., Heller, S., Horvath, K., Houwen, B., Nordin, G., Sarkani, E., Skitek, M., Van Blerk, M., Libeer, J.C.: Haematology working group of the european external committee for external quality assurance programmes in laboratory medicine, guidelines for blood smear preparation and staining procedure for setting up an external quality assessment scheme for blood smear interpretation. Part I: Control Material. Clin. Chem. Labor. Med. **42**, 922–926 (2004)

6. Yildirim, M., Çinar, A.: Classification of white blood cells by deep learning methods for diagnosing disease. Revue d'Intelligence Artificielle **33**(5), 335–340 (2019)
7. Labati, R.D., Piuri, V., Scotti, F.: IEEE International Conference and Image Processing. ALL-IDB: The Acute Lymphoblastic Leukemia Image Database for Image Processing, Universit'a degli Studi di Milano, Department of Information Technology. IEEE International Conference on Image Processing, pp. 2089–2092 (2011)
8. Acharya, V., Kumar, P.: Detection of acute lymphoblastic Leukemia using image segmentation and data mining algorithms. Med. Biol. Eng. Comput. **57**(8), 1783–1811 (2019). ISSN 17410444. https://doi.org/10.1007/s11517-019-01984-1
9. Alsalem, M.A., Zaidan, A.A., Zaidan, B.B., Hashim, M., Madhloom, H.T., Azeez, N.D., Alsyisuf, S.: A review of the automated detection and classification of acute leukaemia: coherent taxonomy, datasets, validation and performance measurements, motivation, open challenges and recommendations. Comput. Methods Programs Biomed. **158**, 93–112 (2018)
10. Agaian, S., Madhukar, M., Chronopoulos, A.T.: Automated screening system for acute myelogenous Leukemia detection in blood microscopic images. IEEE Syst. J. **8**(3), 995–1004 (2014)
11. Rezatofighi, S.H., Soltanian-Zadeh, H.: Automatic recognition of five types of white blood cells in peripheral blood. Comput. Med. Imaging Graph. **35**(4), 333–343 (2011). ISSN 08956111. https://doi.org/10.1016/j.compmedimag.2011
12. Livieris, I.E.: Identification of blood cell subtypes from images using an improved SSL algorithm. Biomed. J. Sci. Techn. Res. **9**(1) (2018). https://doi.org/10.26717/bjstr.2018.09.001755
13. Abbasi, M., Kermani, S., Tajebib, A., Amin, M.M., Abbasi, M.: Automatic detection of acute lymphoblastic leukaemia based on extending the multifractal features. IET Image Process. **14**(1), 132–137 (2019)
14. Bhavnani, L.A., Jaliya, U.K., Joshi, M.J.: Segmentation and counting of WBCs and RBCs from micro-scopic blood sample images. Int. J. Image Graph. Signal Process. **8**(11), 32–40 (2016). ISSN 20749074. https://doi.org/10.5815/ijigsp.2016.11.05
15. Anilkumar, K.K., Manoj, V.J., Sagi, T.M.: A survey on image segmentation of blood and bone marrow smear images with emphasis to automated detection of Leukemia. Biocybern. Biomed. Eng. (2020)
16. Bani Baker, Q., Alsmirat, M.A., Balhaf, K., Shehab, M.A.: Accelerating white blood cells image segmentation using GPUs. Concurr. Comput. Pract. Exp. e5133 (2019)
17. Xing, F., Yang, L.: Robust nucleus/cell detection and segmentation in digital pathology and microscopy images: a comprehensive review. IEEE Rev. Biomed. Eng. **9**, 234–263 (2016)
18. Salem, N., Sobhy, N.M., Dosoky, M.E.: A comparative study of white blood cells segmentation using Otsu threshold and watershed transformation. J. Biomed. Eng. Med. Imag. **3**(3), 15–15 (2016)
19. Razzak, M.I., Naz, S.: Microscopic blood smear segmentation and classification using deep contour aware CNN and extreme machine learning. In: 2017 IEEE Conference on Computer Vision and Pattern Recognition Workshops (CVPRW), pp. 801–807. IEEE (2017)
20. Liu, Y., Cao, F., Zhao, J., Chu, J.: Segmentation of white blood cells image using adaptive location and iteration. IEEE J. Biomed. Health Inform. **21**(6), 1644–1655 (2016)
21. Al-Hafiz, F., Al-Megren, S., Kurdi, H.: Red blood cell segmentation by thresholding and Canny detector. Proc. Comput. Sci. **141**, 327–334 (2018)
22. Prinyakupt, J., Pluempitiwiriyawej, C.: Segmentation of white blood cells and comparison of cell morphology by linear and naïve Bayes classifiers. Biomed. Eng. Online **14**(1), 63 (2015)
23. Zhong, Z., Wang, T., Zeng, K., Zhou, X., Li, Z.: White blood cell segmentation via sparsity and geometry constraints. IEEE Access **7**, 167593–167604 (2019)
24. Chaudhary, A.H., Ikhlaq, J., Iftikhar, M.A., Alvi, M.: Blood cell counting and segmentation using image processing techniques. In: Applications of Intelligent Technologies in Healthcare, pp. 87–98. Springer, Cham (2019)
25. Sajjad, M., Khan, S., Jan, Z., Muhammad, K., Moon, H., Kwak, J.T., Rho, S., Baik, S.W., Mehmood, I.: Leukocytes classification and segmentation in microscopic blood smear: a resource-aware healthcare service in smart cities." IEEE Access **5**, 3475–3489 (2016)

26. Biji, G., Hariharan, S.: White blood cell segmentation techniques in microscopic images for Leukemia detection. IONS J. Dent. Med. Sci. **15**, 45–51 (2016)
27. Mohamed, S.T., Ebeid, H.M., Hassanien, A.E., Tolba, M.F.: Optimized feed forward neural network for microscopic white blood cell images classification. In: International Conference on Advanced Machine Learning Technologies and Applications, pp. 758–767. Springer, Cham (2019)
28. Abbas, S.: Microscopic images dataset for automation of RBCs counting. Data Brief **5**, 35–40 (2015). ISSN 23523409. https://doi.org/10.1016/j.dib.2015.08.006
29. Miao, H., Xiao, C.: Simultaneous segmentation of leukocyte and erythrocyte in microscopic images using a marker-controlled watershed algorithm. Comput. Math. Methods Med. (2018). ISSN 17486718.https://doi.org/10.1155/2018/7235795
30. Bills, M.V., Nguyen, B.T., Yoon, J.-Y.: Simplified white blood cell differential: an inexpensive, smartphone-and paper-based blood cell count. IEEE Sens. J. **19**(18), 7822–7828 (2019)
31. Bala, S., Doegar, A.: Automatic detection of sickle cell in red blood cell using watershed segmentation **4**(6), 488–491 (2015). https://doi.org/10.17148/IJARCCE.2015.46105
32. Elsalamony, H.A.: Healthy and unhealthy red blood cell detection in human blood smears using neural networks. Micron **83**, 32–41 (2016). ISSN 09684328. https://doi.org/10.1016/j.micron.2016.01
33. Alotaibi, K.: Sickle Blood Cell Detection Based on Image Segmentation (2016)
34. Javidi, B., Markman, A., Rawat, S., O'Connor, T., Anand, A., & Andemariam, B.: Sickle cell disease diagnosis based on spatio-temporal cell dynamics analysis using 3D printed shearing digital holographic microscopy. Opt. Exp. **26**(10), 13614 (2018). ISSN 1094-4087. https://doi.org/10.1364/oe.26.013614
35. Lavanya, T.H., Gubbi, T., Sushritha, S.: Detection of sickle cell anemia and thalassemia using image processing techniques
36. Poostchi, M., Silamut, K., Maude, R.J., Jaeger, S., Thoma, G.: Image analysis and machine learning for detecting malaria. Transl. Res. **194**, 36–55 (2018). ISSN 18781810. https://doi.org/10.1016/j.trsl.2017.12.004
37. Duan, Y., Wang, J., Menghan, Hu., Zhou, M., Li, Q., Sun, Li., Qiu, S., Wang, Y.: Leukocyte classification based on spatial and spectral features of microscopic hyperspectral images. Opt. Laser Technol. **112**, 530–538 (2019)
38. Tantikitti, S., Tumswadi, S., Premchaiswadi, W.: Image processing for detection of dengue virus based on WBC classification and decision tree. In: 2015 13th International Conference on ICT and Knowledge Engineering (ICT & Knowledge Engineering 2015), pp. 84–89. IEEE (2015)
39. Poostchi, M., Ersoy, I., McMenamin, K., Gordon, E., Palaniappan, N., Pierce, S., Maude, R.J., et al.: Malaria parasite detection and cell counting for human and mouse using thin blood smear microscopy. J. Med. Imag. **5**(4), 044506 (2018)
40. Rashid, Noor, N.Z., Mashor, M.Y., Hassan, R.: Unsupervised color image segmentation of red blood cell for thalassemia disease. In: 2015 2nd International Conference on Biomedical Engineering (ICoBE), pp. 1–6. IEEE (2015)
41. Sandanayake, T.C., Thalewela, A.T.P.M.N., Thilakesooriya, H.P., Rathnayake, R.M.A.U., Wimalasooriya, S.A.Y.A.: Automated thalassemia identifier using image processing (2016)
42. Tyas, D.A., Ratnaningsih, T., Harjoko, A., Hartati, S.: The classification of abnormal red blood cell on the minor thalassemia case using artificial neural network and convolutional neural network. In: Proceedings of the International Conference on Video and Image Processing, pp. 228–233 (2017)
43. Patel, N., Mishra, A.: Automated leukaemia detection using microscopic images. Proc. Comput. Sci. **58**, 635–642 (2015). ISSN 18770509. https://doi.org/10.1016/j.procs.2015.08.082
44. Negm, A.S., Hassan, O.A., Kandil, A.H.: A decision support system for Acute Leukaemia classification based on digital microscopic images. Alexandria Eng. J. **57**(4), 2319–2332 (2018). ISSN 11100168. https://doi.org/10.1016/j.aej.2017.08.025
45. Neoh, S.C., Srisukkham, W., Zhang, L., Todryk, S., Greystoke, B., Lim, C.P., Hossain, M.A., Aslam, N.: An intelligent decision support system for leukaemia diagnosis using microscopic blood images. Sci. Rep. **5**, 1–14 (2015). ISSN 20452322. https://doi.org/10.1038/srep14938

46. Singh, H., Kaur, G.: Automatic detection of blood cancer in microscopic images: a review. Balkrishan Int. J. Innov. Adv. Comput. Sci. **6**(4), 40–43 (2017)
47. Shafique, S., Tehsin, S., Anas, S., Masud, F.: Computer-assisted acute lymphoblastic leukemia detection and diagnosis. In: 2019 2nd International Conference on Com- munication, Computing and Digital Systems, C-CODE 2019, pp. 184–189 (2019). https://doi.org/10.1109/C-CODE.2019.8680972
48. Putzu, L., Di Ruberto, C.: White blood cells identification and counting from microscopic blood image. World Acad. Sci. Eng. Technol. **7**(1), 363–370 (2013)
49. Jha, K.K., Dutta, H.S.: Mutual Information based hybrid model and deep learning for acute lymphocytic Leukemia detection in single cell blood smear images. Comput. Methods Progr. Biomed. **179**, 104987 (2019). ISSN 18727565. https://doi.org/10.1016/j.cmpb.2019.104987
50. Moshavash, Z., Danyali, H., Helfroush, M.S.: An automatic and robust decision support system for accurate acute Leukemia diagnosis from blood microscopic images. J. Dig. Imaging **31**(5), 702–717 (2018). ISSN 1618727X. https://doi.org/10.1007/s10278-018-0074-y
51. Mishra, S., Majhi, B., Sa, P.K.: Texture feature based classification on microscopic blood smear for acute lymphoblastic Leukemia detection. Biomed. Signal Process. Control **47**, 303–311 (2019)
52. Labati, R.D., Piuri, V., Scotti, F.: All-IDB: the acute lymphoblastic Leukemia image database for image processing. In: 2011 18th IEEE International Conference on Image Processing, pp. 2045–2048. IEEE (2011)
53. Ahmed, N., Yigit, A., Isik, Z., Alpkocak, A.: Identification of Leukemia subtypes from microscopic images using convolutional neural network. Diagnostics **9**(3), 104 (2019)
54. Sahlol, A.T., Kollmannsberger, P., Ewees, A.A.: Efficient classification of white blood cell Leukemia with improved swarm optimization of deep features. Sci. Rep. **10**(1), 1–11 (2020)
55. Deshpande, N.M., Gite, S.S., Aluvalu, R.: A brief bibliometric survey of Leukemia detection by machine learning and deep learning approaches (2020)
56. Deshpande, N.M., Gite, S., Aluvalu, R.: A review of microscopic analysis of blood cells for disease detection with AI perspective. PeerJ Comput Sci **7**, e460 (2021)
57. Kaur, M.P.: A normal blood cells. Significant analysis of leukemic cells extraction and detection using KNN and Hough transform algorithm **3**(1), 27–33 (2015)
58. Matek, C., Schwarz, S., Spiekermann, K., Marr, C.: Human-level recognition of blast cells in acute myeloid leukaemia with convolutional neural networks. Nature Mach. Intell. **1**(11), 538–544 (2019)
59. Gupta, A., Gupta, R.: ALL challenge dataset of ISBI 2019 [data set]. Cancer Imag. Arch. (2019).https://doi.org/10.7937/tcia.2019.dc64i46r
60. Belekar, S.J., Chougule, S.R.: WBC segmentation using morphological operation and SMMT operator—a review, pp. 434–440 (2015)
61. Patel, N., Mishra, A.: Automated leukaemia detection using microscopic images. Proc. Comput. Sci. **58**, 635–642 (2015)
62. Bhanushali, A., Katale, A., Bandal, K., Barsopiya, V., Potey, M.: Automated disease diagnosis using image microscopy **02**, 2–6 (2016)
63. Chougale, M.B., Mohite-patil, T.B.: Automated red blood cells counting using image processing techniques **3**(12), 748–750 (2016)
64. Australian national parks service and wildlife. Special issue. Australian Ranger Bull. **4**(1), 9–10 (1986). ISSN 0159-978X
65. Thiruvinal, V.J., Ram, S.P.: Automated blood cell counting and classification using image processing, pp. 74–82 (2017). https://doi.org/10.15662/IJAREEIE.2017.0601010
66. Bhagavathi, S.L., Thomas Niba, S.: An automatic system for detecting and counting RBC and WBC using fuzzy logic. ARPN J. Eng. Appl. Sci. **11**(11), 6891–6894 (2016). ISSN 18196608
67. Biswas, S., Ghoshal, D.: Blood cell detection using thresholding estimation based watershed transformation with Sobel filter in frequency domain. Proc. Comput. Sci. **89**, 651–657 (2016). ISSN 18770509. https://doi.org/10.1016/j.procs.2016.06.029

68. Sukhia, K.N., Ghafoor, A., Riaz, M.M., Iltaf, N.: Automated acute lymphoblastic leukaemia detection system using microscopic images. IET Image Process. **13**(13), 2548–2553 (2019)
69. Al-Tahhan, F.E., Sakr, A.A., Aladle, D.A., Fares, M.E.: Improved image segmentation algorithms for detecting types of acute lymphatic leukaemia. IET Image Process. **13**(13), 2595–2603 (2019)

Chapter 7
Investigating Clinical Named Entity Recognition Approaches for Information Extraction from EMR

Pranita Mahajan and Dipti Rana

Abstract Electronic Medical Record (EMR) contains much information used in various applications, such as identifying similar patients, keeping track of follow-ups, etc. An essential feature of EMR is that it is rich in context and may lead to ambiguity during analysis if undetected in the initial stages and could result in wrong interpretation. The chapter includes a detailed literature review of recent clinical Named Entity techniques. The chapter demonstrates comparative results of Clinical Named Entity Classification using rule-based, deep learning-based, and hybrid approaches. The chapter expresses the efficacy of clinical Named Entity Recognition (NER) techniques for Information Extraction. Our experimentation validates state-of-art recitation about the high accuracy of combined Deep Learning (DL) models with a sequential model. The experiment appraises the need for improved clinical word embeddings for efficient entity identification.

Keywords Information Extraction (IR) · Electronic Medical Record (EMR) · (Clinical Named Entity Recognition) cNER · Deep Learning (DL) · Long and Short Term Memory (LSTM) · Bidirectional LSTM (BLSTM)

7.1 Introduction

In healthcare, data mining intensively is becoming essential. Information Technology can benefit doctors in making better data-driven decisions [1]. Healthcare Technology emerged intending to automate the critical task and provide deeper insights. Research shows several areas in which Information Technology has provided solutions to Healthcare industries, for example, diagnosis treatment recommendations, patient communication care, and coordination. A significant feature provided by Information Technology to hospitals in the development of Electronics Medical Records (EMR). EMR stores a tremendous amount of patient information, yet not all information is useful, which gave rise to EMR analysis. It is essential to extract meaningful

P. Mahajan (✉) · D. Rana
SIESGST, Navi Mumbai, India

SVNIT, Surat, India

M. Mehta et al. (eds.), *Tracking and Preventing Diseases with Artificial Intelligence*, Intelligent Systems Reference Library 206,
https://doi.org/10.1007/978-3-030-76732-7_7

data from the patient's records. Named entity recognition is an important subtask of information extraction.

MUC-6, the sixth in a series of Message Understanding Conference first introduced the term "Named Entity". NER maps extracted keywords to predefined categories such as disease names, medical codes, etc. In the health domain, entities of interest are disease, symptom, gene, gene products, etc. A named entity recognizes the keywords having similar properties from a collection of features. NER is a classification problem, where tokens get classified into more abstract level named entities. The chapter concentrates on the state-of-the-art clinical Named Entity Recognition and empirical study of sequential and Deep Learning models.

This study majorly concentrates on retrieving precise information from unstructured data such as patient's discharge summary. Discharge summaries would be processed and analyzed, using Natural Language Processing (NLP) techniques such as Part of Speech, Named Entity Recognition (NER). The scope of this chapter is an empirical study of various NER techniques for clinical information extraction. The chapter answers, Which clinical NER technique performs better on unstructured data such as Electronic Medical Records (EMR)?, What is the impact of applying machine learning and NLP to extract NERs from EMR?, Which Deep Learning algorithm is suitable for NERs classification? Chapter showcases both theoretical as well as empirical analysis of techniques of clinical Named Entity Recognition.

Chapter Content
The remaining chapter is organized as follows; Sect. 7.2 covers a state-of-the-art survey of clinical NER elicitation and specification approaches/tools. Section 7.3 is the experimental evaluation, and the results are discussed in Sect. 7.4. Finally, we summarize our work in conclusion with open challenges in information extraction in the healthcare domain, which may help the researchers.

7.2 Clinical Named Entity Recognition

MUC-6, the sixth in a series of Message Understanding Conference first introduced the term "Named Entity". NER maps extracted keywords to predefined categories such as disease names, medical codes, etc.; for example, in the health domain, entities of interest are disease, symptom, gene, gene products, etc. A named entity recognizes the keywords having similar properties from a collection of features. Literature shows a high impact of NER on information extraction [2–6]. This Section is a detailed study of Clinical NER techniques.

A formal definition of NER [3, 4], see Fig. 7.1, for input sequence of words, $S = \{w_1 \ldots w_N\}$ with output as three terms NER $\{I_s, I_e, t\}$.

Where I_s and I_e are the starts and the end indexes of a named entity mention. t is the entity type from a predefined category set. Figure 7.1 is an example of four predefined named entities. Clinical NER is a critical NLP task to extract named entities, e.g., problem, symptom, treatment, form, dosage, etc., from clinical narratives such as

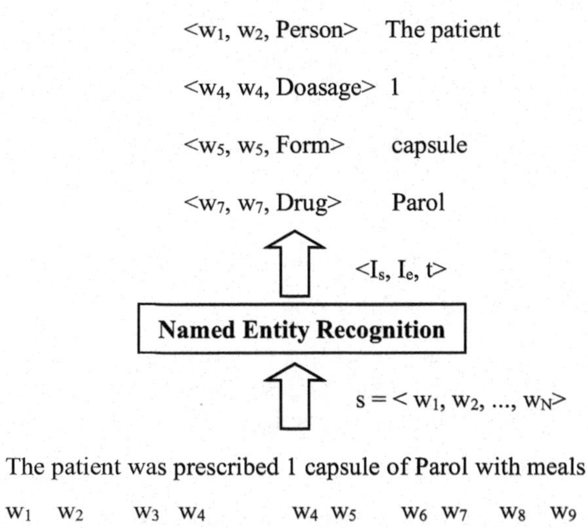

Fig. 7.1 Formal representation of NER

Electronic Medical Records, Pathological reports etc. This section is a literature review of clinical NER research to date.

NER is a classification problem, where tokens get classified into more abstract level named entities. This chapter concentrates on clinical Named Entity Recognition. Figure 7.2 shows the recent NER classification approaches preferred by researchers.

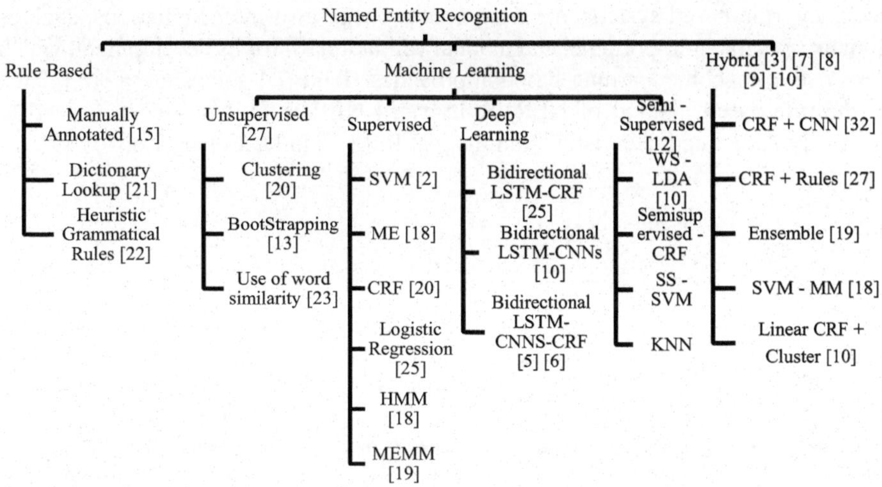

Fig. 7.2 NER classification techniques

According to the literature, the broad classification of NER is Rule-based approaches, Machine learning-based approaches, and hybrid approaches. Broad categories further evolved as follows.

1. Rule-based Approaches
   a. Manually Annotated
   b. Heuristic Grammatical Rules
   c. Dictionary Lookup
2. Machine Learning Approaches
   a. Unsupervised Classification
   b. Deep Learning Classification
   c. Supervised Classification
3. Hybrid Approaches

The following sections explain the above summarization in detail.

7.2.1 Rule-Based Approach

Traditional systems were most often based on manual rules [7]. These systems are hand-crafted rules-based to identify named entities. Rule-based NER systems are highly efficient due to the use of context-based knowledge [6]. Even with high accuracy, rule-based systems are not popular because of their limitations, such as domain-specific, non-portable, and manual validation [8] for its development, which gave rise to machine learning-based approaches. Table 7.1 shows literature related to the rule-based method of NER. Rule-based NER is further subcategorized to Manually Annotated, Heuristic Grammatical Rules, and Dictionary Lookup [9].

Table 7.1 Rule-based NER system

Authors and year	Language/Domain	NEs found	Technique used	Dataset used	Observations
Munoz et al. 2016 [10]	Medical	Diagnosis, treatment	Dictionary-based approach for stemming	The i2b2 heart disease risk NLP dataset	Similarity-based approach to identify entities
Rahem and Omar 2015 [11]	Drug-related crime news documents	Types, price, amount of drug, nationality	POS tagging, Rule-based contextual, heuristics, and grammatical rules	Crime news documents retrieved from two local Malay websites	Simple rule-based approach useful in specific morbidity entity identification

7.2.1.1 Manually Annotated

This method, manually creates a list of domain-specific words with the respective named entity, and calculates semantic similarity to classify NER. This method is a traditional way of data annotation [12]. The process takes raw text as an input, and manually, entity spans are highlighted by domain experts. At the end of the process, this corpus is used for training new datasets. The author has discussed the semantic similarity approach for calculating the semantic similarity approach for identifying entities [10]. This process needs a domain expert for the validation of annotated entities. The annotations are domain-specific and context-dependent, which limits the generalization of these systems.

7.2.1.2 Heuristic Grammatical Rules

Hand-crafted grammatical rules are written to identify entities. These rules are domain-specific. Rules are written considering the position of a word in a sentence, PoS tagger output, etc. [9, 10]. Uzuner et al. [13] discussed heuristic rule-based named entity approaches such as; lexicon-based approaches. A list of excluding lexicons is used to calculate the probability of words being named entity, suffixes, and co-occurrence statistics used to detect multiword entities. Datla et al. [14], discussed Viterbi implementation to maintain previous word's output tags and calculate rule-based probabilities to label each word, "Number", "Measure" and "Time". The literature emphasizes the use of heuristics rules for augmenting and pruning identified named entities. These techniques can be used along with machine learning approaches to increase the accuracy of identified entities, especially context-based identification of named entities.

7.2.1.3 Dictionary Lookup

Dictionary lookup based algorithms use external knowledge source, for example, domain-specific dictionary. Rahem and Omar [11] classified dictionary lookup into the general list, list of entities, and entity cues. General list resolves disambiguation such as capital words, ambiguous position of words in the sentence. List of entities used to determine nouns such as organization names etc. The third list, entity cues used to detect context-specific entities. The process of entity identification requires at least one candidate word to exact match with the maintained dictionary. The similarity between tokens extracted after pre-processing and dictionary words is measured, and then respective NERs are identified [10, 15, 16].

Table 7.1 is a literature work related to the above-discussed methods in a rule-based system.

7.2.2 Machine Learning-Based Approaches

Machine learning is a method of learning hidden patterns to make efficient decisions or predictions. Machine Learning-based approaches can be divided into the following categories:

7.2.2.1 Unsupervised Learning

Unsupervised learning is an algorithm that learns from implicit features of unlabeled data. The unsupervised learning makes its decisions considering the linguistic, positional features of data to learn more about the data. Clustering and association rules-based approaches are classic examples of unsupervised learning. In case of unavailability of the annotated datasets, the clustering-based systems are used [9]. These methods use distributional features of data to find similarities between elements to combine or distribute them in clusters. Association rule-based approaches learn by finding an association between items from large datasets.

7.2.2.2 Supervised Learning

Supervised learning-based approaches use labeled training data and construct adaptive feature correlations. Appropriate learning algorithms use these features and identify similar information from unseen samples [6]. The supervised model needs labeled corpus for training, for named entities identification; these labeled instances are annotated manually and validated with domain experts. A NER system uses raw text data, and every word in the data is a feature. Identifying relevant features is an important task in supervised learning NER systems. Features in the NER task are well explained by Chen et al. [17]. Author classified features into namely, list lookup features, corpus features, and word-based features. These features are scrutinized using language knowledge to determine association with one of the categories.

Other feature selection techniques such as context, location, structure etc. of document or corpus consider language knowledge such as orthographical, contextual, and morphological properties to identify word-level features. The next important task is to identify best learning algorithm. We have discussed, Hidden Markov Model (HMM) [18], Support Vector Machine (SVM) [19], Conditional Random Field (CRF) [15], Maximum Entropy Markov Model (MEMM) [17], etc. Named Entity models along with ensemble classifier techniques [15, 20, 21]. Table 7.2 is a literature review of machine learning-based NER.

Table 7.2 Machine learning-based NER system

Authors and year	Language/Domain	NEs found	Technique used	Dataset used	Observation
Lin et al. 2020 [22]	General domain	Persons, organizations, locations	Improved CRF with hidden information using, word position, n-gram features, POS	CHEMDNER, CONLL02	Good work in sequence labeling task to replace handcrafted CRF layer with latent variable CRF to capture sentence level features
Abulaish et al. 2019 [23]	Biomedical domain	Disease–symptom	Graph-based technique to construct disease–symptom knowledgebase	Documents crawled from PubMed database containing 107,302 biomedical abstracts	Communicable disease knowledgebase constructed
Zhang and Elhadad 2013 [9]	Biomedical domain	Cell_type, DNA, RNA, treatment, test and protein	Unsupervised classifier using lexical based similarity	Documents crawled from PubMed database	Reduces annotation time, rule based approach affects scalability
Wang et al. 2014 [18]	Clinical records	Symptom names	SVM and HMM classification techniques from large volume of clinical records	Chinese structured clinical records	Need of structured labeled dataset
Savova 2010 [16]	Clinical records	Disease, symptom	Use of dictionary look up algorithm along with OpenNLP toolkit	Built own gold standard datasets for named entity	Does not solve entity ambiguity
Lim et al. [12]	Biomedical Domain	Disease name	Supervised CRF+ handcrafted dictionary for fuzzy classification	NCBI disease corpus and BioCreative corpus	Along with CRF fuzzy string matching algorithm is used to identify rare diseases

(continued)

Table 7.2 (continued)

Authors and year	Language/Domain	NEs found	Technique used	Dataset used	Observation
Uzuner et al. 2010 [13]	Healthcare domain	Medications Dosages Modes Frequencies Durations Reasons list/narrative	I2b2 challenge data is annotated using SVM and CRF algorithms	i2b2 Dataset	Data annotation is performed, this structured information can be used further for relation identification

7.2.2.3 Deep Learning

Lately, deep learning (DL, also named deep neural network) has shown success in various domains. The most appreciated work by Li et al. [19] is DL-based NER systems; the author proposed a minimal feature engineering method. State-of-the-art shows a considerable number of studies based on deep learning NER, to mention a few are [2, 6]. Table 7.2 is a literature review of machine learning methods of Named Entity Recognition.

Machine learning algorithm reviews highly recommend deep learning-based approaches for the best results. Table 7.3 is a literature review of deep learning methods of Named Entity Recognition.

7.2.2.4 Semi-supervised Learning

Supervised classifiers are data and computation hungry classifiers. Labeling done by human annotator can be expensive, biased, complicated, and time consuming process [6]. Semi-supervised methods uses small amount of labeled corpus, called "seed" data. More data samples are labeled using this seed corpus. This process is continued to generate a label dataset.

7.2.3 Hybrid Approaches

Hybrid approaches combines advantages of learning-based and rule-based techniques. It combines learning with hand-crafted rules for efficient results. This section shades a light on research in NER using hybrid approach. Xu et al. [4] combines CRF (rule-based learning) with, transformation-based learning for NER. Work by Pasca [15] is an ensemble classifier to extract named entities from Biomedical Data. Li et al. [8] is Turkish Named Entity Recognition, which combines lexical resources and patterns and rule-based learning. Linear CRF and Cluster-based approaches are combined to recognizing entities from English tweets. State of art shows improved results with better accuracy by using hybrid systems, refer Table 7.4.

This section is the study of literature for clinical NER techniques. Study parameters are mentioned in respective method tables. Literature shows that hybrid approaches are beneficial over rule-based and machine learning-based approaches. Comparison between rule-based and machine learning-based methods of NER highlights that the research scope is open in the area of identification of linguistic entity features, computation, and scaling of big data [29].

Table 7.3 Deep learning-based NER system

Authors and year	Domain	NEs found	Technique used	Embeddings	Parameter tuning	Dataset used	Observation
Lin et al. 2020 [24]	Sequence labeling using character level and word level information	Persons, organizations, locations	Hierarchical attention neural semi-Markov conditional random fields (semi-CRF) model	Sequence labeling using character level and word level information	Learning rate is set as 0.01 dropout rate of 0.5, BiLSTM level 50, hidden layer size of 256, and output vector size equal to output labels in each dataset	CONLL02, BC2GM, JNLPBA	Hierarchical LSTM-CRF attention based model to combine word level feature model and BERTs
Jason and Nichol 2020 [25]	News	Location, organization, person, and miscellaneous	Bidirectional LSTM-CNN	Glove, Word2vec	Additional layer of Collobert features, 80 epochs, 0.68 Dropout, 1 LSTM layer, CNN output vector size 53	OntoNotes 5.0 CoNLL 2003	The need for a transfer learning approach
Hofer et al. 2018 [26]	Electronic medical records	Problem, test, treatment	Bidirectional LSTM-CRF	Custom word embedding (BioNLP2016, MIMIC-III and UK CRIS)	Dropout 0.5, pool_size 52, 200 LSTM state	The 2012 i2b2/VA	Good results due to the combined approach, ordering of entities also taken into consideration
Ma and Hovy 2016 [27]	News	Person, location, organization	Bidirectional LSTM-CNN-CRF	Character embedding and glove word embedding layers	50 Epochs, 0.5 dropout, LSTM state size 200, 30 CNN output vector size	CoNLL 2003	Annotation customization can help in the medical domain

Table 7.4 Hybrid NER systems

Authors and year	Domain	NEs found	Technique used	Dataset used	Evaluation measures
Keretna et al. 2014 [28]	Biomedical	Drug named entities	Rule-based and lexicon based model	i2b2 2009 dataset	Low precision, increased FP with language-based rules
Basaldella et al. 2017 [3]	Biomedical	Biological processes, species, molecular functions, cellular component	CRF + CNN	CRAFT corpus	Dictionary-based pre-annotator results are improved by pipelining it to a machine-learning classifier
Xu et al. 2018 [4]	Clinical records	Disease names	Semantic BLSTM and CRF	PubMed, NCBI corpus	Results are good, data is manually annotated by 14 medical professionals
Ji et al. 2019 [5]	Clinical records	Disease, symptom, drug names	Attention BLSTM + CRF	Chinese medical records	Results are improved by adding embedding layer

7.3 Experimental Evaluation

Challenge in Information Extraction from Electronic Health Records (EHR) is the need for careful examination of facts. Extracted information need to be significant and meaningful. Health mining is evolved from structured to unstructured data analysis. Clinical relation extraction like Disease–Symptom relations, temporal relations, etc. is the key areas that have contributed a lot to developing many useful medical text information processing systems. Section 7.2 literature review of clinical Named Entity Systems (NER) emphasizes the outperformance of the Deep Learning (DL) techniques. To persuade this theory we experimented on three datasets, i2b2 corpus of clinical summaries, UMLS dataset and CONLL 2003 dataset [22, 24, 26]. This section is the details of the experimental study.

This section's experimental method is mainly the comparative analysis method, which reflects the experimental results by comparing the entity recognition efficiency under different model under the same dataset.

Dataset Description

We used datasets described in Table 7.5 for the experimental evaluation. These are popularly used datasets for the NLP NER task. Datasets are publically available, and can be accessed after signing an agreement of use.

The dataset used for experiments is the 2009 i2b2 challenge corpus composed of 1249 patient discharge summaries provided by Partners Health-care [30]. The available dataset is non annotated discharge summaries of patients. Included details are

Table 7.5 Dataset description

Dataset	Description
i2b2 2009 [30]	Electronic medical records, for clinical entity recognition publicly available i2b2 challenge dataset. This dataset is a collection of discharge summaries obtained from Partners Healthcare, Beth Israel Deaconess Medical Center, and the University of Pittsburgh Medical Center
CONLL 2003 [4, 5]	We performed the named entity recognition classification using the English data from a CoNLL 2003 shared task. This dataset contained four different types of named entities, i.e., person, location, organization, and misc
UMLS [31, 32]	The Unified Medical Language System (UMLS) is a medical repository that provides licensed based access to the individual researcher. We have constructed disease–symptom knowledge base using UMLS

Gender, Past History, Present illness, Family History, etc. This dataset is available for researchers and can be accessed by signing an agreement of use. For experimenting with information extraction we have used medication information corpus.

We have used the CoNLL-2003 English dataset. The CoNLL-2003 data files contain four columns, word, a part-of-speech (POS) tag, chunk tag and the named entity tag. We have used NER tags for our experiment. NER tags are in IOB format [4, 5].

The Unified Medical Language System (UMLS) is a medical repository that provides licensed based access to the individual researcher. We have constructed Disease–Symptom Knowledge Base using UMLS [2], which is the metathesaurus, a repository of interrelated biomedical concepts. We scrapped Disease–Symptom data from the UMLS database [32]. We pre-processed data in two columns. The first column is the disease, the second the number of discharge summaries containing a positive and current mention of the disease and the associated symptom.

The following section explains developed models using Python, Spacy NER model, and then Conditional Random Field (CRF) followed by Long and Short Term Memory (LSTM) technique for NER. Experiment details are discussed in each section with model selection criteria and extracted details; Fig. 7.11 shows the comparative results of implemented models.

7.3.1 spaCy NER Model

This section explains the Python package, spaCy, an open-source library for advanced Natural Language Processing (NLP) task. spaCy provides a model which identifies wide variety of named entities such as, company name, location, organization, product-name, etc. spaCy provides platform for customization of user defined entity-recognition model, by training the model to update it with newer trained examples. Figure 7.3 is the spaCy NER model; it shows the spaCy NER classification process. Input raw text is tokenized and processed to return named entities.

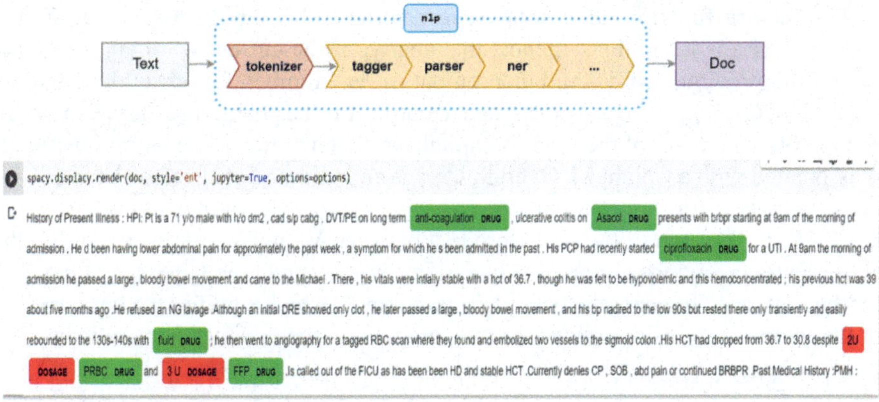

Fig. 7.3 spaCy model and NER output

The figure shows a result of simple spaCy NER model which is trained on clinical data. The result has false predictions; many of the entities are not recognized. There was no improvement in the results even after pre-processing.

7.3.2 *Conditional Random Field NER Model*

Literature shows a plethora of work using CRF [4, 13, 26, 33]; we referred to these papers for implementing CRF with custom features for clinical NER.

7.3.2.1 Conditional Random Fields with Synthetic Feature Selection

Conditional Random Field (CRF) is a sequence modeling algorithm. CRF learns patterns by observing future features with the basic assumption that features are dependent on each other. It combines feature selection methods of both, Hidden Markov Model (HMM) and Maximum Entropy Markov Model (MEMM) [4, 17]. CRF weighs each word in the training set based on predetermined features like Capitalization, word length and assign a label to each word based on its features and features of the neighboring terms too. CRF uses feature functions to quantify different characteristics of each word in a sentence. For examples whether the word is capitalized etc. Below is CRF formula where y is the output variable and X is the input sequence. CRF models output sequence as the normalized product of the feature function.

$$p(y|X, \lambda) = \frac{1}{Z(X)} \exp \sum_{i=1}^{n} \sum_{j} \lambda j f j(X, i, yi - 1, yi)$$

The feature function takes four basic parameters as input—a sentence X, the position i of a word in the sentence, the label y_i of the current word, and the label y_{i-1} of the previous word—and then outputs a real number. For example, a feature function $f(X, i, y_i, y_{i-1})$ can be designed to output 1 if the position $i = 1$, y_i = NAME, and the first character of the word is capitalized. Each output of a feature function fj is associated with a weight λj. Overall, there is an n-dimensional vector Vf = {f1(x), f2(x), …, fn(x)} for a word input x and a n-dimensional vector of weights $V_\lambda = \{\lambda_1, \lambda_2, \ldots, \lambda_n\}$. The sum of the scalar product between V_f and V_λ for each word in the entire sentence outputs a score that quantifies how accurate the sequence of labels is for the sentence. To generate the highest possible X, which signifies the best labeling of the sentence, V_λ is optimized through gradient descent, which is as follows:

1. For each feature function f_j, randomly assign a value between 0 and 1 to associated weight λ_j.
2. Carry out the following iterations until a certain number of iterations is performed (the stopping condition):

   a. Calculate the gradient of the log of probability for the score s (obtained through exponentiation and normalization) for λ_j.
   b. Move λ_j in the direction of the gradient with a constant learning rate α.

The vector of weights V_λ is optimized.

Broadly, there are two components to the CRF formula: Normalization and Weight features. Normalization factor normalizes the output label probabilities and weight feature assigns weights to the features by likelihood estimation.

Implementation Details

We synthesized our vocabulary by maintaining domain-specific words and their Part of Speech (PoS). Figure 7.4 depicts this process. Pre-processing includes tokenization, removal of stop words, etc. We modified the traditional method by adding Part of Speech of each word, which helped us building feature selection rules.

Features are assigned weights with respect to rules learned by CRF model. Figure 7.5 shows the result of CRF with synthesized feature vectors.

Figure 7.5 shows CRF model performance metric. Though the f1-scores are suitable for each label but the model did not generalize well when exposed to real data. The false-positive ratio was too high to be considered as a good model. The reason could be a limited vocabulary to generalize the model. Limited feature selection rules also added in not generalizing well.

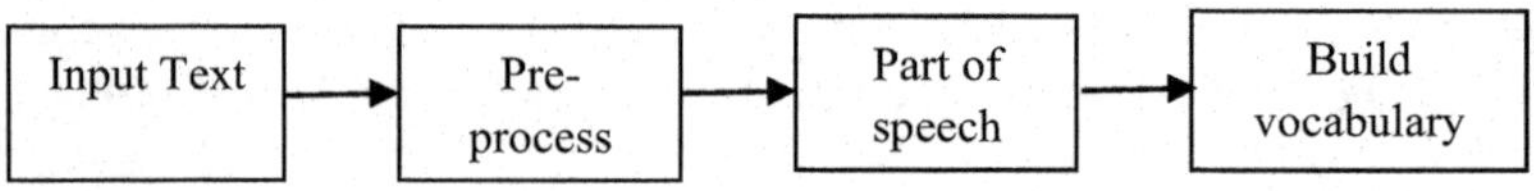

Fig. 7.4 Vocabulary construction process

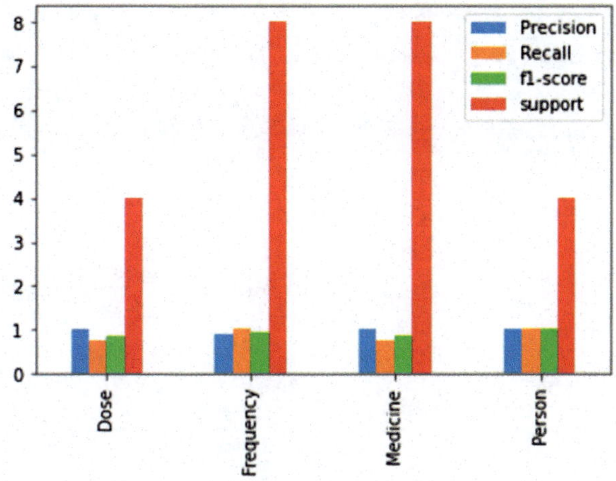

Fig. 7.5 CRF model result

7.3.2.2 Clinical Language Annotation, Modeling, and Processing (CLAMP)

We studied the CRF based model invented in year 2016 for clinical Named entity recognizer: CLAMP [34]. CLAMP is implemented with three approaches; NER using the machine learning conditional random fields (CRF) algorithm, a dictionary-based NER system, a UMLS lexicon dictionary, and a regular expression-based NER for common entities such as dates and phone numbers. Figure 7.6 shows a CLAMP result.

7.3.3 BLSTM NER

Long short-term memory (LSTM) cells are the building block of recurrent neural networks (RNNs). While plain LSTM cells in a feed-forward neural network process text from left to right, BLSTMs also consider the opposite direction. This technique allows the model to considers the sequence of tokens ahead of a word and tokens before the word as BLSTM learns in both directions. Figure 7.7 shows BLSTM architecture where x is input sequence, h is output sequence; y is concatenated output sequence where σ concatenates the forward and backward output elements [33].

We implemented a deep learning model using a bidirectional long and short-term memory algorithm on open-source dataset. We have used trained clinical word embeddings (details are given in the model implementation section) to identify domain-specific entities like disease names, treatment, dosage, etc.

Feature Selection
We used hand-crafted features with rules such as one sentence with five words, Casing features (lowercase, Title case), Character Embedding (for spelling correction). The

Results.xml Results.txt

Location_Start	Location_End	Semantics	CUI	Assertion	Entity
64	67	problem	null	present	dm2
70	73	problem	null	present	cad
78	82	treatment	null	present	cabg
85	91	problem	null	present	DVT/PE
95	121	treatment	null	present	long term anti-coagulation
124	142	problem	null	present	ulcerative colitis
146	152	drug		null	Asacol
167	172	problem	null	present	brbpr
185	203	temporal	null	null	9am of the morning
238	258	problem	null	present	lower abdominal pain
238	253	BDL	null	null	lower abdominal
277	290	temporal	null	null	the past week
293	302	problem	null	present	a symptom

Fig. 7.6 CLAMP result

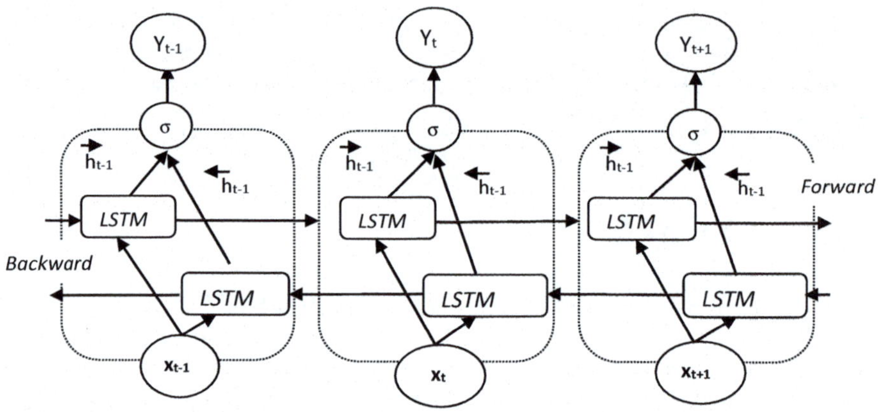

Fig. 7.7 BLSTM architecture

target tags used for annotations are Inside-outside-beginning (IOB) (short for inside, outside, beginning) tagging format. The IOB format is a standard annotation format for tagging tokens with named entity recognition. Here we are explaining each tag concerning an example of Symptom entity, 'cough and cold':

1. B-indications: It indicates the beginning of a symptom entity word (for example, 'cough').
2. I-indications: It identifies if word is inside an entity (for example, 'cold').

3. O: Words that are outside of a symptom entity are tagged as 'O'.

Therefore, any word which does not represent the symptom name is tagged as "O" tag. Similarly, the first word of symptom name has to be classified as "B-Symptom" and the following words of symptom name as "I-Symptom".

Word Embedding

Word embedding of unique words from the documents is used. The Patients discharge summaries were pre-processed to remove stop words and tokenize. The Vocabulary of uncommon words is created using Word2Vec model. This vocabulary is used to generate an embedding matrix. Embedding matrix is then fed to BiLSTM model, which is explained in following section.

Model Implementation

Bidirectional deep learning model was trained with a softmax output layer. It is a popular approach for sequence labeling. Input to the model is the whole sentence and compute tags as output. The first layer learns local features from each token of sentence. It maintains contextual information with word embedding in next layer. Feature layer initialize word embedding with pre-trained vector representation of tokens. The output of layer is a sequence of vector for each token. This vector output is fed to a softmax classifier to produce an NER tag per token.

To avoid overfitting, drop out layers are added in forward and backward directions. LSTM model is optimized by running it on Tensorflow Product Unit (TPU). This trained model cannot be used for identifying NER from unstructured data as pre-processing would be needed to convert the unstructured data into a model supported format (IOB format). Figure 7.8 shows model summary after fine-tuning the BLSTM model. The annotated test data model achieved f1-score of 77.8%.

```
[ ] model.summary()

Model: "model_1"
_________________________________________________________________
Layer (type)                 Output Shape              Param #
=================================================================
input_1 (InputLayer)         (None, 180)               0
_________________________________________________________________
embedding_1 (Embedding)      (None, 180, 180)          5513220
_________________________________________________________________
dropout_1 (Dropout)          (None, 180, 180)          0
_________________________________________________________________
bidirectional_1 (Bidirection (None, 180, 300)          397200
_________________________________________________________________
time_distributed_1 (TimeDist (None, 180, 4)            1204
=================================================================
Total params: 5,911,624
Trainable params: 5,911,624
Non-trainable params: 0
_________________________________________________________________
```

Fig. 7.8 Trained model summary of BLSTM result

7.3.4 BLSTM with CRF

This study classified clinical NER into three major entities: Problem, Treatment, and Test. These are the most practical entities being used in healthcare analytics, and we trained this model using i2b2 dataset—a part of Challenges in NLP for Clinical Data. The goal is to extract clinical entities (Problem, Treatment, and Test) from highly unstructured clinical text. We have used a pre-trained Spark NLP clinical NER model, which uses BLSTM + CRF algorithm [4, 5].

Implementation Details
Pre-trained Deep Learning NER (DL-NER) model implementation is based on the Deep Learning architecture shown in Fig. 7.9. Implementation pipeline takes clinical summaries as input and uses BERT Embeddings for feature selection; each word is translated to a 768-dimensional vector. Vector outputs are input to deep learning NER layer, forward LSTM, and backward LSTM layers. The output of backward LSTM is fed to the CRF layer for final classification.

As shown in Fig. 7.9, the word embeddings generated by the embedding layer are given as an input to the character level component. The individual character of

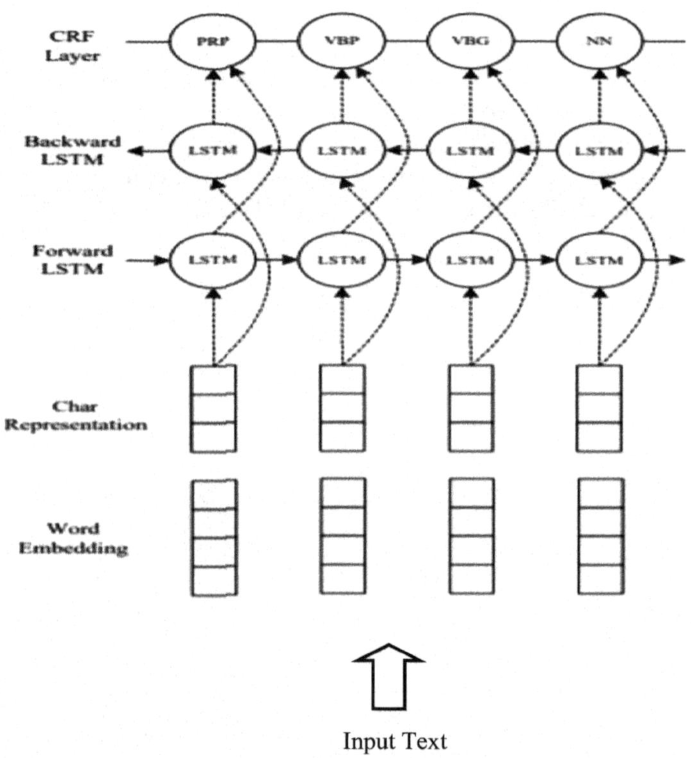

Fig. 7.9 DL architecture for BLSTM + CRF Model

the words is mapped to character embedding and is encoded by a forward LSTM. After pre-training the forward and backward learning model separately, we remove the top layer softmax and concatenate the forward and backward encoding to the traditional linguistic feature-based CRF model. This improves the NER labeling by adding the feature engineering layer before the final output layer. Given below are the algorithmic steps used for training process of BLSTM CRF model.

1. The first forward pass of the bidirectional LSTM-CRF model encompasses one forward pass for forward state LSTM and one forward pass for backward state LSTM.
2. In the next pass, feature engineering is performed with the CRF Layer in forward and backward pass.
3. The next forward pass of the bidirectional LSTM-CRF model encompasses one backward pass for forward state LSTM and one backward pass for backward state LSTM.
4. Parameters are fine tuned to improve the model accuracy.

We have trained the model on Spark NLP version 2.4.4 and Apache Spark version 2.4.5. We ran the NER BERT model for 50 epochs, a 0.5 dropout with 200 LSTM state size and three convolutional width. Figure 7.10 shows the LSTM + CRF model result with identified NER on unstructured clinical summary of a patient.

7.4 Result Discussion

Presented work is an empirical study of clinical NER approaches. For experiment, we have categorized our implementation into three significant categories, sequential (CRF, CLAMP), deep learning (BLSTM), and Deep learning combined with sequential (BLSTM with CRF) model with word embedding. This comparative study highlights the need for clinical word embedding and highlights that the sequential model combined with the deep learning model can be used for precise information extraction from unstructured electronic health records. Figure 7.11 shows the comparative result of implemented algorithms for clinical NER classification.

Figure graphically represents the comparison of the above explained four methods of NER. Graph X-axis shows identified NER mentions, and Y-axis reflects the count of respective NER. The following points summarize the analysis of results to the best of our knowledge.

- The spacy result had false predictions; many of the entities are not recognized. There was no improvement in the results even after pre-processing.
- Though the graph shows CRF performance as highest for the identified entities, we noted the high false-positive ratio after manual analysis because of dependency on feature engineering. CRF needs high-quality word embeddings or combined with a dictionary-based approach (CLAMP) to improve performance. CRF has

	chunks	begin	end	sentence_id	entities
0	dm2	64	66	0	PROBLEM
1	cad	70	72	0	PROBLEM
2	cabg	78	81	0	PROBLEM
3	DVT/PE	85	90	0	PROBLEM
4	long term anti-coagulation	95	120	0	TREATMENT
5	ulcerative colitis	124	141	0	PROBLEM
6	Asacol	146	151	0	TREATMENT
7	brbpr	167	171	0	TREATMENT
8	lower abdominal pain	237	256	1	PROBLEM
9	a symptom	292	300	1	PROBLEM
10	ciprofloxacin	375	387	2	TREATMENT
11	a UTI	393	397	2	PROBLEM
12	a large , bloody bowel movement	444	474	3	PROBLEM
13	his vitals	510	519	4	TEST
14	a hct	547	551	4	TEST
15	hypovolemic	588	598	4	PROBLEM
16	this hemoconcentrated	604	624	4	PROBLEM
17	his previous hct	628	643	5	TEST
18	an NG lavage	687	698	6	TREATMENT
19	an initial DRE	711	724	7	TEST

Fig. 7.10 BLSTM + CRF result

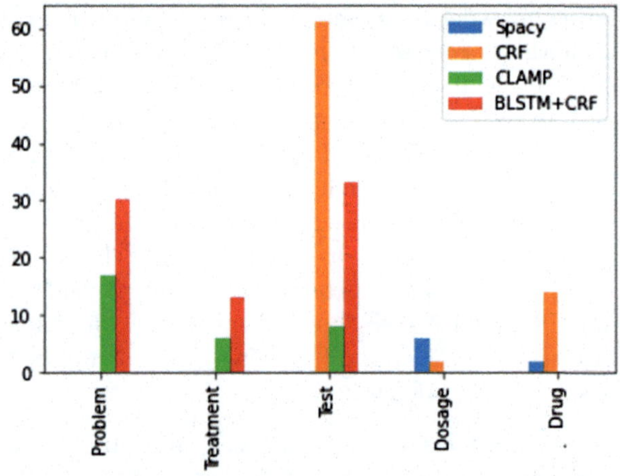

Fig. 7.11 Clinical NER classification result

a short term memory. It is a linear model, so it cannot describe more complex relations.

- BLSTM + CRF model overcomes the drawback of CRF and takes advantage of its sequential approach to consider the ordering of words in the sentence. Hence the model shows improved performance in identifying the accurate entities from the input text.

This experiment proves the better performance of BLSTM + CRF over other traditional models implemented in this chapter. As can be seen in the graph, the existing model does not fulfill the requirements of the proposed architecture, as even the best model is not identifying all entities due to insufficient word embedding. The possible reason, to the best of our knowledge, can be insufficient embeddings for out of vocabulary words. We have used character embeddings that lack context information, and word embedding fails to capture morphological information. We will upgrade the existing DL + Sequential model by improving the word embeddings to identify necessary entities with minimum possible error. Researchers can refer to recent work in sequential labeling and attention-based mechanisms for better feature engineering to improve results [24, 22].

7.5 Conclusion and Future Scope

A Preliminary step in information extraction from unstructured data is identifying essential keywords, popularly known as NER. The following Section explains Clinical NER methods and implementation details with comparative analysis of the state of the art algorithms. The Chapter reviewed clinical feature extraction task literature and discussed issues that make it quite challenging, such as nested entities, ambiguity in the text, availability of resources, etc. These challenges need careful handling. NER techniques literature promises outperformance of Deep Learning (DL) approaches. The Chapter evaluates sequential and DL NER discussed the results of the implementation of the NER module.

The Deep-learning approach, combined with sequential learning (BLSTM + CRF), conferred precise clinical NER results from discharge summaries. This is because the sentences in clinical narratives show related influences on each other. The improved clinical word embeddings to detect context-based entities can improve the studied NER techniques.

References

1. Mahajan, P., Rana, D.P.: Text mining in healthcare. Int. J. Innov. Technol. Exploring Eng. (IJITEE) **9**(2) (2019)

2. Ma, X., Hovy, E.: End-to-end sequence labeling via bi-directional LSTM-CNNs-CRF. In: Proceedings of the 54th Annual Meeting of the Association for Computational Linguistics, vol. 1 (2016)
3. Basaldella, M., Furrer, L., Tasso, C.: Entity recognition in the biomedical domain using a hybrid approach. J. Biomed. Semant. **8**(51) (2017)
4. Xu, K., Zhou, Z., Gong, T.: SBLC: a hybrid model for disease named entity recognition based on semantic bidirectional LSTMs and conditional random fields. BMC Med. Inf. Decis. Mak. **18**(114) (2018)
5. Ji, B., Liu, R., Li, S.: A hybrid approach for named entity recognition in Chinese electronic medical record. BMC Med. Inf. Decis. Mak. **19**(64) (2019)
6. Goyal, A., Gupta, V., Kumar, M.: Recent named entity recognition and classification techniques: a systematic review. Comput. Sci. Rev. **29** (2018)
7. Shaalan, K.: Rule-based approach in Arabic natural language processing. Int. J. Inf. Commun. Technol. **3**(3) (2010)
8. Li, Z., Liu, F., Antieau, L., Cao, Y., Yu, H.: Lancet: a high precision medication event extraction system for clinical. J. Am. Med. Inf. Assoc. **17**(5) (2010)
9. Zhang, S., Elhadad, N.: Unsupervised biomedical named entity recognition: experiments with clinical and biological texts. J. Biomed. Inf. **46**(6) (2013)
10. Munoz, O.M., Quimbaya, A.P., Sierra, A.: Named entity recognition over electronic health records through a combined dictionary-based approach. In: International Conference on Health and Social Care Information Systems and Technologies, vol. 100 (2016)
11. Rahem, K.R., Omar, N.: Rule-based named entity recognition for drug-related crime news documents. J. Theoret. Appl. Inf. Technol. **77**(2) (2015)
12. Lim, E.H.Y., Liu, J.N.K., Lee, R.S.T.: Knowledge discovery from text learning for ontology modelling. In: Sixth International Conference on Fuzzy Systems and Knowledge Discovery, vol. 7, pp. 227–231 (2009). https://doi.org/10.1109/FSKD.2009.669
13. Uzuner, Ö., Solti, I., Cadag, E.: Extracting medication information from clinical text. J. Am. Med. Inf. Assoc. **17**(5) (2010)
14. Datla, V., Lin, K., Louwerse, M.: Capturing disease-symptom relations using higher-order co-occurrence algorithms. In: 2012 IEEE International Conference on Bioinformatics and Biomedicine Workshops, Philadelphia, PA, pp. 816–821 (2012)
15. Pasca, M.: Weakly-supervised discovery of named entities using web search queries. In: CIKM'07, pp. 683–690 (2007)
16. Savova, G.K: Mayo clinical text analysis and knowledge extraction system (cTAKES): architecture, component evaluation and applications. J. Am. Med. Inf. Assoc. **17** (2010)
17. Chen, Y., Lasko, T.A., Mei, Q., Denny, J.C., Xu, H.: A study of active learning methods for named entity recognition in clinical text. J. Biomed. Inf. **58** (2015)
18. Wang, Y., Yu, Z., Chen, L., Chen, Y., Liu, Y., Hu, X., Jiang, Y.: Supervised methods for symptom name recognition in free-text clinical records of traditional Chinese medicine: an empirical study. J. Biomed. Inf. **47** (2014)
19. Li, J., Sun, A., Han, J., Li, C.: A survey on deep learning for named entity recognition. IEEE Trans. Knowl. Data Eng. (2020). https://doi.org/10.1109/TKDE.2020.2981314
20. Funde, K., Kuffner, R., Zimmer, R.: RelEx—relation extraction using dependency parse trees. Bioinformatics **23**(3) (2007)
21. Roberts, K., Rink, B., Harabagiu, S.: Extraction of medical concepts, assertions, and relations from discharge summaries for the fourth i2b2/va shared task. In: Proceedings of the 2010 i2b2/VA Workshop on Challenges in Natural Language Processing for Clinical Data, i2b2, Boston, MA, USA (2010)
22. Lin, J.C.-W., Shao, Y., Zhang, J., Yun, U.: Enhanced sequence labeling based on latent variable conditional random fields. Neuro Comput. **403**, 431–440 (2020). https://doi.org/10.1016/j.neucom.2020.04.102
23. Abulaish, M., Parwez, M.A., Jahiruddin: DiseaSE: a biomedical text analytics system for disease symptom extraction and characterization. J. Biomed. Inf. **100** (2019)

24. Lin, J.C.-W., Shao, Y., Djenouri, Y.: ASRNN: a recurrent neural network with an attention model for sequence labeling. Knowl. Based Syst. (2020) 106548, https://doi.org/10.1016/j.knosys.2020.106548
25. Chiu, J.P.C., Nichols, E.: Named entity recognition with bidirectional LSTM-CNNs. Trans. Assoc. Comput. Linguist. **4**, 357–370 (2016)
26. Hofer, M., Kormilitzin, A., Goldberg, P., Nevado, A.J.: Few-shot learning for named entity recognition in medical text (2018). arXiv preprint arXiv:1811.05468
27. Ma, X., Hovy, E.: End-to-end sequence labeling via bi-directional LSTM-CNNs-CRF (2016). arXiv:1603.01354
28. Keretna, S., Lim, C.P., Creighton, D.: A hybrid model for named entity recognition using unstructured medical text. In: 2014 9th International Conference on System of Systems Engineering (SOSE), Adelade, SA (2014)
29. Kormilitzin, A., Vaci, N., Liu, Q., Nevado-Holgado, A.: Med7: a transferable clinical natural language processing model for electronic health records (2020). ArXiv Prepr arXiv:2003.01271, 01271
30. https://www.i2b2.org/NLP/DataSets/Main.php. Accessed June 2020
31. Bodenreider, O.: The unified medical language system (UMLS): integrating biomedical terminology. Nucleic Acids Res. (2004). PubMed Central PMCID: PMC308795
32. https://www.nlm.nih.gov/research/umls/licensedcontent/umlsknowledgesources.html. Accessed June 2020
33. Saad, F., Aras, H., Hackl-Sommer, R.: Improving named entity recognition for biomedical and patent data using bi-LSTM deep neural network models. In: Natural Language Processing and Information Systems, vol. 10 (2020)
34. Soysal, E., Wang, J., Jiang, M., Wu, Y., Pakhomov, S., Liu, H., Xu, H.: CLAMP—a toolkit for efficiently building customized clinical natural language processing pipeline. J. Am. Med. Inf. Assoc. **25**(3) (2018)

Chapter 8
Application of Fuzzy Convolutional Neural Network for Disease Diagnosis: A Case of Covid-19 Diagnosis Through CT Scanned Lung Images

Priti Srinivas Sajja

Abstract The medical diagnosis can be enhanced by intelligent and automatic diagnosis through advances in Information and Communication Technologies (ICT) during the Covid-19 pandemic. This chapter discusses the current scenario, fundamental concepts, and existing solutions for diagnosing corona based diseases and their limitations. The chapter presents a generic and hybrid intelligent architecture for disease diagnosis. The architecture considers CT scanned images along with other fuzzy parameters and classifies the images into various disease categories using a convolutional neural network. The fuzzy convolutional neural network has experimented on 100 CT scanned images of lungs with additional fuzzy symptoms to prove the architecture's utility. The working of the convolutional layer, pooling layer, fully connected layer, fuzzy membership functions, and training data sets used in the experiment are discussed in detail in this chapter. The results are analyzed and presented graphically with improvement in accuracy, sensitivity, and precision. The chapter concludes with applications of the architecture for other disease diagnoses using radiology images and also discusses limitations and future work enhancement.

Keywords Covid-19 · Convolutional neural network · Fuzzy logic · Neuro-fuzzy system · CT scanned lungs images · Disease diagnosis

8.1 Introduction

Conventional and modern Artificial Intelligence (AI)/Machine Learning (ML) techniques have been efficient disease diagnosis instruments since their inceptions. Traditional and symbolic artificial intelligence has many limitations and depends on symptomatic data and inputs through manually extracted features. Modern and new AI/ML techniques have overcome such limitations, and the diagnosis's effectiveness has increased significantly. In this chapter, a generic and novel architecture for a fuzzy

P. S. Sajja (✉)
Sardar Patel University, Vallabh Vidyanagar, Anand, India
e-mail: priti@pritisajja.info

M. Mehta et al. (eds.), *Tracking and Preventing Diseases with Artificial Intelligence*,
Intelligent Systems Reference Library 206,
https://doi.org/10.1007/978-3-030-76732-7_8

convolutional neural network is presented along with an experiment on diagnosing viral diseases such as Covid-19.

The Covid-19 is the corona virus-based infectious disease that has affected the whole world in late 2019. The World Health Organization (WHO) has declared it as an international public health emergency. As per the December 2020 data, more than seventy million people have infected, and approximately one million people lost their lives due to this novel virus [37]. With the highly spreading nature and without specific vaccines and medicine, it is challenging to control the disease's spreading. The only possible remedy is preventing such disease by maintaining the social distance and healthy practices in the given situation.

Early prediction of the disease is also beneficial in such cases, especially for asymptomatic infections. Artificial neural networks have been an effective instrument in such disease detection [18]. However, the traditional artificial neural network requires extensive data relating to the problem's essential features before deciding. It may not be efficient when it comes to computer vision and image processing. Further, the tedious task of feature extraction remains manual. Here, the deep convolutional neural network helps in many ways, which is a model of an in-depth learning approach using a feed-forward neural network with multiple hidden layers [30]. In the case of the disease diagnosis such as Covid-19 based viral disease, diagnosis is made based on the lungs' CT scanned images. CT scanned images are computer-aided tomographical images that are taken by a computerized rotating X-ray machine. CT scanned images show more information than typical X-ray images. When plenty of such lung images are available, it is complicated to extract significant features from the images and feed them to the neural network.

The convolutional neural network extracts the CT scanned lung images' features automatically, which are sometimes not possible or feasible manually. Still, there is a challenge while employing the convolutional neural networks for the problem. The operators and experts are from medical and paramedical fields. It is complicated for non-computer professionals to deal with such crisp and normalized technology. Further, additional symptoms are also needed, which are vague in nature. This leads to the use of fuzzy logic, along with the convolutional neural network. Both the artificial neural networks and the convolutional neural network cannot handle uncertainty and computing with words as the knowledge in a neural network is stored indirectly in its connections. Hybridization of the convolutional neural network with fuzzy logic helps achieve dual advantages of feature extraction and handling uncertainty and computing with words for the benefit of non-computer professionals and increases the effectiveness and efficiency of the solution.

Many researchers have used CT scanned images for disease diagnosis through the use of machine learning techniques. Latest work in this stride includes the diagnosis of chest diseases [1, 13], skin cancer diagnosis [5], and diagnosis of liver disorders [19]. Many other researchers [15, 17] have also applied fuzzy logic for various applications, including disease diagnosis. Researchers [6, 31, 35] discuss the latest work done in the domain of disease diagnosis. Some common observations made from such existing solutions are (i) utilization of machine learning techniques restricted to a specific domain, (ii) non-utilization of machine learning techniques, and (iii) application of deep learning techniques without hybridization, etc. Refer Sect. 8.3 for the detailed discussion on the related work. A generic and hybrid model is provided in the chapter for the effective disease diagnosis and experimented for the Covid-19 diagnosis. The output of the research work is compared with similar existing solutions. The work documented in the chapter is as follows.

This chapter describes the novel hybrid fuzzy convolutional neural network for the classification of patients for the diagnosis of various virus-based diseases. Section 8.2 of the chapter documents a brief description of fundamental technologies such as fuzzy logic, convolutional neural networks, and hybridization. Section 8.3 of this chapter enlists the available solutions and work done so far in the domain, along with the common observations and limitations. Section 8.4 presents a domain-independent architecture for the disease diagnosis using scanned radiology images and briefly discusses its components. Section 8.5 demonstrates an experiment using the fuzzy convolutional neural network to classify the CT scanned images in various categories to diagnose diseases, if any. The section also provides necessary technical details about the image data sets, working of the convolutional neural network, fuzzy membership functions, sample training data of the underlying flatten neural network, and the experimental result analysis. Section 8.6 concludes the chapter by providing applications in other domains, limitations, and possible future enhancements.

8.2 Background Technologies

This section discusses the required background technologies and fundamental concepts related to the artificial neural networks, convolutional neural network and fuzzy logic, and their possible hybridization. The section starts with a brief note on the fundamental concepts of fuzzy logic, fuzzy membership functions, fuzzy rules, and a fuzzy rule-based system structure. Later, the section discusses artificial neural networks' limitations and establishes the convolutional neural network's need, particularly while working with feature extractions of images. The section also presents the fundamentals of the convolutional neural network, the feature extraction process, Rectified Linear Unit (ReLU), max pooling, and average pooling methods. At the end of the section, the need for possible hybridization is discussed.

8.2.1 Fuzzy Logic

Fuzzy logic introduced by Zadeh [44] is based on fuzzy sets, allowing an individual's graded membership to a set having no boundaries. Traditionally, all the sets have rigid boundaries, and a given individual has crisp or binary membership to the set. That is, the given element is either totally a member of the set or not. However, human beings are used to categorize the things into sets without boundaries. Thus, a member can have simultaneous graded memberships in different sets. For example, a tall person

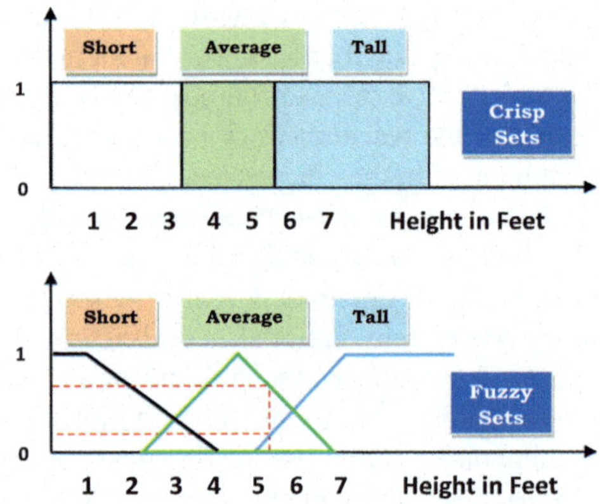

Fig. 8.1 Fuzzy membership functions for tall, average, and short people

can be a member of a *Tall* people class with graded membership of 0.8 degrees and a member of *Short* people class with a 0.2 membership degree.

Further, the fuzzy logic has the capability to work with the linguistic variables in conjunction with membership degrees. The machine needs appropriate membership functions to map the membership degree to the crisp values for further computing. The membership functions are also useful for converting the crisp output calculated by the machine into the linguistic values that are easy to handle for human beings. The procedure for converting the crisp values into their equivalent fuzzy values is known as the fuzzification, which is typically done with the help of membership functions. The process of converting the linguistic values into their appropriate crisp values is known as defuzzification. There are various defuzzification techniques available such as Centroid, Area under the curve, Adaptive integration, Bisector of the area, etc., [2]. Figure 8.1 displays the crisp sets and fuzzy membership functions for *Tall, Average,* and *Short* people.

As per the crisp sets defined in Fig. 8.1, all the sets have well-defined boundaries. For belonging to the *Short* class people, as per the definition, a person's height must be less than or equal to 3.2 ft. People with a height of 3.21–5.6 ft belong to the set of *Average* class people. To be a *Tall* class member, a person must have a height greater than 5.6 ft. In this classification approach, a person having 5.6 ft is treated at par with a person with 3.21 ft height, as they fall into *Average* class people height.

On the other hand, persons with heights of 5.6 and 5.61 ft are treated as two different classes. While the first one is *Average* class member, the second one is *Tall* class member. As per the fuzzy membership functions defined in Fig. 8.1, a person with a height 5.2 is *Average* class member with a membership degree of 0.75. The person is also a *Tall* class member with a membership degree of 0.20.

Figure 8.1 shows a graphical representation of fuzzy sets for *Short* class people and *Tall* class people with trapezoidal membership functions. The Figure also illustrates the *Average* people set definition with the triangular membership function. These are also called types of membership functions. There are many types of membership functions possible such as Normal, Gaussian, Trapezoidal, etc., depending on the membership function's characteristics and the shape of the graph represented by the membership function. Operations such as union, intersection, and complementation are also possible on the fuzzy membership functions.

The linguistic variables mentioned above can also be used in conjunction with fuzzy rules. Such fuzzy rules may incorporate the uncertainty factors or membership degree also. Traditionally, in fuzzy rule-based systems, such rules are compiled from the domain experts and other resources providing the domain knowledge. Later, such knowledge is validated and represented in the form of fuzzy rules into a knowledge base of the system. Since most of the content of such a knowledge base is in the form of rules, it is also called a fuzzy rule base.

Besides the rule base, other components are also needed to make the system work totally. These components are inference engine, repositories of membership functions, defuzzification methods, and user interface. These components are illustrated in Fig. 8.2, showing their relationships with each other.

Figure 8.2 illustrates the generic architecture of the fuzzy rule-based system. As shown in Fig. 8.2, a fuzzy rule-based system traditionally encompasses the rule base. To understand, interpret, and use the linguistic variables within the fuzzy rules by machine, their appropriate values need to be extracted through fuzzy membership functions. These fuzzy membership functions are also stored in the system. Besides these, the inference mechanism using forward, backward, or hybrid inference technique is also needed [2]. An inference engine's primary work is to infer knowledge/conclusion from the data provided (forward inference mechanism), or

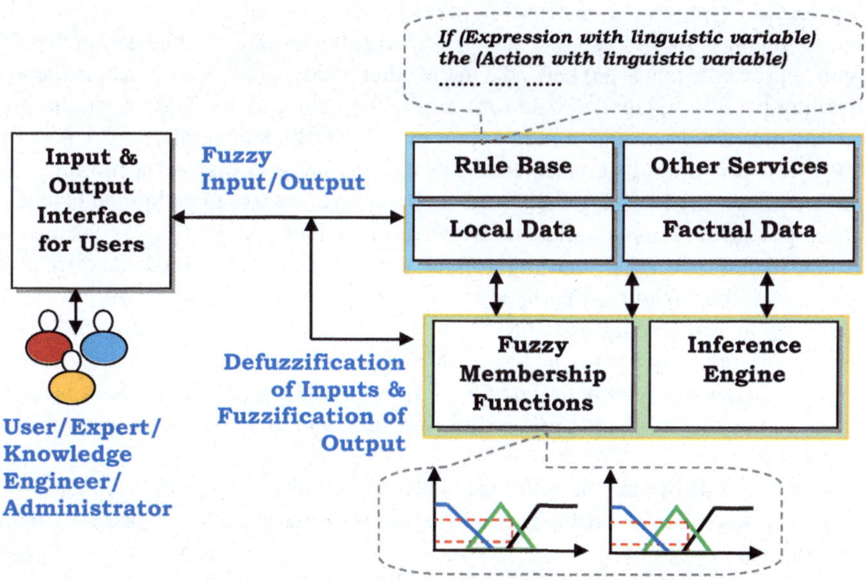

Fig. 8.2 Fuzzy rule-based system

set a hypothesis or goal, and test it in light of the existing data (backward inference mechanism). Above this, components such as approximate reasoning, detailed explanation, self-learning, and user interface are also needed to be designed.

Such fuzzy logic-based systems can provide reasoning and explanation, due to the ability to document the knowledge in the form of rules, which is not available with popular machine learning techniques such as neural network and genetic algorithms. In addition to this, a fuzzy logic-based system will have advantages such as handling vagueness through linguistic variables and using documented knowledge for further training.

8.2.2 Convolutional Neural Network

An Artificial Neural Network (ANN) is a tool that tries to simulate the human nervous system by implementing a set of neurons connected in a specific topology and learns in parallel, asynchronous, and distributed way. It is to be noted that a neuron generally does not contribute much towards the required solution. However, a large number of neurons connected with each other work in a parallel fashion together contribute in a significant manner. Such a network also exhibits a virtue of fault tolerance. The artificial neural network's popular topologies are Hopfield network, single perceptron for the linearly separable problems, support vector machines, multi-layer perceptron for non-linearly separable problems, Kohonen self-organizing map, etc. These topologies or structures of neural networks work in a supervised, unsupervised, or hybrid manner. In the case of a supervised learning paradigm, the training data sets with appropriate labels are needed. On the other hand, in the case of the unsupervised learning paradigm, labelled data for training the neural network are not required.

The most common and popular topology of the artificial neural network is a multi-layer feed-forward fully connected artificial neural network that learns in a supervised manner. Such networks typically have an input layer, two to three hidden layers, and an output layer. This is called shallow learning. If the hidden layers are many, it is considered as a deep learning architecture. Often, the artificial neural network does not extract features that need to be provided to the neural network as inputs. In a deep learning neural network, such features are extracted automatically by the neural network [30]. For image handling, particularly for feature extractions from the images, a convolutional neural network, which is a deep learning model, is considered. Yamashita et al. [41] present an overview and applications of convolutional networks.

Figure 8.3 illustrates the difference between a multi-layer perceptron and a deep learning model of the artificial neural network in terms of the process of feature extractions.

The reasons why such traditional shallow learning fails in image processing include the size of the image and its resolution. As per the conventional multi-layer perceptron model, each pixel of an image is considered as input. For a colour image with 224×224 size, the number of input nodes would be $224 \times 224 \times 3$ (for

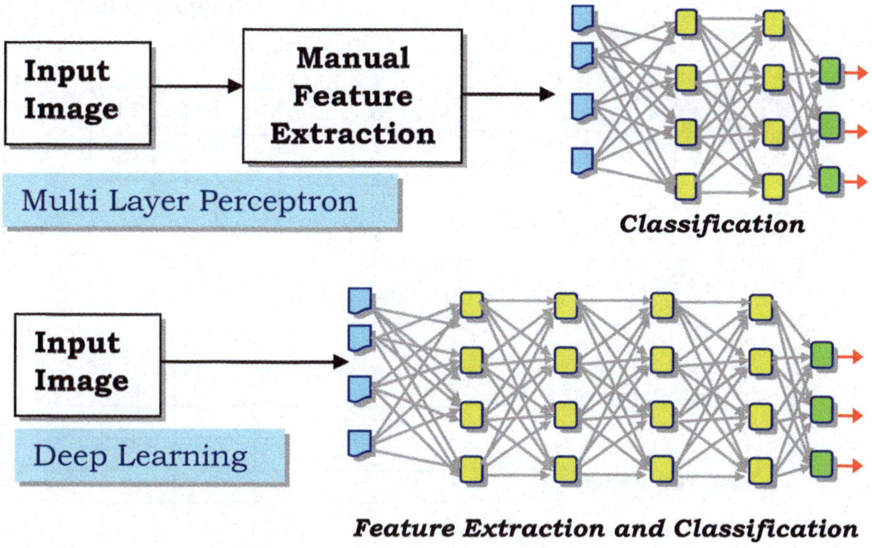

Fig. 8.3 Multi-layer perceptron and a deep learning ANN

three colour channels—RGB), which are significantly high. Further, the size of the weight matrix would also increase. In this case, the number of entries in the weight matrix exceeds 0.15 million. This leads to a tremendous increase in computation effort and time. Another primary reason is that, when an image is flattened into a shallow neural network, some spatial information may be lost. In such a case, there is a need to highlight pixels that are strongly related to each other using virtue of a filter or kernel of size 2×2, 3×3, or 5×5 depending on the size of the main image. Initially, the values of the features are random, and gradually they would be updated. The filter hovers on the image and generates a feature map, as shown in Fig. 8.4.

One can have more than one feature map also. Unlike the fully connected multi-layer perceptron, the convolutional neural networks are not fully connected. Sometimes, full or partial padding is also used to avoid downsampling that usually happens at the image's edges.

As an output function of the network, the convolutional neural network, Rectified Linear Activation (ReLU) function is used. The output of the function is the input value, if the provided input is positive. The output of the function is zero, if the provided input is negative. After the generation of the feature map, pooling is done. There are many methods of pooling. The average pooling, max pooling, and global pooling are some popular pooling methods [11]. The overall structure of a complete convolutional ANN is shown in Fig. 8.5.

In the end, the pooled set of extracted features undergoes flattening processes generating a long vector of data. This vector is used as an input layer of a traditional multi-layer artificial neural network.

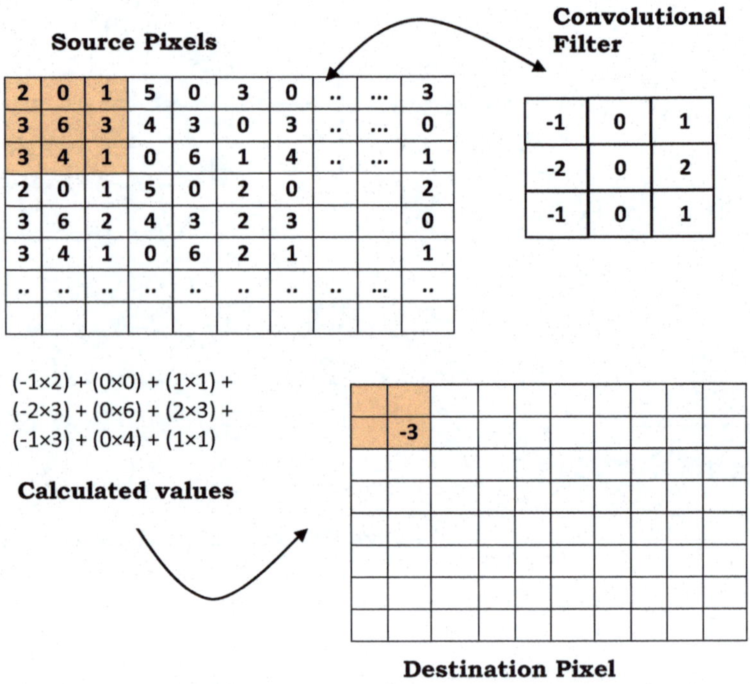

Fig. 8.4 Process of convolution

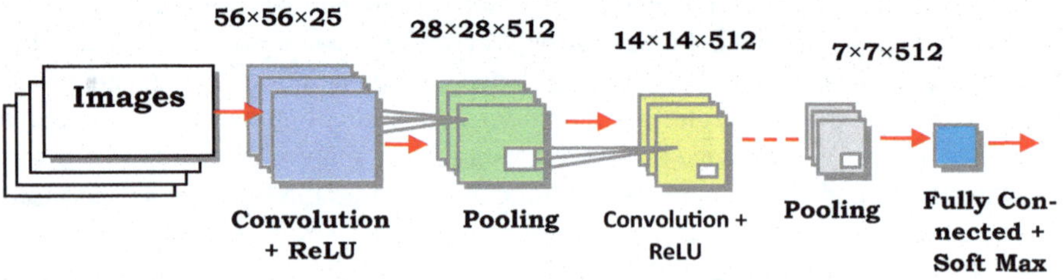

Fig. 8.5 Convolutional artificial neural network

8.2.3 Adding Fuzziness in the Neural Network

The inputs as well as outputs of a neural network are crisp and normalized. As the neural network-based architectures cannot document knowledge, it is not possible to provide inference, explanation, and reasoning knowledge. Above this, to fine-tune the output of a neural network further and present them in the friendliest manner, the use of fuzzy logic is suggested. The fuzzy logic has the virtue of handling uncertainty, approximate reasoning, and fuzzy inference in addition to the human-like reasoning and working with linguistic variables. A neuro-fuzzy hybridization is considered to take the dual advantages of the neural network and the fuzzy logic. Such hybridization can be done in many different ways. Two hybridization methods have become very popular nowadays; the first one is the linear approach, and the second one is the concurrent approach. In the linear approach, either the fuzzy logic or neural network completes its tasks and passes outputs to the other system or technique. Fuzzy logic can be used as input and the output interface to interact with users in a friendly way and deal with partial, uncertain, and loose (natural) linguistic inputs. The fuzzy membership functions convert these vague inputs into their most possible normalized form and send them to the neural network. The neural network processes the normalized input values and outputs equivalent crisp values, which further can be considered by the fuzzy logic. In the concurrent neuro-fuzzy system, the fuzzy logic and the neural network components work concurrently and improve the other components' workings. Here, one can consider the pure hybridization or fusion of both the technologies. A few examples of such pure hybridization are fuzzy activation functions, fuzzy weights, and fuzzy learning mechanisms within a neural network, self-learning of fuzzy rules, and learning parameters related to fuzzy logic-based systems [15, 29].

8.3 Related Work

There are plenty of tools, techniques, and solutions available in medical diagnosis using traditional as well as modern artificial intelligence techniques. Conventional artificial intelligence includes mainly the expert system based applications, which are historically pioneered knowledge based systems. Examples of such applications are the expert systems developed by the Stanford University such as Dendarl [10] and Mycin [20]. Traditional artificial intelligence has many limitations, such as the abstract nature of knowledge, the voluminous knowledge requirement, lack of knowledge acquisition, knowledge representation, and knowledge based system development methods and guidelines. Further, a high degree of effort is required to develop the knowledge base and the other utilities. Such utility includes an inference engine, self-learning, explanation, and reasoning facilities. After giving such a high amount of effort, cost, and time for the development of such a traditional artificial intelligence based system, there are chances that the system becomes obsolete due to new type of diseases, symptoms, medicine/drugs, vaccines, and clinical methods. In this situation, modern, bio-inspired techniques or machine learning based systems are most useful. This section discusses disease diagnosing with traditional intelligence based techniques and machine learning based techniques focusing on the novel Covid-19 disease diagnosis and prediction.

The use of the simple neural network in medical diagnosis is presented by a few researchers [3–5, 40]. The work presented by Qeethara [4] discusses how artificial neural networks are used in various disease diagnosis. The work also examines the acute nephritic cases with normalized data regarding symptoms and disease diagnosis with image data. Work presented by Samir [40] also discusses machine learning approaches to diagnose heart diseases. Aleksander and Morton [3] have also introduced artificial neural networks in disease diagnosis. Work presented by Orhan [9] demonstrates traditional artificial intelligence techniques for chest disease diagnosis. A similar type of review on artificial intelligence applications in the Covid-19 pandemic is presented in the work of Raju [33]. Demonstration of how the neural network based techniques can be used for the novel coronavirus detection is presented by [31]. The authors have used the X-ray images classified into two categories: infected and non-infected, with the support vector machine's help.

These works use simple artificial neural networks and do not extract features automatically. All the symptoms and the features related to the medical images need to be provided manually.

Convolutional neural networks and other machine learning techniques are also being used for various disease diagnoses through image processing. Some of the convolutional neural networks' uses are demonstrated by Jiuxiand [11]. The use of the deep convolutional neural network for disease diagnosis in general is presented in the work of [39].

The work of Pim [21] highlights the use of the convolutional neural network for brain disease diagnosis. Work presented by Manjit [14] discusses various e-health applications using random forest techniques. The convolutional neural network, which considers radiological images, is discussed by Venugopal et al. [34] to diagnose Covid-19 virus based diseases. Similarly, the diagnosis of various plant diseases through the plant images is experimented by Ramcharan et al. [26] to show applications of the technology in the agricultural domain. Sharada [22] also did similar work using the plant images. How the deep convolutional neural networks can be used to classify various images on the Image Net is demonstrated in the work of Alex [16]. In the year 2018, a convolutional neural network was used to detect diseases related to chest. The convolutional neural network, in this work, first considers x-ray images of chests and tries to classify the images [1]. Tulin [24] also did similar work.

The use of various filters and hierarchal threshold values is demonstrated in Pattrapisetwong and WChiracharit [25] work. The authors try to achieve a satisfactory level of automation in the segmentation of the chest radiography images. Machine learning techniques, particularly artificial neural network based models, have been popular in medical image processing, classification, and disease diagnosing. Geert [18] document a brief survey on how the deep learning artificial neural network can be useful in the domain. Islam et al. [13] also dealt with chest x-ray images to identify the level and location of the possible abnormalities within the images.

The latest work in the domain of the diagnosis of the disease using the machine learning and image processing include works of a deep learning model for corona related radiology image classification [17, 35, 42]. Recent experiments on the coronavirus using the CT scanned images are carried out by a few researchers [6, 32, 36, 38]. An experiment considering the CT scanned images of a possible Covid-19 patients conducted in China [6]. They have considered more than 100 patients and evaluated their CT scanned images at regular time interval for the possibility of the disease. However, they have not used the hybrid machine learning approach. In another experiment [32], CT scanned images are experimented for exploring the possibilities of early prediction of disease. It is to be noted that they have used only 21 images. Further, researchers [35, 38] experimented the deep learning technology to diagnose corona virus disease using CT scanned images. Inception transfer-learning model, which is a pre-trained neural network based model along with validations to diagnose the disease via CT scanned images is extended [35]. Three dimensional convolutional neural network for the diagnosing is particularly used in experiment [38].

These are non-hybrid convolutional neural networks experimented for a specific problem domain. Hybrid evolutionary convolutional neural networks are also used by a few researchers [8, 27, 43]. Fuzzy logic is also experimented along with the deep convolutional neural network [12, 15, 19, 23].

It is to be noted that the convolutional neural networks are being used for various disease diagnosis using medical images. The most popular are heart diseases, diabetic retinopathy, arthritis and osteoporosis, tuberculosis (TB), other tumours, and various organs' cancer. As the domain of interest is the chest diseases and virus-based infectious diseases such as novel Covid-19, the research work done for the chest related images through the convolutional neural network is considered.

The experiments and research work cited above carried out in the medical diagnosis domain using the simple neural network, convolutional network, and hybrid neural network have certain limitations. The significant observations on the research work reviewed and the article surveyed are as follows.

- In the initial period, the disease diagnosis used to be done with the traditional and symbolic artificial intelligence methods. The most popular among such techniques is the rule-based expert system. Such systems require a lot of effort, cost, and time for development and not adaptive enough in novel diseases, symptoms, and drugs/vaccines. Further, the rule-based expert system deals with data related to symptoms instead of radiology and medical images;
- Later, artificial neural networks are utilized for disease diagnosis. The simple and traditional artificial neural networks do not extract features from the images in an automatic manner;
- To overcome the limitations of automatic feature extraction, the use of deep convolutional neural networks started. Such convolutional neural networks extract features; however, they cannot handle uncertainty and cannot exhibit the ability to compute with words;
- Knowledge based advantages such as explanation, reasoning, use of the documented knowledge for future training and learning, etc., cannot be achieved with a simple or complex convolutional neural network. The fundamental reason behind this is that there is no symbolic representation of learned knowledge in the system. The neural network based systems implicitly store the knowledge into its weights matrix and connections;
- The convolutional neural network is used for a specific disease. Minimal work is done for diagnosis of the corona virus-based diseases using CT scanned images;

- The convolutional neural network developed for the Covid-19 analysis use manual inputs and statistical techniques instead of CT scanned images; and
- Use of fuzzy logic is not much seen for ease of use and other fuzzy logic related advantages for Covid-19 related CT scanned images.

An architecture involving a fuzzy deep convolutional neural network is proposed and experimented in this chapter to overcome these limitations. The next section represents the domain-independent generic structure and a brief discussion on the components' working.

8.4 Generic Architecture of the Disease Diagnosis System Based on Fuzzy Convolutional Neural Network

A fuzzy logic-based convolutional neural network is used instead of the classical deep convolutional neural network here to overcome various limitations of the existing approaches. The deep convolutional neural network is proposed to extract features of the given image by applying frequent convolution and pooling operations. Later, these features and other symptoms are provided to the flattened neural network to diagnose the disease possibilities. With appropriate fuzzy membership functions and other components, the system's fuzzy logic component enables handling the vague symptoms entry into the flattened neural network. The fuzzy component also works as an interface in co-operation with the deep convolutional layer and facilitates the human-like interface to the system. The hybridization is illustrated in Fig. 8.6.

Figure 8.6 demonstrates three layers. These layers are given as (i) an interface layer, (ii) a fuzzy layer, and (iii) a convolutional layer. The interface layer interacts with the system's major stakeholders such as users, experts, and administrators for collecting their information, symptoms, queries, feedback, and training datasets. These data may be in natural language, partial, and fuzzy. The interface layer selectively stores all these data into various repositories. The significant repositories are user profile repository, feedback repository, frequently asked questions (FAQs), and training sets repositories. These repositories can encompass fuzzy linguistic variables. The user profile can be fuzzy to present vague but useful information about the users, as suggested by [28]. The fuzzy user profile helps in increasing customized presentation and interaction with the system.

The fuzzy layer incorporates fuzzy membership functions and fuzzy rules along with the standard fuzzy inference and reasoning. The fuzzy queries and fuzzy training data sets undergo defuzzification with membership functions and various defuzzification techniques encoded into the system within the fuzzy layer. The result of the convolutional neural network classification and diagnosis can also be fine-tuned with fuzzy rules, which elaborate and extend the results achieved along with an interactive explanation in user-friendly manner. As the neural network cannot explain and handle vague input/output, the fuzzy logic component overcomes the limitations.

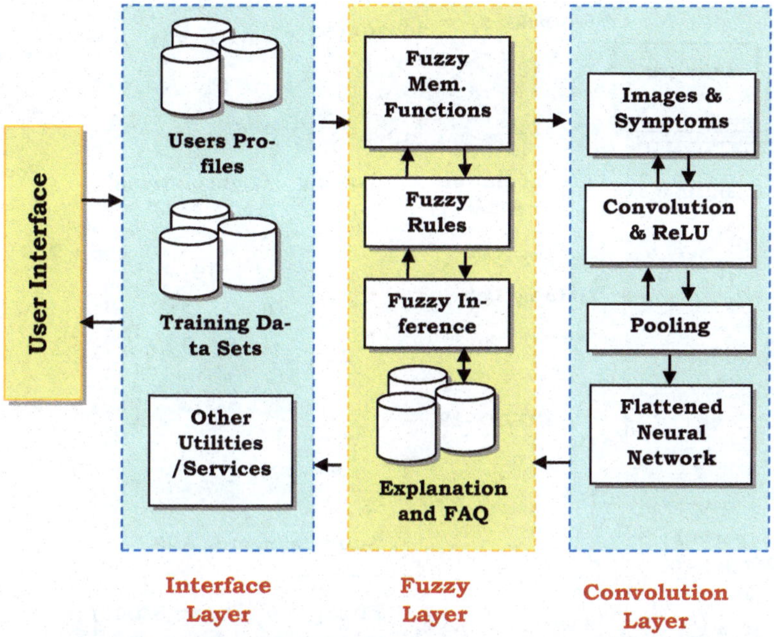

Fig. 8.6 Architecture of the fuzzy convolutional neural network for disease diagnosis

The convolutional neural network uses the scanned images and generates a convolved matrix/feature. After undergoing different convolutional and pooling layers, features are provided to a flattened feed-forward neural network with other inputs. The network is trained with necessary data and able to classify the scanned lung images into various categories. The neural network usually works with crisp and normalized data. The convolved feature provides crisp and normalized data; however, some symptoms are also needed for a final and complete diagnosis with images. Hence, with the convolved features, some more variables related to the patients' symptoms and experts observations on the patients are provided to the network. These symptoms are fuzzy; hence, they are converted first into their appropriate crisp values through the fuzzy layer and sent to the neural network along with the convolved feature matrix.

An experiment to demonstrate the working of the architecture is discussed in the next section.

8.5 Detailed Method, Experiment, and Results

To experiment with the generic architecture demonstrated in the previous Section, the CT scanned images of lungs are considered for classifying the patients into the

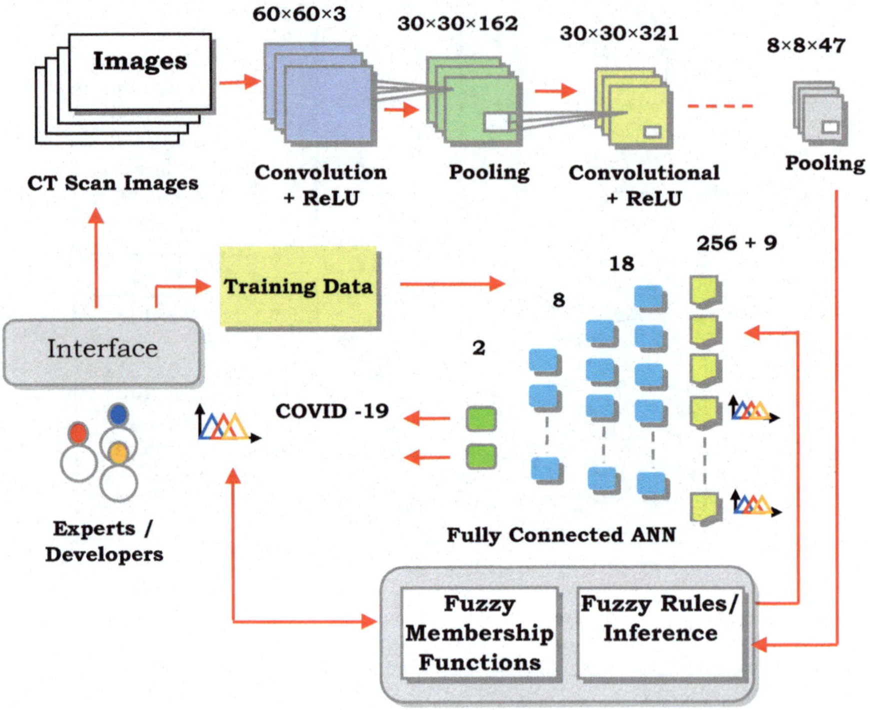

Fig. 8.7 Working of the experimental system

categories given as Covid-19 and other diseases. Figure 8.7 demonstrates use of the convolutional neural network with the pooling and convolution operations along with the flattened neural network with the fuzzy interface layer.

The dataset containing the CT scanned images of lungs is available as mentioned by Paul [7]. The dataset contains approximately 50 normal and 50 Covid-19 positive cases. Few other datasets are also available at Kaggle,[1] Github,[2] Visible human project by the USA,[3] and Neuro-archive,[4] which provide related data in standard image formats.

As denoted in Fig. 8.7, a selected image, after necessary pre-processing, goes to a convolutional neural network for feature extractions. The convolutional neural network works in three phases, namely (i) convolution, (ii) pooling, and (iii) flattened a fully connected neural network.

Convolution: The first phase considers the colour image; it needs to be divided into red, green, and blue (RGB) colour values (planes). The kernel value (k) is generally

[1]https://www.kaggle.com.

[2]https://github.com/.

[3]https://www.nlm.nih.gov/research/visible/getting_data.html.

[4]https://neurohive.io/en/popular-networks/u-net/.

taken as 3 with stride 2 or 2 with stride 2. The kernel is like a hovering flashlight from the top left corner of the array moving through the image to calculate convolved features. Each time a mathematical operation, such as matrix multiplication, is carried out and a new kernel size (3 × 3 × 1) matrix is generated. The newly generated matrix is called a matrix of the convolved values or convolved feature map.

The mathematical formula to output the feature matrix or feature map is given below.

$$H_j^n = Maximum\left(0, \sum^{k}\left(H_k^{(n-1)} \times W_{kj}^n\right)\right),$$

$$where\ H^n\ is\ the\ output\ feature\ map\ and$$

$$H^{(n-1)}\ is\ the\ input\ feature\ map. W\ is\ a\ kernel.$$

The convolved feature map is applied to a ReLU function to generate a rectified feature map. The ReLU application offers local linear approximation and non-linearity. Occasionally, contrast normalization is also done at this point.

Pooling: After the featured map is generated by convolutional and application of ReLU to get the feature map, the pooling layer comes into the picture. The pooling further reduces the convolved feature's spatial size with the Max pooling or Average pooling technique. Such pooling tries to reduce the over fitting and extracts important representative features from the image. Because of less size, it often reduces computation and increases computational efficiency.

The max-pooling and the average pooling are defined as:

$$H_j^n(x, y) = Maximum\ of\ H_j^{(n-1)}(x', y')\ and$$

$$H_j^n(x, y) = 1/K\left(\sum_{x,y} H_j^{(n-1)}(x', y')\right)$$

Figure 8.8 illustrates an example of max pooling and average pooling.

The pooling layer is independent for every convolved matrix and generally uses 2 × 2 or 3 × 3 size filters with stride 2, as shown in Fig. 8.8. By such down-sampling, about 75% of the activations can be reduced, and prepare a summarized version of the feature map. Instead of the max pooling, average pooling, L2 norm score pooling, or fractional max pooling can also be used.

Fully Connected Layer: The features pooled out from the previous phases are provided to a flattened neural network. A standard back-propagation algorithm is used to train such a flattened and fully connected neural network. This layer adds some symptoms extracted through medical professionals and patients in a fuzzy manner in this proposed research work. That makes the flattened fully connected neural network a little more comprehensive in terms of input neurons. The fully

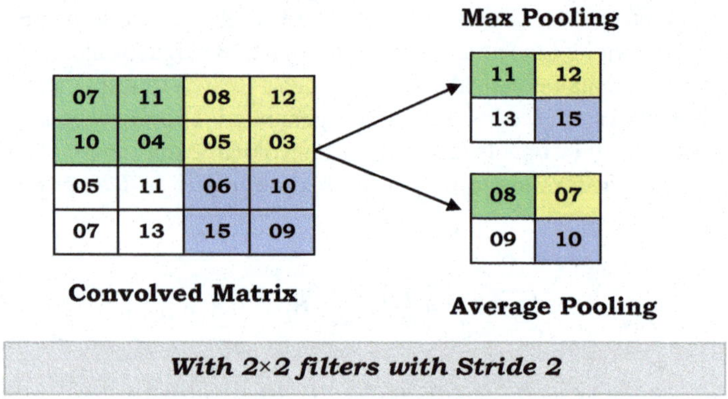

Fig. 8.8 Max pooling and average pooling in convolutional neural network

connected neural network now classifies the image into a required category. The neural network must be trained and tested with a sufficient amount of data.

Training Data to Fully Connected Layer: Besides the CT scanned image features, the other major input to the system is the data related to the symptoms. These data are fuzzy, which needs to be converted into their equivalent crisp values using the fuzzy membership functions. The fuzzy linguistic variables that play critical roles in identifying the Covid-19 are Fever, Dry cough, Short breath, Fatigue, Joint pain, Sore throat, Chills, Nausea, Diarrhea, etc. These data are collected from professionals and patients [29]. Table 8.1 enlists the fuzzy sample data related to the symptoms.

Fuzzification of the Training Data: As stated, the system works with the fuzzy data, which need to be converted into their equivalent crisp values using the fuzzy membership functions. The fuzzy linguistic variables that play critical roles in identifying the novel coronavirus disease named Covid-19 are mainly Fever, Dry cough, Short breath, Fatigue, Joint pain, Sore throat, Chills, Nausea, and Diarrhea. Some of

Table 8.1 Sample fuzzy data collected for input to the system

No.	S_1	S_2	S_3	S_4	S_5	S_6	S_7	S_8	S_9
1	Severe	Severe	High	Mod	V. Low	Low	Low	V. Low	V. Low
2	Severe	High	High	Mod	V. Low	V. Low	Low	Low	Low
3	Mod	V. Low	V. Low	High	High	V. Low	Low	Mod	Severe
4	Severe	V. Low	V. Low	Mod	Mod	V. Low	Low	V. Low	V. Low
5	Severe	Severe	Severe	Severe	High	Mod	V. Low	Low	Low
6	V. Low	V. Low	V. Low	High	Low	V. Low	Mod	Low	Severe
...	...	...	...	...	...	...	...	...	...

The Encoded Symptoms are: No.: Sr. no. of patient, S_1: Fever, S_2: Dry Cough, S_3: Short Breath, S_4: Fatigue, S_5: Joint Pain, S_6: Sore Throat, S_7: Chills, S_8: Nausea, S_9: Diarrhea

these functions are demonstrated in Fig. 8.9. Other fuzzy functions can be developed similarly.

It is to be noted that the first membership function demonstrated in Fig. 8.9 uses the degree Fahrenheit values on X-Axis as the measures for the entities within the universe of discourse. In the case of the other membership functions illustrated in

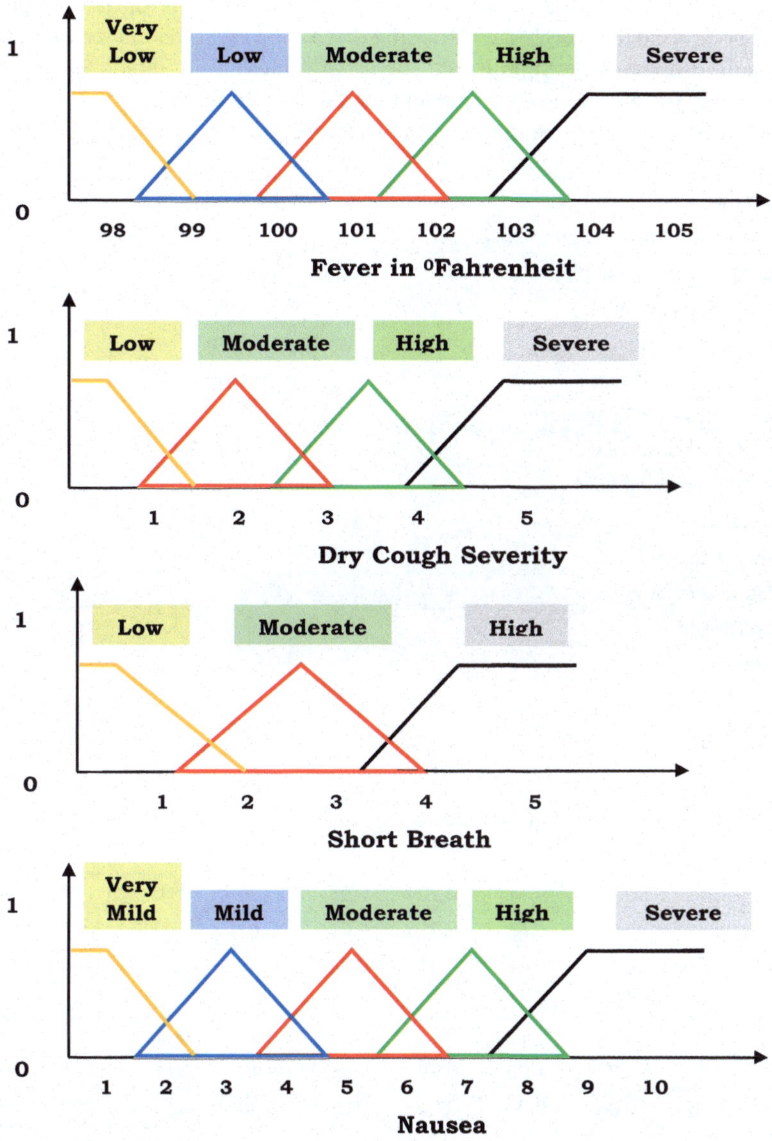

Fig. 8.9 Example fuzzy membership functions

Fig. 8.9, scale of 5 or a scale of 10 are used as measures on X-Axis. Here, type-2 fuzzy membership functions can also be used. In such cases, an additional type reducer mechanism is needed to convert the type-2 fuzzy membership functions into traditional membership functions.

Diagnosis is made based on the CT scanned image along with the symptoms defuzzified. However, the defuzzified values of the symptoms are not directly used. Each symptom value, after defuzzification, multiplies with appropriate weights. The experts determine the weights, considering that all the input parameters/variables are not equally important or significant. The crisp values of the inputs and the weight factors taken for the experiment are demonstrated in Table 8.2.

The statistic related to the convolutional, ReLU, pooling, and training is as follows.

Image size	**60 × 60 × 3**
Average forward time for an example	8 ms
Backdrop time per example	7 ms
Loss for classification	1.6018–2.408
1st conv. activation range	– 2.7 to 1.9
2nd conv. activation range	– 3.2 to 2.9
3rd conv. activation range	– 1.7 to 1.2
1st ReLU activation range	– 0.6 to 2.4,
2nd ReLU activation range	– 0.3 to 1.6
Pooling stride	2 × 2 (all pooling)
Training accuracy	0.5
Validation accuracy	0.39
Examples seen	100
Learning rate	0.01

Table 8.2 Defuzzified inputs to the flattened network

No.	S_1	S_2	S_3	S_4	S_5	S_6	S_7	S_8	S_9
Wts.	0.15	0.16	0.11	0.10	0.08	0.2	0.08	0.04	0.08
1	0.81	0.79	0.62	0.42	0.21	0.19	0.21	0.00	0.00
2	0.71	0.56	0.61	0.43	0.25	0.17	0.00	0.00	0.00
3	0.39	0.11	0.11	0.63	0.61	0.07	0.27	0.41	0.81
4	0.81	0.01	0.01	0.41	0.41	0.01	0.00	0.00	0.00
5	1.00	0.77	1.00	0.79	0.61	0.40	0.00	0.21	0.20
6	0.21	0.00	0.00	0.61	0.23	0.00	0.42	0.22	0.79
.	.	.	.	.	.	.	.	.	.

The Encoded Symptoms are: No.:Sr. no. of patient, S_1*: Fever,* S_2*: Dry Cough,* S_3*: Short Breath,* S_4*: Fatigue,* S_5*: Joint Pain,* S_6*: Sore Throat,* S_7*: Chills,* S_8*: Nausea,* S_9*: Diarrhea*

Table 8.3 Performance of the experiment

		Batch size	10	20	50	75	100
1	Sensitivity	Correctly identifying disease	0.8620	0.8167	0.8923	0.9233	0.9438
2	Specificity	Actual negatives identified	0.8912	0.8739	0.9134	0.9411	0.9567
3	Precision	Correct Prediction on total positive predictions	0.8710	0.8529	0.8876	0.9017	0.9108
4	False positive	Wrongly predicted as having the disease	0.0120	0.0180	0.0090	0.0030	0.0020
5	False negative	Wrongly predicted as not having the disease	0.0561	0.0540	0.0659	0.0700	0.0690

The data sets used has 100 images at the time of execution. The dataset is a kind of collaborative and can have a new addition of the images from time to time. The experiment is done with images with different batch sizes. In the beginning, only 10 images are taken. Later, the experiment is carried out with image sizes 20, 50, 75, and 100. In each experiment, approximately 70–75% of the images are used for training, and the remaining 25–30% of the images are used for the validations. Such a model's success is often measured by considering how effectively and correctly the images are identified and classified into appropriate data sets. The patients who have the disease need to be recognized as positive, and those who do not have the disease must be classified as negative. These facts can be measured in terms of sensitivity, specificity, precision, false-positive, false-negative rate, and accuracy using the literature's standard formulas. The results are enlisted in Table 8.3.

Some of the results are graphically presented in Fig. 8.10.

The results can be compared with the existing approaches of Wang and Wong [35], and Sethy and Bahera [31]. The results achieved from the proposed method align with their results with added benefits of fuzzy logic. Further, the results will become better with bigger data sets, as shown in Table 8.3. For comparison of the results, refer to Table 8.4. From Table 8.4, it may be observed that overall results are improved even though sensitivity for the proposed approach is slightly lower.

8.6 Conclusion

The proposed system architecture involving a fuzzy convolutional neural network is domain-independent and can be used for many disease diagnosis, image processing, and computer vision solutions. An experiment is carried out using approximately 100 CT scanned images of lungs to diagnose the novel Covid-19 virus-based disease with fuzzy symptoms to demonstrate the proposed architecture's utility. Other possible applications of this architecture include any disease diagnosis that deals with radiology images such as CT scanned images, x-ray images, etc. The skin diseases,

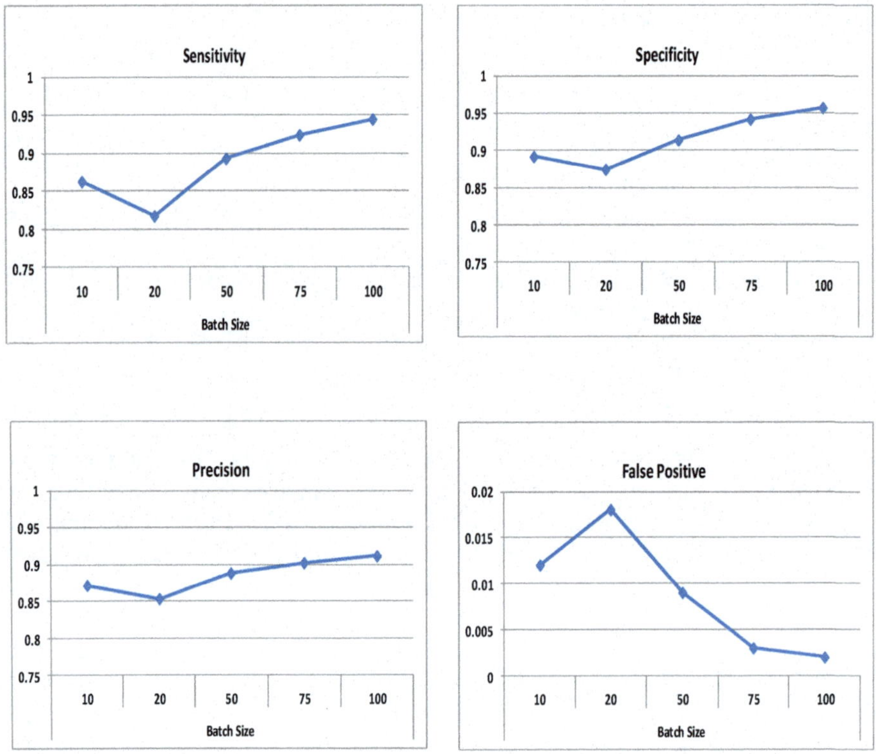

Fig. 8.10 Analysis of results

Table 8.4 Comparison of the results with the other approaches

		Wang and Wong [35]	Sethy and Bahera [31]	Proposed approach
1	Sensitivity	0.80	0.95	0.9438
2	Specificity	0.92	0.70	0.9567
3	Precision	–	0.90	0.9108
4	False-positive	–	–	0.0020
5	False-negative	–	–	0.0690

abdomen diseases (from endoscopy images), osteoporosis, retina, and fundus (eye) diseases, plant and crop diseases, etc., can be diagnosed effectively with machine learning techniques. Particularly for infectious diseases with high risk such as Covid-19, such systems help the medical professionals and non-professionals in effective diagnosis in the absence of symptoms (asymptomatic patients) and vague symptoms with some degree of uncertainty.

In the future, this system can be extended by adding a facility to handle the big data related to the health informatics and can provide analytics on the trend of virus-based infectious diseases, patients' statistics, hospital resource planning, work management of medical and paramedical staff, and other supporting information such as health insurance and claims. A full-fledged and web based ERP system can be planned with the proposed architecture as a primary facility for various disease diagnosis, advisory, analytical support, and other facilities.

References

1. Abiyev, R.H., Ma'aitah, M.K.: Deep convolutional neural networks for chest diseases detection. J. Healthcare Eng. (2018)
2. Akerkar, R.A., Sajja, P.: Knowledge-Based Systems. Jones & Bartlett Publishers, Sudbury, MA, USA (2010)
3. Aleksander, I., Morton, H.: An Introduction to Neural Computing. International Thomson Computer Press, Boston (1995)
4. Al-Shayea, Q.: Artificial neural networks in medical diagnosis. Int. J. Comput. Sci. Issues **8**(2), 150–154 (2011)
5. Andre, E., Brett, K., Roberto A.N., Ko, J., Swetter, A.M., Balu, H.M., Sebastian, T.: Dermatologist-level classification of skin cancer with deep neural networks. Nature **542**(7639), 115–118 (2017)
6. Bernheim, A., Huang, M., et. al.: Chest CT findings in coronavirus disease-19 (COVID-19): relationship to duration of infection. Relationship to duration of infection (2020)
7. Cohen, J.P., Morrison, P., Dao, L.: COVID-19 image data collection. Retrieved from https://arxiv.org/abs/2003.11597 (2020)
8. da Silva, G.L., da Silva Neto, O.P., Silva, A.C.: Lung nodules diagnosis based on evolutionary convolutional neural network. Multimed. Tools Appl. **76**, 19039–19055 (2017)
9. Er, O., Yumusak, N., Temurtas, F.: Chest diseases diagnosis using artificial neural networks. Expert Syst. Appl. **37**(12), 7648–7655 (2010)
10. Feigenbaum, E.A., Buchanan, B.G.: DENDRAL and Meta-DENDRAL: roots of knowledge systems and expert system applications. Artif. Intell. **59**, 233–240 (1993)
11. Gu, J., Wang, Z., Kuen, J., Ma, L., Shahroudy, A., Shuai, B., Liu, T., Wang, X., Wang, G., Cai, J., Chen, T.: Recent advances in convolutional neural networks. Pattern Recogn. **77**, 354–377 (2018)
12. Hsu, M., Chien, Y., Wang, W.: A convolutional fuzzy neural network architecture for object classification with small training database. Int. J. Fuzzy Syst. **22**, 1–10 (2020)
13. Islam, M.T., Aowal, M.A., Minhaz, A.T., Ashraf, K.: Abnormality detection and localization in chest x-rays using deep convolutional neural networks. https://arxiv.org/abs/1705.09850 (2017)
14. Kaur, M., Gianey, H., Sabharwal, M., Singh, D.: Multiobjective differential evolution based random forest for e-health applications. Mod. Phys. Lett. B **33**(5), 1950022 (2019)
15. Korshunova, K.P.: A convolutional fuzzy neural network for image classification. In: 3rd Russian-Pacific Conference on Computer Technology and Applications (RPC), pp. 1–4. Vladivostok (2018)
16. Krizhevsky, A., Sutskever, I., Hinton, G.: ImageNet classifcation with deep convolutional neural networks. Commun. ACM **60**(6), 84–90 (2017)
17. Lalmuanawma, S., Hussain, J., Chhakchhuak, L.: Applications of machine learning and artificial intelligence for Covid-19 (SARS-CoV-2) pandemic: a review. Chaos Solitons Fractals **139** (2020)
18. Litjens, G., Kooi, T., Bejnordi, E.B.: A survey on deep learning in medical image analysis. Med. Image Anal. **42**, 60–88 (2017)
19. Ma'aitah, M.K., Abiyev, R., Bus, I.J.: Intelligent classification of liver disorder using fuzzy neural system. Int. J. Adv. Comput. Sci. Appl. **8**(12), 25–31 (2017)
20. Melle, W.: MYCIN: a knowledge-based consultation program for infectious disease diagnosis. Int. J. Man Mach. Stud. **10**(3), 313–322 (1978)

21. Moeskops, P., Viergever, M., Mendrik, A., De Vries, L., Benders, M., Isqum, I.: Automatic segmentation of MR brain images with a convolutional neural network. IEEE Trans. Med. Imaging **35**(5), 1252–1261 (2016)
22. Mohanty, S., Hughes, D., Salathé, M.: Using deep learning for image-based plant disease detection. Front. Plant Sci. **7**, 1419 (2016)
23. Nguyen, T.-L., Kavuri, S., Lee, M.: A fuzzy convolutional neural network for text sentiment analysis. J. Intell. Fuzzy Syst. Special Section Green Human Inf. Technol. **35**(6), 6025–6034 (2018)
24. Ozturk, T., Talo, M., Yildirim, E.A., Baloglu, U., Yildirim, O., Acharya, R.: Automated detection of COVID-19 cases using deep neural networks with X-ray images. Comput. Biol. Med. **121** (2020)
25. Pattrapisetwong, P., Chiracharit, W.: Automatic lung segmentation in chest radiographs using shadow filter and multilevel thresholding. In: Proceedings of 2016 IEEE Conference on Computational Intelligence in Bioinformatics and Computational Biology (CIBCB). Manchester, UK: IEEE. Retrieved 2020 (2016)
26. Ramcharan, A., Baranowsk, K., McCloskey, P., Ahmed, B., Legg, J., Hughes, D.: Deep learning for image-based cassava disease detection. Front. Plant Sci. **8**, 1852 (2017)
27. Ravi, K.S., Heang-Ping, C., Lubomir, M.H., Mark, A.H., Caleb, R., Kenny, C.: Evolutionary pruning of transfer learned deep convolutional neural network for breast cancer diagnosis in digital breast tomosynthesis. Phys. Med. Biol. **63**(9) (2018)
28. Sajja, P.: Application of fuzzy user's profile for mining reusable e-learning repositories on web through lightweight mobile agent. In: Yan, L. (ed.) Handbook of Research on Innovative Database Query Processing Techniques, pp. 500–521. IGI Global, Hershey, PA, USA (2015)
29. Sajja, P.S.: Illustrated Computational Intelligence. Springer, Singapore, Singapore (2020)
30. Sajja, P., Akerkar, R.: Deep learning for big data analytics. In: Banati, H., Mehta, S., Kaur, P. (eds.) Nature-Inspired Algorithms for Big Data Frameworks, pp. 1–21. IGI Global Book Publishing, Hershey, PA, USA (2018)
31. Sethy, P., Behera, S.: Detection of coronavirus disease (COVID-19) based on deep features. Preprints **2020**, 2020030300 (2020)
32. Shan, F., Gao, Y., Wang, J., Shi, W., Shi, N., Han, M., Xue, Z., Shen, D., Shi, Y.: Lung infection quantification of COVID-19 in CT images with deep learning. Retrieved from https://arxiv.org/abs/2003.04655 (2020)
33. Vaishya, R., Javaid, M., Khan, I., Haleem, A.: Artificial Intelligence (AI) applications for COVID-19 pandemic. Diabetes Metab. Syndr. **14**(4), 337–339 (2020)
34. Venugopal, V.K., Mahajan, V., Rajan, S., Agarwal, V.K., Rajan, R., Syed, S., Mahajan, H.: A systematic meta-analysis of CT features of COVID-19: lessons from radiology. Retrieved from https://www.medrxiv.org/content/10.1101/2020.04.04.20052241v1 (2020)
35. Wang, L., Wong, A.: COVID-net: a tailored deep convolutional neural network design for detection of COVID-19. Cases from chest radiography images. In Arxiv (2020)
36. Wang, S., Kang, B., Ma, J., Zeng, X., Xiao, M., Guo, J., Cai, M., Yang, J., Li, Y., Meng, X., Xu, B.: A deep learning algorithm using CT images to screen for corona virus disease (COVID-19). Retrieved from https://www.medrxiv.org/content/10.1101/2020.02.14.20023028v5 (2020)
37. WHO.: Retrieved December 24, 2020, from https://www.who.int/emergencies/diseases/novel-coronavirus-2019/situation-reports (2020)
38. Xu, X., Jiang, X., Ma, C., Du, P., Li, X., Lv, S., Yu, L., Ni, Q., Chen, Y., Su, J., Lang, G., Wu, W.: Deep learning system to screen coronavirus disease 2019 pneumonia. Retrieved from https://arxiv.org/ftp/arxiv/papers/2002/2002.09334.pdf (2020)
39. Yadav, S.S., Jadhav, S.M.: Deep convolutional neural network based medical image classification for disease diagnosis. J. Big Data **6**(113) (2019)
40. Yadav, S., Jadhav, S., Nagrale, S., Patil, N.: Application of machine learning for the detection of heart disease. In: 2nd International Conference on Innovative Mechanisms for Industry Applications. IEEE, Bengaluru (2020)
41. Yamashita, R., Nishio, M., Do, R.: Convolutional neural networks: an overview and application in radiology. Insights Imaging **9**, 611–629 (2018)

42. Yoon, S.H., Lee, K.H., Kim, J.Y., Lee, Y.K., Ko, H., Kim, K.H., Park, C.M., Kim, Y.H.: Chest radiographic and CT findings of the 2019 novel coronavirus disease (COVID-19): analysis of nine patients treated in Korea. Korean J. Radiol. **21**(4), 494–500 (2020)
43. Yosuke, T., Fumio, O.: How convolutional neural networks diagnose plant disease. Plant Phenom. **9237136**, 1–14 (2019)
44. Zadeh, A.: Fuzzy sets. Inf. Control **8**, 338–353 (1965)

Chapter 9
Computer Aided Skin Disease (CASD) Classification Using Machine Learning Techniques for iOS Platform

C. Alvino Rock, E. Bijolin Edwin, C. Arvinthan, B. Kevin Joseph Paul, Richard Jayaraj, and R. J. S. Jeba Kumar

Abstract Skin lesions is caused due to thermal exposure, genetic disorder, accumulation of dead cells etc. This is due to the abnormal growth in skin tissue, defined as cancer. This disease plagues are found in more than 14.1 million patients which results in 8.2 million deaths worldwide. Therefore the classification construction model for three lesions is proposed in a cost effective way. Once users submit a picture of a skin lesion they will receive back instant prediction. This model classifies skin lesions into three classes. Create ML technique of iOS systems is opted for creatng machine learning models which will be deployed in the app. Create ML technique of iOS systems is used along with tools including Swift and mac-OS playgrounds for creating and training custom machine learning models on the Mac system. Designing an online tool which intimates doctors and lab technologists about three types of highest probability diagnostic diseases for a given skin lesion observation. This helps quickly detect high priority patients in less than three seconds. Ensuring privacy the images go-through pre-processing and gets stored locally and never gets uploaded to an outsourced server, thereby maintaining the medical privacy. CASD system include working of skin lesion analysis in a robust and secured way.

Keywords Image processing · Data science · Skin lesion · Artificial intelligence · Biomedical image analysis · Automated diagnosis

C. Alvino Rock · E. Bijolin Edwin · C. Arvinthan · B. Kevin Joseph Paul · R. Jayaraj
Department of Computer Science and Engineering, Karunya Institute of Technology and Sciences, Coimbatore, Tamil Nadu, India
e-mail: alvinorockc@karunya.edu.in

E. Bijolin Edwin
e-mail: bijolin@karunya.edu

B. Kevin Joseph Paul
e-mail: kevinjosephpaulb@karunya.edu.in

R. J. S. Jeba Kumar (✉)
Department of Electronics and Communication Engineering, Karunya Institute of Technology and Sciences, Coimbatore, Tamil Nadu, India

M. Mehta et al. (eds.), *Tracking and Preventing Diseases with Artificial Intelligence*, Intelligent Systems Reference Library 206,
https://doi.org/10.1007/978-3-030-76732-7_9

9.1 Introduction

9.1.1 Background on Existing System

Dermatology is an important sector of medicine, with various skin ailment due to hypertension and external factors which ultimately leads to dermatology cancer. At present skin cancer is a health, economic and societal problem faced by all community, that for many years have been considered with the similiar ideology by the dermatology sector [1]. This is problamatic when we inspect that for the past 30 years the count of cases investigated with skin cancer has been doubled considerably [2]. A major portion of penny gets spent by individual for inspection of diagnosis through conventional lab based reports for skin lesion identification. The doctor inspects the lesion and takes immediate action on the pieces of statement report elucidates. By reducing the conventional steps of report based confirmation, it will reduce the expenditure for the whole dermatology treant process. Skin diseases create issue worldwide which ranks 18th globally in health problems generated. Medical imaging is valued highly, as dermatology holds more diseases to treat. Additionally, the opted way to identify early skin diseases is to know beforehand of new or upcoming skin growths [3]. Cross sectioning using the naked eye is the first tool used by specialists together with technique ABCDE, which involves scanning the skin area [4]. By this method, the cross section from medical images is similiar to the cross section with the naked eye and thus would apply the similiar methodologies which supports the conventional concept that skin cancer can be identified through photography, but this approach fails to predict in early stage [5].

9.2 System Analysis

9.2.1 Literature Survey on Existing System

There are algorithms and tools to help professionals detect diseases in various fields which provides more confidence to doctors as they hold more data to treat patients. History points out many trials had been made over the years,dealing with algorithms like KNearest Neighbors and Support Vector Machines brought good output, but tiresome to build applications of such methodologies. Hamblin et al. [1] proposed an approach to classify melanoma, seborrheic keratosis, and nevocellular nevus, using dermoscopic images. Their work showed an ensemble solution with two binary classifiers obtained from age and sex information of the patients. Armstrong et al. [2] used image processing before training which results in normalization and noise reduction on the dataset as non dermoscopic images possess noise. Khan et al. [3] brought a major breakthrough in the research by comparing the result of the their learning model infront of 21 board certified dermatologists and proved accurate. Mete et al. [4] proposed work for ISIC Challenge where it uses deep learning models on ImageNet. Models used were ResNet-101 and Inception-v4 by experimenting them with several configurations of the dataset and data-augmentation with transformations.Critical success of the project was obtained by the volume of data, normalization. The latter is an SVM layer acts as the final output of the deep-learning models, that map the outputs to the three classes that were proposed in the ISIC Challenge where it secured a prize. Lundervold et al. [5] classified the skin lesions into unique but not meta-classes like benign and malignant. It uses ResNet-152 model trained on the ImageNet model. The images were collected from Korean hospitals. Ethnic differences in the context was the reason for bad results when we tried out with different datasets, for collecting data from different ethnics and ages to show it as real world problem.All above works have been observed that they one common problem which is data scarcity with very few publicly available data. The researchers tried to collect data from private hospitals. Furthermore, data collection did not fully hold the problem, rather it was still mandatory to use methodologies such as transfer-learning and data-augmentation [6, 7].

9.2.2 Proposed CASD System

Skin lesions are detected with Testing output of 84% in training and validation accuracy. Since skin lesion disease have very less stint features to classify them. properly so try to obtain large datasets from the hospitals and then train the model.

9.2.3 Requirements for CASD System

Functional requirements include the behaviors like registration, login and picture clicking. Non functional requirements include attributes like first name, lastname, email-id, password, username, password. Hardware requirements include Iphone, Macbook pro, Docker cable.Software requirements include XCode, iOS Simulator, Create ML.

9.3 System Design

9.3.1 Create ML Model

Create ML creates machine learning models which get deployed in an app. Create ML uses applications like Swift and macOS playgrounds to create and train default

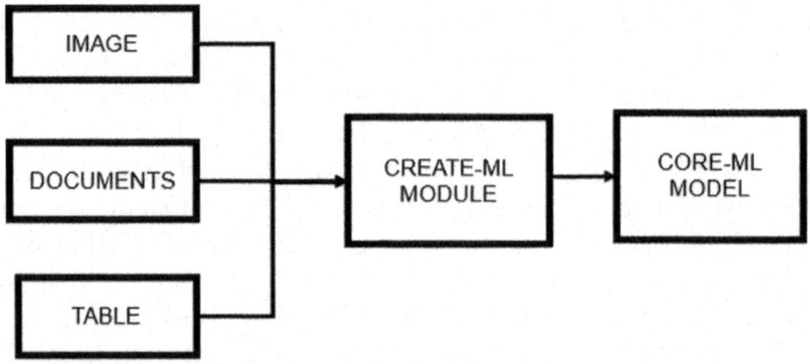

Fig. 9.1 Simplified Create-ML and Core-ML architecture

machine learning models on Apple laptop. Model gets trained to do works like image recognization, extraction meaning from text and relate numericals.

Figure 9.1 depicts models which get trained to recognize patterns by giving sample test images. For example model is trained for recognizing cats by giving cat images as training data and testing with new data.

9.3.2 Core ML Model

"Core ML" is used to integrate machine learning models for the given application. Core ML is a combined representation for all models. Core ML APIs and user information are required to make predictions, train and fine tune models.

Figure 9.2 depicts a model that is created by applying a machine learning algorithm for training data to make predictions based on new inputs. For example model gets trained to classify photos, or detect specific objects within a photo. Built and trained a model with the Create ML app together with Xcode. Models trained using Create ML are in the Core ML model format and the application is ready to use. Contrastingly other machine learning libraries and Core ML Tools hold the model in Core ML format. Model found in user's device uses Core ML to retrain models using user's data found in device. Core ML adjusts device performance by rising the CPU, GPU, and Neural Engine but reducing its memory footprint and power consumption. Running the model on device removes the need for internet connection, thereby increasing data privacy and app responsiveness (Fig. 9.3).

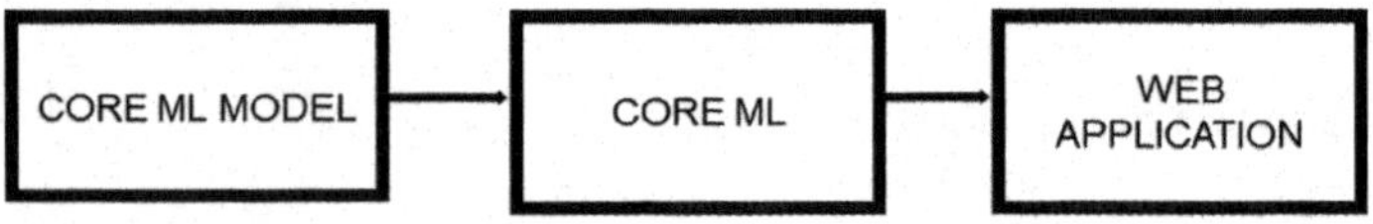

Fig. 9.2 Core ML workflow of CASD system

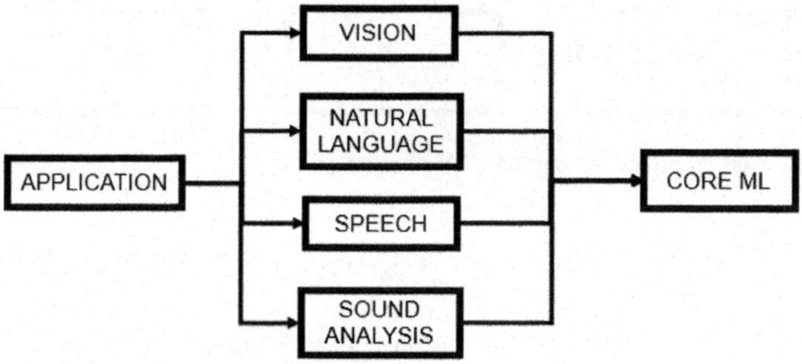

Fig. 9.3 Create-ML application module topology

9.3.3 Apple iOS Architecture for CASD Machine Learning System

The iOS is the operating system for mobiles introduced by apple. Different mobile types include iPhone, iPod, iPad etc. The iOS is costly but Android is popular. The iOS architecture is layered and contains an intermediate layer between the applications and the hardware does not have communication. The lower layers give basic level services and the higher layers provide user interface and complicated graphics (Fig. 9.4).

Fig. 9.4 Apple iOS architecture

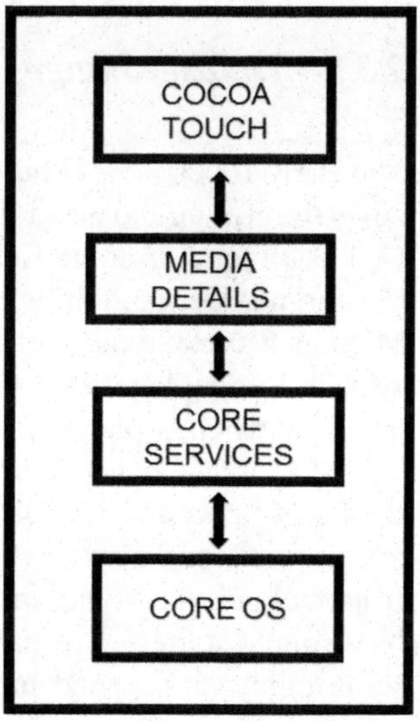

Fig. 9.5 Firebase cloud messaging for CASD system

9.3.4 Firebase Architecture for CASD Database System

Firebase is a essential platform for mobile and web app development to build high-quality applications, grow user base, and increase profit. The Firebase Realtime Database is a cloud based NoSQL database which syncs data in realtime. The Real-time Database is really just one big JSON object that the developers can manage in realtime. With single API, the Firebase database provides both current value and any updates to the data. While offline, use local cache to store changes but while online, the local data gets automatically synchronized. The Realtime Database joins together with Firebase Authentication to provide a simple authentication.

Figure 9.5 depicts Firebase Cloud Messaging (FCM) provides a battery-efficient connection between server and devices to send and receive messages at zero cost. Either send target messages using predefined code segments or create it on your own. It sends messages to a group of devices that are subscribed or to a single device. It sends messages immediately, or at a future time and no coding is involved. It is integrated with Firebase Analytics which helps in tracking.

9.3.5 Dataset Images for CASD System

The HAM10000 dataset holds images needed for designing the model. It includes a collection of ten thousand labelled images. Types of skin lesions are dataset-Basal Cell Carcinoma, Actinic Keratoses, Vascular Lesions, Benign Keratosis, Malignant Melanoma, Dermatofibroma, and Melanocytic Nevi. Three kinds of diseases like Malignant melanoma, melanocytic nevi, benign keratosis are used in this CASD system. Benign keratosis is a generic class that includes seborrheic keratoses, solar lentigo and lichen-planus like keratoses which appear different dermatoscopically but similar biologically. By dermatoscopically lichen planus-like keratoses are difficult to identify because they display morphologic features like melanoma and so they are biopsied for diagnosing diseases [8, 9]. Melanoma is a harmful neoplasm arising from melanocytes found in different forms. If it is found in initial stage it is cured by a simple surgery. Melanomas can be harmful or not but usually not harmful and some highly vary depend on anatomic areas. Melanocytic nevi are harmful neoplasms arising from melanocytes and have different forms which differ dermatoscopically. In contrast to melanoma they are symmetric with respect to color and structure [10] (Figs. 9.6, 9.7 and 9.8).

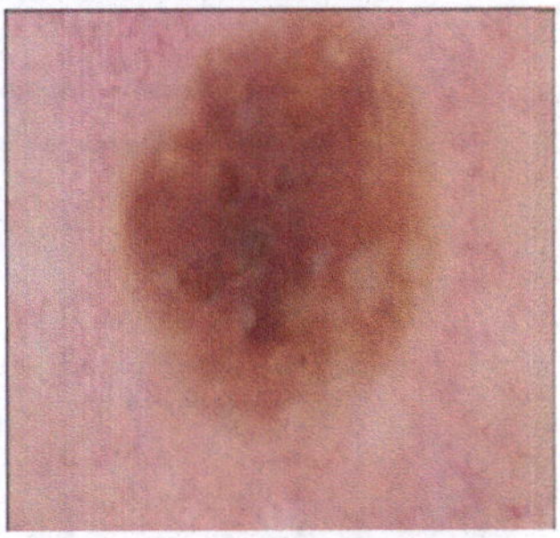

Fig. 9.6 Sample image tile of Benign Keratosis

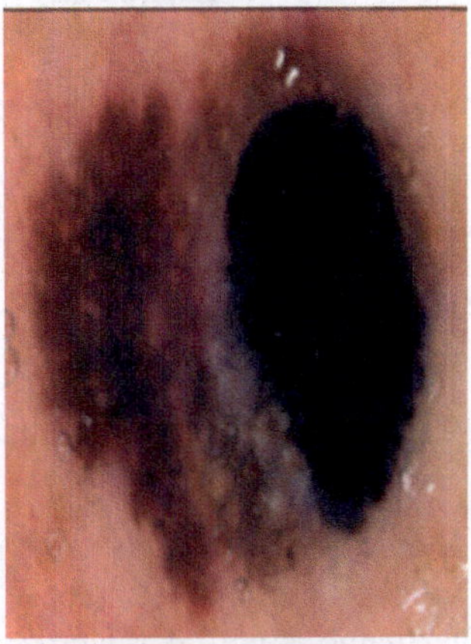

Fig. 9.7 Sample image tile of Malignant Melanoma

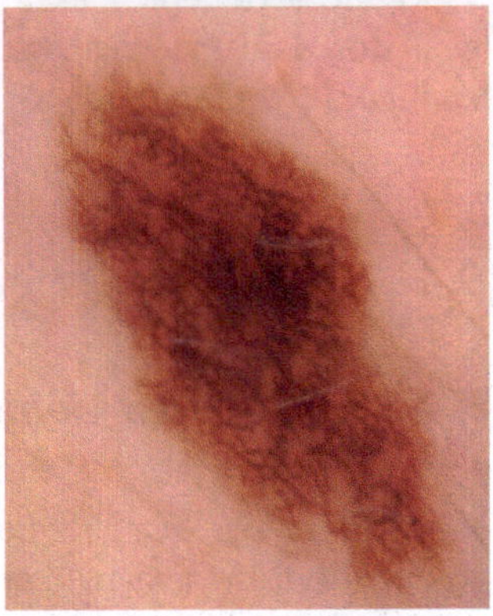

Fig. 9.8 Sample image tile of Melanocytic Nevi

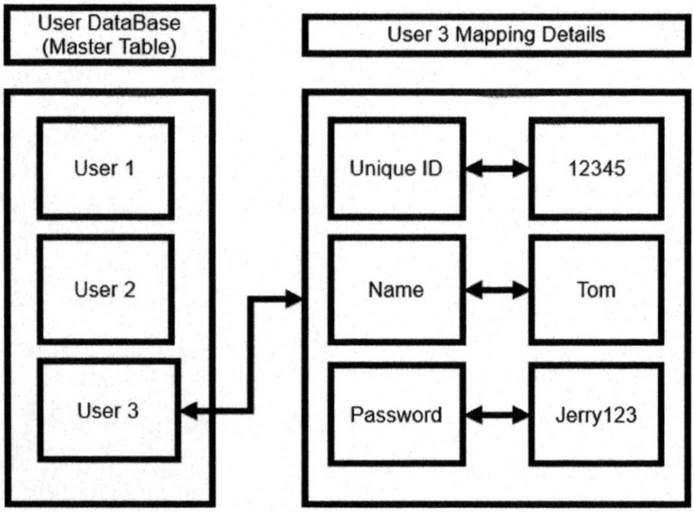

Fig. 9.9 Database structure of CASD system

9.3.6 Database Structure of CASD System

Figure 9.9 depicts this database holding registration details entered by the user which will be stored in the database while signup. The registration details include firstname, lastname, email, password.

Figure 9.10 depicts user authentication details like emailid and password which gets stored in the database while app login. The app reacts to the UI and gets configured with the firebase store [11, 12].

9.4 System Implementation

9.4.1 Four View Controllers of CASD System

Sign-Up-View-Controller holds text fields like filename, last-name, email and password to style the text-field and buttons. The fields are checked by shortening whitespace and newlines which references the fields filled. Password security is checked with the help of regular expression format which holds characters, number and special character.Above fields once stored in the firebase datastore sends an acknowledgement which makes the app page to move from Sign-Up-View-Controller to the home-view-controller. Login-View-Controller supports login of the user using app which holds email, password, text fields with a login button and an error label. After user data entry analyses the presence in the database then if present login page wlll redirect to the home-view-controller. In Home-View-Controller click picture button

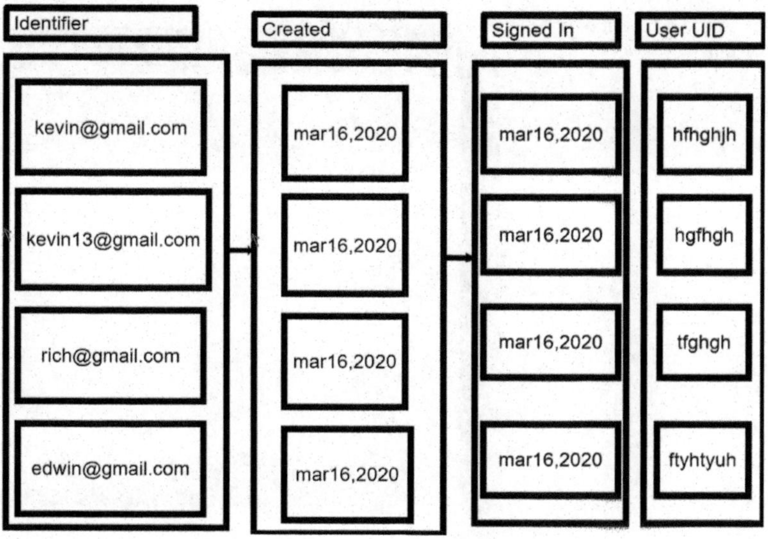

Fig. 9.10 User authentication details in database

creates alert function and UI-Image picker controller delegate thereby gets the mobile application's images. View Controller holds the front display page showcasing the welcome page holding both signup and login buttons.

Figure 9.11 depicts Flow navigation layout which includes login, sign-up, click picture pages. Signup page receives input on first-name, last-name, email, passwords

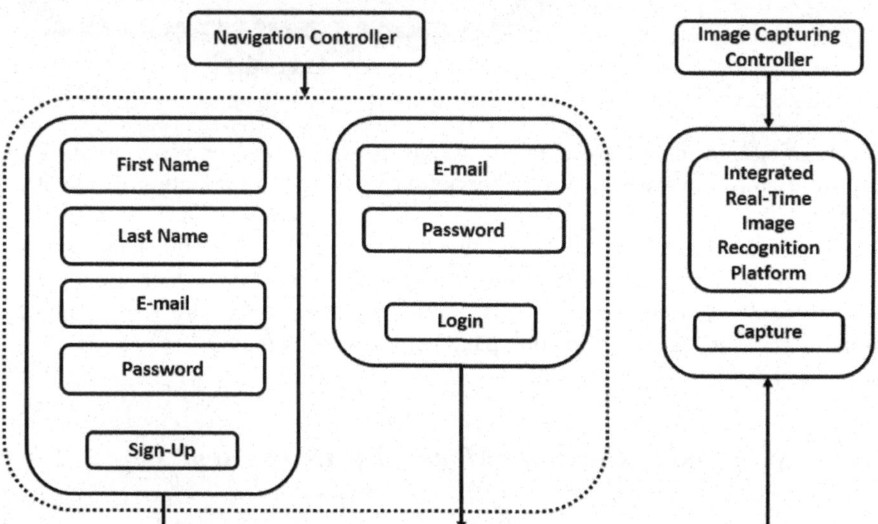

Fig. 9.11 Flow navigation of CASD system

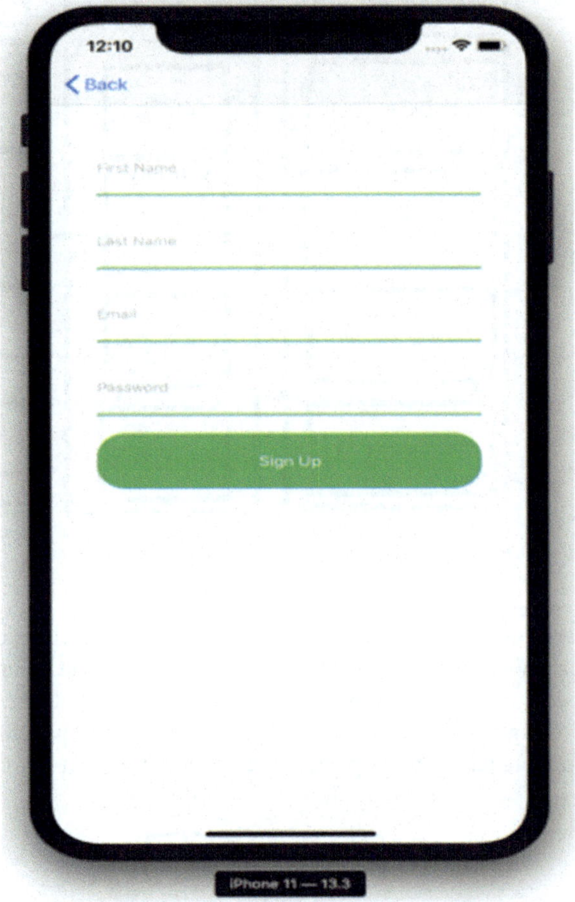

Fig. 9.12 App UI layout of CASD system

which get stored in firebase database. Login page receives input on email-id and password stored in the database. Click picture page gives the option to select the picture.

Figure 9.12 depicts App layout holding login, signup, click picture pages. Signup page receiving input of first-name, last-name, email, password which are stored in firebase database. Login page receives input on email-id and password which are found in the database. Click picture page helps choosing the picture.

9.4.2 *Skin Lesion Classifier Model of CASD System*

Skin-Lesion-Classifier model file is a core ML machine learning file. Three different skin lesion images namely Malignant melanoma, Melanocytic nevi and Benign

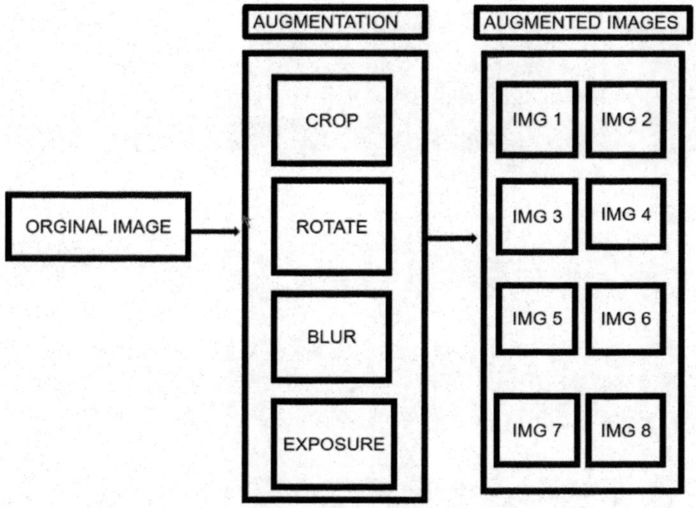

Fig. 9.13 Improving model's validation accuracy of CASD system

Keratosis are used for classification. Training data and testing data are presented in the ratio 80:20. Other models took more than 20 h of time. Receiving less model accuracy leads to augmentations namely cropping the image.

Low model training accuracy leads to current configuration of the model which will not capture the data complexity therefore training parameters should be altered. When dealing with information on images, we double the maximum iteration number where the standard value is 10. While the model is trained accuracy on the validation set is low or constantly switches between low and high each time, gives more data. Using examples we produce more input information which are collected by a methodology called as data augmentation. For images, unite techniques like cropping, rotation, blurring, and exposure adjustment which make images as many examples.

Figure 9.13 depicts There is possibility to have more amount of data and validation accuracy which is lesser than the training accuracy. This indicates an overfitting model, which shows over learning on the training set which will not apply to other examples so reduce the amount of training iterations to stop the model from learning so much on the training dataset.

9.4.3 Steps of Model Creation

Training and testing dataset hold input type, classification, Model prediction and structured images. Model gets availability for mac OS, iOS, tv-OS. Input image to be classified as color (kcvpixeltype_32GRA) image buffer,299 pixels wide by 299 pixels high. Image feature value is given back using function ml feature value. The model prediction output type is similar as MAC OS, tv-OS, IOS, image categories

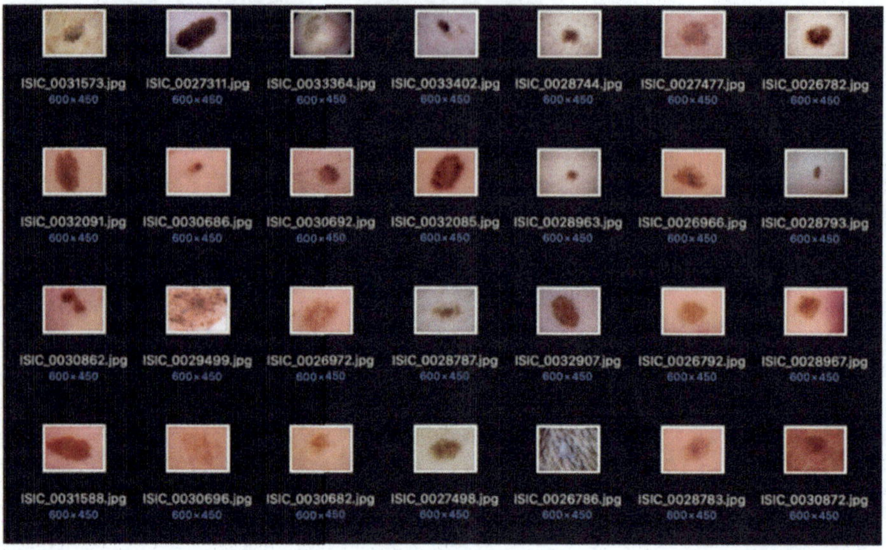

Fig. 9.14 Skin Lesion Classifier Dataset of CASD system for memory storage in its image database

which are considered to be string values. The feature name and feature value returns with label later model is loaded and prediction is calculated. Since prediction is done with the help of structured interface which would create ml model which gets saved as a core ML [13, 14].

Figure 9.14 depicts Skin lesion images dataset holding three types of skin lesions. View Controller holds both UI image view [15, 16] and camera functionalities necessary for taking the picture. Photo captured will get stored inside the UI image view. Function named detect loads the ML model [17, 18]. The model name is SkinLesionClassifier3().model. Console's top result holds confidence and identified gets displayed. If the top result's identifier has any one kind of disease used with the dataset with top result's confidence higher than 60% then corresponding disease gets printed. In the navigation bar if name is found out by the ML model which shows lack of recognition which gets visible in the navigation bar.

9.4.4 Lıve Testing Output Using the Developed CASD System in iOS Platform

Figure 9.15 depicts Melanocytic nevi [19, 20] image was is the output obtained by detecting skin lesion from 3 varieties of skin diseases like malignant melanoma, melanocytic nevi, benign keratosis which is taken for the observation.

Table 9.1 depicts Input, training and testing data providing accuracy and attribute details which holds the metric rate and dataset information.

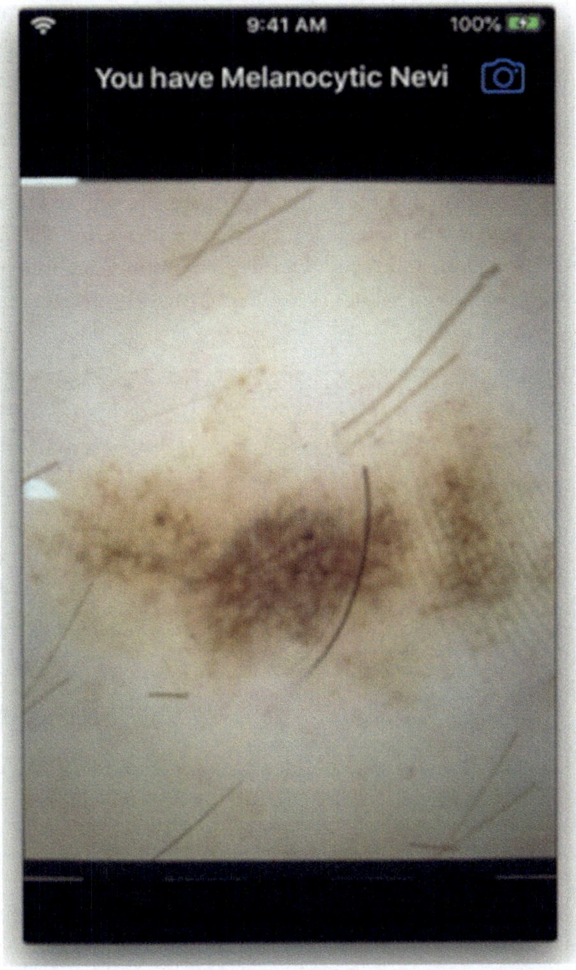

Fig. 9.15 Output of CASD system experiment

Table 9.1 Input, training and testing data of CASD system

Guidelines		Information
Metrics rate	Training rate	84%
	Validation rate	84%
	Testing rate	78%
Dataset information	Training data count	7134 data's
	Validation data count	Automatic
	Testing data count	1783 data's

Figure 9.16 depicts Graph shows the result of training and validation accuracy. Training accuracy achieved at 84% and testing accuracy achieved at 84% in a maximum of 25 iterations.

Table 9.2 depicts Training accuracy shows the accuracy percentage received in the precision and recall details for the three types of diseases under consideration.

Table 9.3 depicts Validation accuracy shows the accuracy percentage obtained in the precision and recall details for the three kinds of diseases under consideration.

Table 9.4 depicts Testing accuracy providing the accuracy percentage received in the precision and recall details. The output Machine learning file shows the machine learning model which is trained to identify images.

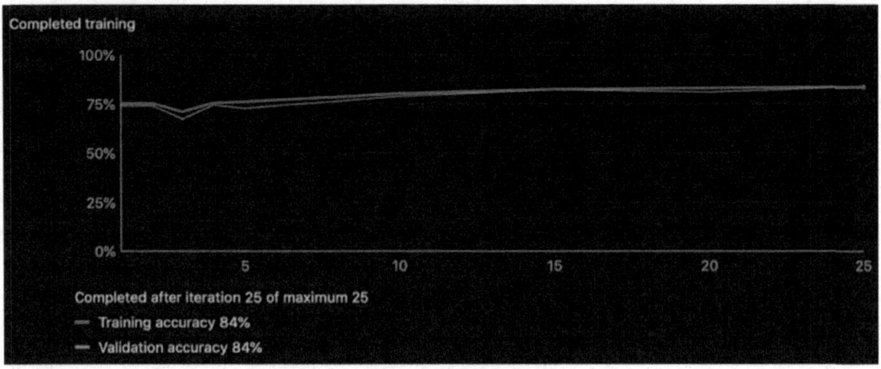

Fig. 9.16 Training and validation accuracy graph of CASD system

Table 9.2 Training accuracy of CASD system

Types of disease	Number of diseases	Precision rate (%)	Recall rate (%)
Benign Keratosis	4105	67	52
Malignant Melanoma	4315	61	43
Melanocytic Nevi	25,600	88	96

Table 9.3 Validation accuracy of CASD system

Types of disease	Number of diseases	Precision rate (%)	Recall rate (%)
Benign Keratosis	59	91	49
Malignant Melanoma	27	54	48
Melanocytic Nevi	244	85	96

Table 9.4 Testing accuracy of CASD system

Types of disease	Number of diseases	Precision rate (%)	Recall rate (%)
Benign Keratosis	219	54	36
Malignant Melanoma	223	41	17
Melanocytic Nevi	1341	83	95

$$\text{Precision} = \frac{(\text{TruePositives\_ 1} + \text{TruePositives\_ 2})}{((\text{TruePositives\_ 1} + \text{TruePositives\_ 2}) + (\text{FalsePositives\_ 1} + \text{FalsePositives\_ 2}))} \tag{9.1}$$

$$\text{Recall} = \frac{\text{TruePositives}}{(\text{TruePositives} + \text{FalseNegatives})} \tag{9.2}$$

$$\text{Validation accuracy} = \frac{\text{Correctly Predicted Class}}{\text{Total Testing Class}} \times 100\% \tag{9.3}$$

When inputting a sample test image as mentioned in the before image, CASD system reacts with a image label of flexible prediction. Training and Testing accuracy is obtained as 84% and Validation accuracy at 78% by implementing in an iOS platform.

9.5 Conclusion and Future Scope

Skin lesion detection with Testing results 78 and 84% are found in training and validation accuracy output. Due to the skin lesions hold stony character it has become tuff to separate them so more datasets are obtained from hospitals and finally the model is trained. The above current technologies advantage is huge. Misdiagnosis in therapy is challenged by Artificial intelligence technology which is the core heart of the CASD system which we have created. The more the data is supplied the more it grows and gains accuracy. The models have gained doctor's ability to see and learn from large number of images. Artificial intelligence technology of CASD system has been rooted in the healthcare division. While adopting mainstream medicine, the quality of treatments improve in marvelous rates. Artificial Iintelligence for diagnosis—stop preventable diseases, and form a future secured and instant diagnosis result.

References

1. Hamblin, M.R., Avci, P., Gupta, G.K. (eds.): Imaging in Dermatology. Academic Press, Cambridge (2016)
2. Armstrong, B.K., Kricker, A.: The epidemiology of UV induced skin cancer. J. Photochem. Photobiol. B **63**(1–3), 8–18 (2001)
3. Khan, M.A., et al.: An implementation of normal distribution based segmentation and entropy controlled features selection for skin lesion detection and classification. BMC Cancer **18**(1), 638 (2018)
4. Mete, M., Kockara, S., Aydin, K.: Fast density-based lesion detection in dermoscopy images. Comput. Med. Imaging Graph. **35**(2), 128–136 (2011)
5. Lundervold, A.S., Lundervold, A.: An overview of deep learning in medical imaging focusing on MRI. Z. Med. Phys. **29**(2), 102–127 (2019)
6. Khan, S.U., et al.: A novel deep learning based framework for the detection and classification of breast cancer using transfer learning. Pattern Recogn. Lett. **125**, 1–6 (2019)
7. Shorten, C., Khoshgoftaar, T.M.: A survey on image data augmentation for deep learning. J. Big Data **6**(1), 60 (2019)
8. Kassem, M.A., Hosny, K.M., Fouad, M.M.: Skin lesions classification into eight classes for ISIC 2019 using deep convolutional neural network and transfer learning. IEEE Access **8**, 114822–114832 (2020)
9. Chaturvedi, S.S., Gupta, K., Prasad, P.S.: Skin lesion analyser: an efficient seven-way multi-class skin cancer classification using MobileNet. In: International Conference on Advanced Machine Learning Technologies and Applications. Springer, Singapore (2020)
10. Sondermann, W., et al.: Prediction of melanoma evolution in melanocytic nevi via artificial intelligence: a call for prospective data. Eur. J. Cancer **119**, 30–34 (2019)
11. Bharti, U., et al.: Android based e-voting mobile app using Google firebase as BaaS. In: International Conference on Sustainable Communication Networks and Application. Springer, Cham (2019)
12. Albertengo, G., et al.: On the performance of web services, google cloud messaging and firebase cloud messaging. Digit. Commun. Networks **6**(1), 31–37 (2020)
13. Thakkar, M.: Introduction to core ML framework. In: Beginning Machine Learning in iOS, pp. 15–49. Apress, Berkeley, CA (2019)
14. Thakkar, M.: Custom core ML models using create ML. In: Beginning Machine Learning in iOS, pp. 95–138. Apress, Berkeley, CA (2019)
15. Malik, Z.H., Munir, T., Ali, M.: UI design patterns for flight reservation websites. In: Future of Information and Communication Conference. Springer, Cham (2020)
16. Alsswey, A., Al-Samarraie, H.: Elderly users' acceptance of mHealth user interface (UI) design-based culture: the moderator role of age. J. Multimodal User Interfaces **14**(1), 49–59 (2020)
17. Shung, D.L., et al.: Validation of a machine learning model that outperforms clinical risk scoring systems for upper gastrointestinal bleeding. Gastroenterology **158**(1), 160–167 (2020)
18. Gupta, R., et al.: Machine learning models for secure data analytics: a taxonomy and threat model. Comput. Commun. **153**, 406–440 (2020)
19. Tronnier, M.: Melanotic spots and melanocytic nevi. In: Braun-Falco's Dermatology, pp. 1–18 (2020)
20. Colebatch, A.J., et al.: Molecular genomic profiling of melanocytic nevi. J. Invest. Dermatol. **139**(8), 1762–1768 (2019)

Chapter 10
A Comprehensive Study of Mammogram Classification Techniques

Parita Oza, Yash Shah, and Marsha Vegda

Abstract Cancer instances have increased in the recent past and are risking many lives. Every kind of cancer is caused due to a malignant (cancerous) tumor which looks pretty similar to a benign (non-cancerous) tumor which makes it difficult to distinguish from one another. Computer-Aided Diagnosis (CAD) is very useful in the early detection of a tumor from its development stage. Many techniques are available in the literature for lesion or mass classification but the challenges faced to train a model are also of great concern. There comes a question about which method to implement for early detection of cancer. In this paper, we have discussed various classification techniques that are categorized based on function, probability, rule and similarity. Analysis of these methods, their drawbacks, and challenges are also discussed. Comparative analysis of existing approaches used to classify mammograms is also presented in the paper. The challenges to train a model also have come from the type of dataset used for classification process. Sometimes the dataset has many anomalies like redundancy, variable size and inconsistency in dimensions that cannot be negotiated. Sometimes the model (architecture) we chose for the training purpose has extensive computation, making it inefficient. These challenges have to be solved during the preprocessing and training phase of a model. This paper mainly focused at various challenges related to mammogram datasets and classification techniques. Methods to handle these challenges are also discussed in the paper.

Keywords Mammograms · Machine learning · Classification · Computer-aided diagnosis

10.1 Introduction

Cancer is the growth of malignant (cancerous) cells in body parts. Breast cancer is a cell dysfunction that composes breast tissue that causes the multiplication of such cells and results in malignant and benign structures [1]. It is an uncontrolled growth of

P. Oza (✉) · Y. Shah · M. Vegda
Nirma University, Ahmedabad, India
e-mail: parita.prajapati@nirmauni.ac.in

M. Mehta et al. (eds.), *Tracking and Preventing Diseases with Artificial Intelligence*, Intelligent Systems Reference Library 206,
https://doi.org/10.1007/978-3-030-76732-7_10

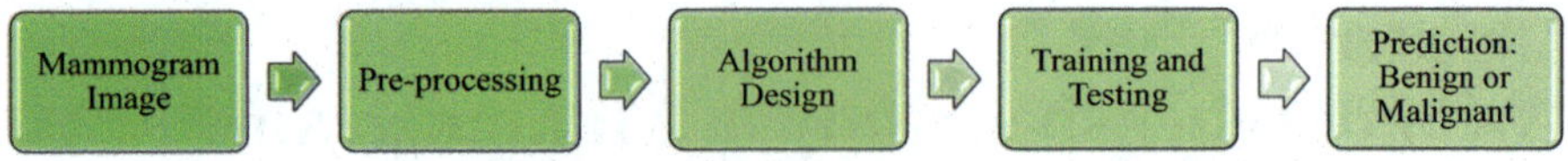

Fig. 10.1 Workflow of lesion classification

pathological cells [2]. Instances of breast cancer have increased in the recent past and risking their lives. It is the second most spread cancer after non-melanoma skin cancer [1]. The main aim of early detection by the computer-aided diagnosis of lesions is to reduce the mortality rate in women. X-ray mammography is best suited to detect lesions accurately and is the standard method [2]. Mammograms are generated in large amounts which consist of X-Ray images of the breast with which we can detect the cancerous tumor if any. There are numerous negative examples (non-cancerous) but they look very similar to the positive (cancerous) examples. Due to this similarity, it becomes very difficult to accurately classify them as benign or malignant. In fact, manual analysis of these X-Rays is extremely difficult [2]. Due to these difficulties in classifying or detecting tumors specifically, Computer-Aided Diagnosis tools have been developed to reduce manual and tiresome work of radiologists. These CAD tools help in classifying tumors easily as compared to manual inspection. There are various algorithms implemented to classify the images and various other tasks such as object detection, regression, interpolation, and many more [2, 3]. Workflow for the computer-aided diagnosis of mammogram classification is shown in Fig. 10.1.

To design a good classifier we need proper dataset and a well defined method to do classification process. Varieties of mammogram datasets and classification techniques are available in the literature. Selection of proper dataset and a proper classification technique is really a challenging task. In this paper we presented an overview of classification techniques that are commonly used to design a classification model. These techniques are also categorized based on their characteristics. Most commonly used mammogram datasets are also discussed in the paper along with few challenges associated with these dataset

The main objective of this study is to give an idea of various classification approaches and their categories. Every method has a different concept of implementation, execution, and performance evaluation parameters. But a method also aims to provide the best result among the state-of-the-art techniques. This paper also provides a statistical comparison between these methods. The paper is structured as follows: In Sect. 10.2, most widely used mammogram datasets are discussed. Related works of state-of-the-art techniques for mammogram classification along with analysis of existing CAD systems are presented in Sect. 10.3. Section 10.4 focused at various classification approaches of machine learning. These approaches are categorized in four different categories: Function based, probability based, rule based and similarity based. Different methods under each of these categories are also presented in this section. These techniques may have some difficulties in terms of datasets or in terms of the technique itself due to its characteristics. Such setbacks that come in way of classification techniques are mentioned in Sect. 10.5. To handle these drawbacks, there are some techniques for producing improved results and help overcome the setbacks. These methods are mentioned in Sect. 10.6. After analyzing, describing, and comparing much about the machine learning techniques, datasets, outcomes, associated procedures, drawbacks and difficulties faced during implementing the Machine Learning Techniques, Sect. 10.7 focuses on an approach towards Deep Learning techniques.

10.2 Mammography and Mammogram Datasets

There are many imaging modalities, used by the radiologist for breast cancer diagnosis. Digital mammography is the most common and widely used screening method for breast diagnostics and imaging. This method is used for the examination of breast cancer. It is basically an X-ray image of a women's breast with a very low dose of X-rays. Breast abnormality has mainly three types; Mass, Micro-calcification, and Architectural Distortion. Masses are also sometimes called tumors or lumps and are described by features such as density, contour, and shape. These features are also called morphological features. Microcalcifications are very small deposits of calcium. Scattered Microcalcification is generally non-cancerous wherein, a tight cluster of these spots may result in malignant mass later on. Architectural Distortion is the structure of the breast with no visible mass and can be seen as a distorted region in the breast tissue. Mammography is recommended to detect Microcalcification in a very efficient way to reduce over-sight errors. Images taken by mammography technique is called a mammogram. This method is the gold standard to detect cancer at a very early stage. Digital mammography is highly sensitive in young women and its specificity is also very high as it can detect Microcalcification accurately [4].

For training a machine learning algorithm, a large amount of properly annotated data are very much essential. There are many public mammogram datasets are available and widely used by the research community. The Mammographic Image Analysis Society (MIAS) [6] is a very old mammogram image repository. This dataset is originated from a UK research group. MIAS has a total of 322 mammograms with all three abnormalities mentioned above. The dataset is supported with a CSV file which states all the metadata about images. Images are available with center and radius coordinates of the abnormal region as ground truth. MIAS has all images with the same size, 1024×1024 pixels. SureMaPP [7] is a recently published mammogram dataset. The dataset is equipped with a total of 343 images annotated by expert clinicians. INBreast Dataset is published in Ref. [8]. The dataset has a total of 410 images of all three breast abnormalities in DICOM format. Dataset is equipped with a pixel-level boundary as a ground truth. Annotations in the dataset were done by an expert in the field and also validated by another expert. A very huge repository of breast cancer dataset is DDSM mammography [9] which has more than 10000 images from a total of 2500 case studies. This dataset is also supported with a pixel-level boundary as ground truth. Another variant of the DDSM dataset family is CBIS-DDSM [5] which is a curated form of DDSM. Images with ambiguous annotation and low resolution

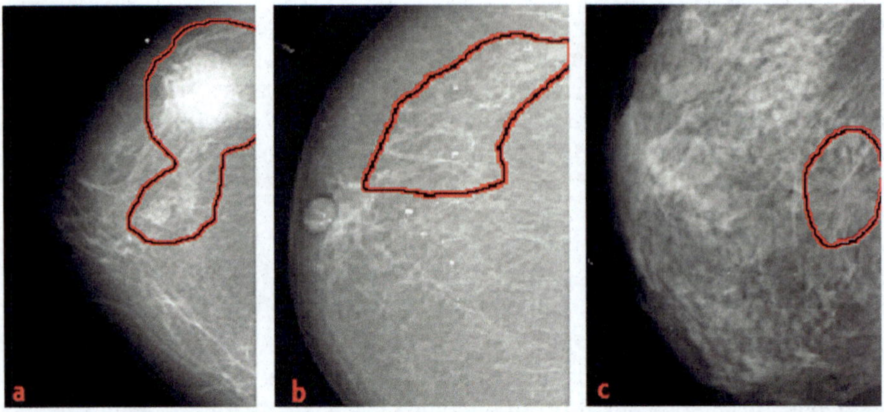

Fig. 10.2 Mammogram images from DDSM dataset [5]

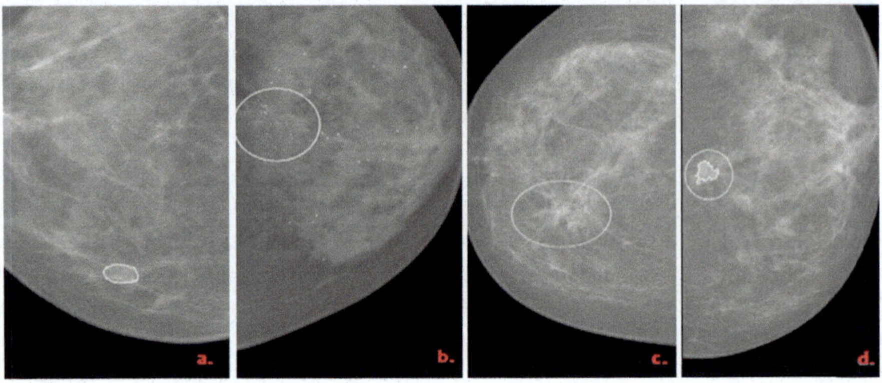

Fig. 10.3 Mammogram images from INBreast Dataset [8]

and quality are filtered out in this dataset. This dataset has a total of 10,239 images. Details of other public and private datasets used by research communal for breast cancer diagnosis are well presented in [8]. Reference mammogram image for the DDSM dataset and INBreast Dataset are given below as Figs. 10.2 and 10.3.

10.3 Related Works in Classification of Lesions Using Data Mining Techniques

Lot of work has been done in the field of medical imaging for breast cancer classification using machine learning approaches. This section discusses the work has been done by the research communal to address the breast mammogram classification. An analysis form various papers on breast classification technique is also presented in this section.

Microcalcification and tiny masses are the most important signs towards the development of malignant tissue in a breast. In [29], a CAD system is developed to detect and classify such abnormalities. Authors also have compared various techniques and stated their advantages and disadvantages. Reference [30] gave an overview of various data mining classification methods for breast cancer classification, diagnosis and prognosis process. Reference [31], summarize and analyze various image mining techniques for breast cancer screening.

Authors in paper [32] showed an automatic Microcalcification detection technique in mammograms using image processing. The method is based on segmentation by thresholding and classifier based on structural featured such as area, image energy, density, and circularity index. In [33], the authors showed a CAD system developed using Artificial Neural Network (ANN) to classify the regions that are suspicious in a mammogram image. A technique to separate two classes having many similarities is to implement the SVM method. SVM introduces or learns a boundary or finding a hyperplane that separates two classes with the maximum distance between the two, this technique is also called Maximum Margin Hyperplane. This is more efficient when classes to be classified have very little difference. Authors in paper [34] have also developed CAD system using support vector machine to classify a mammographic image as normal or abnormal by learning the features extracted. In [35], the author showed a method that used K-means for image segmentation and classifying the segmented portion as benign (normal) or malignant (abnormal).

Many other methodologies have been developed which classifies the mammogram images. Summary of such methodologies used in different researches/review works are presented in Table 10.1.

10.4 Techniques for Classifying Mammogram Images

As stated in previous section, there are various techniques developed for classifying lesions from mammogram images. we have categorized these techniques based on different aspects like based on function, based on probability, based on rules and based on similarity. This categories are taken from Ref. [2]. These categories are defined based on the functions and conditions they use for the classification process [2]. These techniques have their own aims, advantages, and disadvantages. Figure 10.4 shows categorical view of these classification technique.

10.4.1 Function-Based Methods

These methods can be addressed in a mathematical term with some specific set of parameters which are fit into a linear or non-linear function as:

Table 10.1 Summary of machine learning techniques used in various state-of-the-art researches

References	Technique proposed	Dataset used	Parameters/results
[10]	SVM classifier For segmentation, the fuzzy c-means algorithm, For feature extraction and texture property, Local binary pattern (LBP) and Dominant rotated LBP (DRLBP) is used	Mammographic Image Analysis Society (MIAS)	Accuracy = 94.21%
[11]	DSOM (Self organizing map based on distance)	Wisconsin diagnostic breast cancer (WDBC)	Accuracy = 97%
[12]	Fuzzy multiple parameter SVM	Digital database for screening mammography (DDSM)	Accuracy = 93% MCC = 86.16% Sensitivity = 96% Specificity = 90%
[13]	Transfer learning – CNN + RBF based SVM	Private	Sensitivity = 1 Specificity = 0.86 Accuracy = 92%
[14]	Subtraction of temporally sequential mammogram pairs followed by machine learning techniques, k-fold cross-validation technique (k = 4, 5, 10) and leave-one-patient-out (l-o-p-o) technique for validation of training and testing dataset	Private	SVM provided the best results with Accuracy = 99.55%
[15]	Local binary patterns (LBP) feature extraction method with a machine learning model 10-fold cross-validation technique for validation	(DDSM)	Accuracy: Gaussian process = 84.80% SVM = 92.60% kNN = 94.90% Adaboost = 94.60%
[16]	Discrete wavelet transform (DWT) and random forest (RF Features were extracted and used as input to random forest classifier.	DDSM	Sensitivity = 93% Specificity = 97% Accuracy = 95% False Positive Rate = 3% Area under ROC = 0.92

(continued)

Table 10.1 (continued)

References	Technique proposed	Dataset used	Parameters/results
[17]	Adaboost, kNN, DT, SVM Dataset features are fed into Neighborhood Components Analysis (NCA) to reduce the number of features and hence reducing the complexity,	Breast Cancer Wisconsin Dataset	Accuracy SVM = 94.74% kNN = 92.12% Adaboost = 93.86% DT = 89.47%
[18]	SVM, Random forest, Bayesian network Extreme learning machine	Wisconsin breast cancer dataset	In metrics of accuracy, specificity, precision, SVM has the highest performance compared to random forest and Bayesian network
[19]	Random forest classifier for classification	MIAS	Accuracy = 97.32% Sensitivity = 97.45% Specificity = 98.13% Area under curve = 97.28%
[20]	RF with the combination of RF-ELM classifier Feature extraction is done by texture analysis	MIAS	Accuracy: RF = 89% RFELM = 98%
[21]	SVM, KNN,		Machine learning classifiers such as KNN, Extreme learning machine SVM, and Naïve Bayes are compared The best classifier among these is extreme learning machine
[22]	GLCM based feature extraction	Mini-MIAS	Sensitivity = 99.3% Expressness = 100% Precision = 99.4%

(continued)

Table 10.1 (continued)

References	Technique proposed	Dataset used	Parameters/results
[23]	There are 2 proposed CADx systems using Particle swarm optimization (PSO) Gaussian mixture model (GMM) Classification techniques used here are non-linear SVM Validation technique used here is 10-fold cross-validation	Mini-MIAS	Accuracy of classification for PSO = 89.5% Accuracy of classification for GMM = 87.5%
[24]	Adaptive boosting (AB) Validation technique: 5-fold cross-validation	DUKE mammogram database	Sensitivity = 97% Specificity = 56.6% Positive predictive value (PPV) = 55.8%
[25]	Self-regulated multilayer perceptron	MIAS	Accuracy = 90.59% Specificity = 90.67% Sensitivity = 90.53 Area under curve = 0.906
[26]	SVM	Private database of 76 mammograms containing 1120 microcalcification	For evaluation, the free-response receiver Operating characteristic curve (FROC) is used SVM provided better performance
[27]	SVM	DDSM	Average accuracy = 98.88%
[28]	ML classifier: SVM, decision tree (DT), ANN, kNN and Naïve Bayes DL classifiers: CNN with its variants of neuron layers and networks	DDSM, MIAS, INBreast, IRMA	Classifiers showed different behavior with different datasets and networks SVM outperformed other classifiers

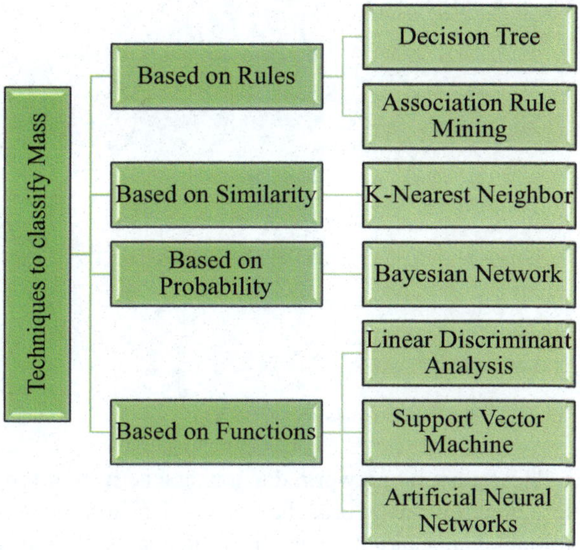

Fig. 10.4 Mass classification techniques [2]

$$y = F(\theta, x) \tag{10.1}$$

where θ is a finite set of parameters, x is a feature vector, y is the prediction for x.

In these methods, the hyperparameters need not be learned but the coefficients are learned which makes these methods fast. Non-linear functions have higher accuracy than linear classifiers. Various function based methods are discussed below.

10.4.1.1 Artificial Neural Networks (ANN)

Due to the good performance of ANN in pattern recognition and classification, its use in detection and classification in mammogram images of the breast for early detection of breast cancer has been increased [2]. Artificial neural networks has fully connected layers and every layer has large number of neurons which are acting as processing elements. The network has large number of weighted connection between these processing elements. For ANN learning Process is implemented to acquire knowledge for a particular domain task. ANNs use non-linear functions for learning the features extracted from mammogram images and that is why the accuracy has an upper hand than the other methods. Learning involves finding the internal weights of the network by artificial neurons just like the biological neuron which receives an input signal and produces output by use of activation functions [36, 37]. The network has many hidden layers which are used to perform non linear transformation of the input image. A basic architecture for ANN with one hidden layer is depicted in Fig. 10.5 [38] and general formula to find output volume is shown in Eq. 10.2. While

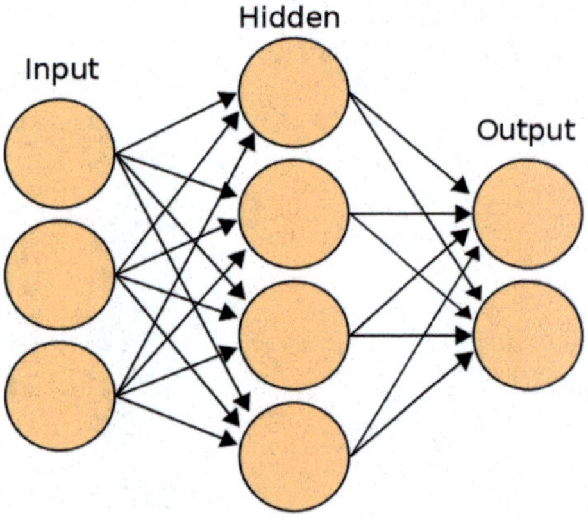

Fig. 10.5 ANN with 1 hidden layer

training a neural network, due to some activation function like sigmoid, gradient of loss function may approaches to zero. Due to small gradients, the weights and bias of initial layers can not be updated effectively. This makes training process very hard. This problem is knows as vanishing gradient. This problem can be solved by replacing activation function or by adding residual connections. ReLu is most commonly used activation function which can weaken the effect of vanishing gradient problem.

$$output = f(bias + \sigma^{n}_{(i=1)} x_i w_i) \tag{10.2}$$

where bias denotes the value b1, b2, and b3 etc. for respective neurons, x is an input and w is weight.

Strength: Neural networks can store information of entire networks. The network can be trained even with incomplete information. Network performance depends on the importance of missing data. This network have capability of parallel processing and fault tolerance. Failure of one or more neurons does not effect training process and does not prevent a network from generating the desired output.

Limitation: Due to fully connected layers, ANN needs processor with parallel computing capabilities. ANNs operates on numerical information hence, there is a need to translate a problem into numerical values. Sometimes network may behave unexpectedly and does not give any clue that why it has happened.

10.4.1.2 Linear Discriminant Analysis (LDA)

It is a very basic and simple method that constructs a decision boundary around the suspicious objects or masses directly by error optimizing criteria [39, 40]. LDA maximizes the ratio of interclass variance to intraclass variance thereby making the maximum separation between the classes. The functions which are used in LDA are linear in nature [39]. The function is optimized in such a way that interclass similarity decreases and intraclass similarity increases. These functions have coefficients that can be optimized based on the feature learning of the training dataset provided to the algorithm [2].

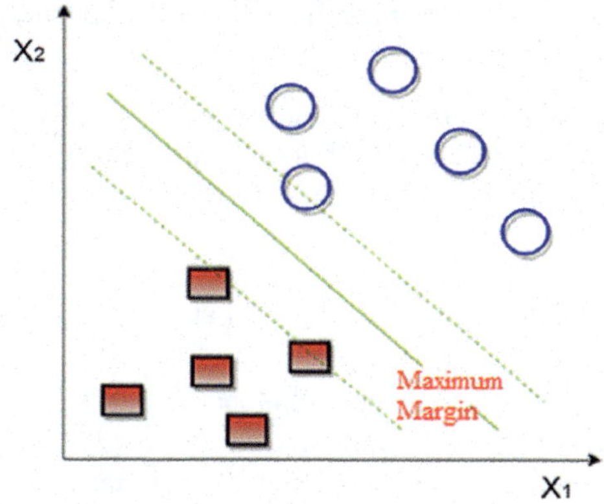

Fig. 10.6 Support vector machine with many hyperplanes

10.4.1.3 Support Vector Machines (SVM)

SVM is a potential classification algorithm that decides the category (class) of an object based on its characteristics and not by probability. Thus making it a non-probabilistic binary classification approach. This is a concept of learning decision planes or boundaries between the two class of data or images [34]. The algorithm learns from training set in the possible multi-dimensional plane and decides (determines) the boundary or plane that can separate two classes. The plane that the algorithm learns or derives is called separating hyperplane where on one side of the hyperplane, are positive data points and negative data points, on the other side [41]. Points near the optimal hyperplane are called support vectors. After finding such a plane, SVM [37] can classify an unlabeled sample (data) [41]. The most common challenge with SVM is to find out exact hyperplane when there are multiple alternatives available as shown in Fig. 10.6 [42]. Maximum margin hyperplane is the plane which has the largest separation with the samples from both the classes. This adds more generalization into a network. SVM is applicable for both, linearly and non-linearly separable data. Kernel trick is most commonly used method when input data are non-linearly separable. Various kernel functions are used to convert linearly non-separable data into linear data such as sigmoid kernel, polynomial kernel and Gaussian kernel. For image processing applications polynomial kernel is mostly used. SVM is a robust binary classifier and most commonly used method by the researcher for mammogram classification task.

Strength: Support vector machines are robust and not easily impacted by noisy data or outliers in the dataset. These networks are used for classification as well as regression task.

Limitation: Training of SVM goes slow when larger sized data are used. The network is memory intensive. The structure of SVM become complex when used with higher dimensional data.

10.4.2 Probability-Based Methods

Classifiers in these methods do not only find out the most probable class but also can give the prediction in the probability distribution over various classes [2]. Below discussed is Bayesian network which is based on the principle of probability.

10.4.2.1 Bayesian Network

As the name depicts, this method uses Bayesian Probability Theory to find the most likely class for that dataset images [31]. It makes a Bayesian Belief Network (BBN) where it creates an acyclic graph denoting joint probability distribution by considering conditional probability [31]. The nodes in the graph denote the features and edges denote the probability (weight) relation or dependency between the features. The assumption taken here is that the weights or relation between the features taken under consideration for classification are independent of each other. By the weighted probabilities extracted from the acyclic graph, these methods predict the class in which the data should belong [40]. The network first convert data points into a frequency table then it creates likelihood table structure with the help of probabilities. At the end naive bayes equations are used to calculate the posterior probability of each of the class in the dataset.
Strength: As compare to other machine learning methods, this network is easily extensible. Adding new data into a network will only need a small calculation of probability and addition of few edges into a graph. To model various cognitive function, these networks are most viable.
Limitation: Designing a Bayesian Network requires more efforts as compared to other ML approaches. This network give very poor result when operating on higher dimensional data.

10.4.3 Similarity-Based Methods

What happens in methods that do not consider or do not know about the probabilities? There comes an idea of calculating distances or similarities [43]. This method predicts the class of unknown data by calculating the distance or similarity of that data with all the other data in the dataset. A way to calculate the distance or similarity between the input data depends on the algorithm (method). The operation of these methods can be divided into two different subtasks, (1) calculating a pairwise similarity between unknown data and the rest of the data and (2) classifying unknown data based on the similarity already calculated [43]. K-Nearest Neighbor approach falls under this category.

10.4.3.1 K-Nearest Neighbor (KNN)

Similarity-based classification methods generally uses k-nearest neighbor where k is user defined parameter which indicates the number of neighbors to look upon to classify the class of the unknown data [43]. Accuracy and performance of this method depend upon the choice of k. If the value of k is very large then the network will take decision regardless of neighborhood. Small value of k may choose noisy data or outlier. There are some strategies to choose the value of k such as;

- Taking square root of number of records in the training set
- Testing several values and decide upon the value which gives better performance
- Take larger value, and then apply weighted voting approach

If a method uses distance as a parameter to calculate similarity, then generally it is Euclidean distance and if the similarity is considered, Cosine similarity is generally used for prediction or classification of mammogram images dataset.
Strength: This approach is very simple and easy to understand. Training process of KNN is very fast as compared to other ML approaches.
Limitation: In KNN, classification is performed based on the training data only. There is no real learning. The model needs large amount of computation space.

10.4.4 Rule-Based Methods

In the above methods, they use the best set of features and obtain the solution in a onestop decision. Rule-based methods cover every aspect (feature) of the data in order to make a decision for the classification process [44]. Thus, better performance is achieved keeping in mind the weightage of every feature of the dataset. The term rule-based can be referred to as the classification scheme that uses the if-else clause for class prediction [44]. The rule-based methods provide ease in understanding, operability, and enhanced feature optimization compared to other methods in this domain [2].

10.4.4.1 Decision Tree Classifier (DT)

This method follows a hierarchical decision structure and forms a tree-like schema while deriving results from a particular attribute. Each node is the split independent attribute and edges or branches from that node denote outcomes of that attribute. The leaf nodes are class labels derived by traversing that path of decisions or outcomes [45]. Due to the consideration of each attribute in the decision making process and giving equal justice or importance to weights of each attribute, DT is used for classifying images in various medical domains [45]. The method is widely used in various applications because of its characteristic to study every attribute and its significance. The formula to calculate entropy of a decision tree is mentioned below (Eq. 10.3).

After splitting process every example in a set should belong to a single class only. Entropy is the measure of impurity of an attribute after the split process.

$$Entropy = \sigma^{n}_{(i=1)} - pilog_2 * pi \tag{10.3}$$

where n is number of different classes and pi is proportion of values falling into ith class in the dataset. Information gain is another parameter which gives the reduction in entropy after the split is performed on data. Pruning is the technique which is used mitigate the effect of overfitting problem. There are various implementation of decision tree such as, C5.0, CART, CHAID and ID3. CART implementation of DT uses GINI Index for splitting process.
Strength: DT can be applied to both numerical as well as categorical data. This methods works well for all size of data, small or large. The method has an ability to provide a definite clue for important features for the classification process.
Limitation: The method is biased towards features which has many levels. It is easily prone to overfitting and underfitting. The method is also prone to errors in classification task if trained using small size dataset.

10.4.4.2 Association Rule Mining (ARM)

It is a very important task in data mining techniques that aims to extract associations or patterns between data in large datasets. It discovers patterns and associations or dependencies between various attributes in the dataset. These algorithms take care of nonlossy handling of medical data by mapping them in transaction format. These algorithms are used in computer aided systems to analyze medical images [46].

10.5 Challenges in Classifying Mammogram Images

The challenges faced in classifying the mammogram images can be widely categorized as (1) challenges related to mammogram images dataset and (2) challenges related to classification techniques. The challenges under each division are depicted in Fig. 10.7. These categories of challenges are taken from Ref. [2].

10.5.1 Dataset Related Challenges

Medical images are captured by various imaging modalities. Mammography is one of them. Images captured by this modality is called mammogram. There are many mammogram datasets are available for the research, some are publicly available and some are private to some organization or hospitals. Images of mammogram dataset may need lot of pre-processing if captured using low resolution device. Pre-

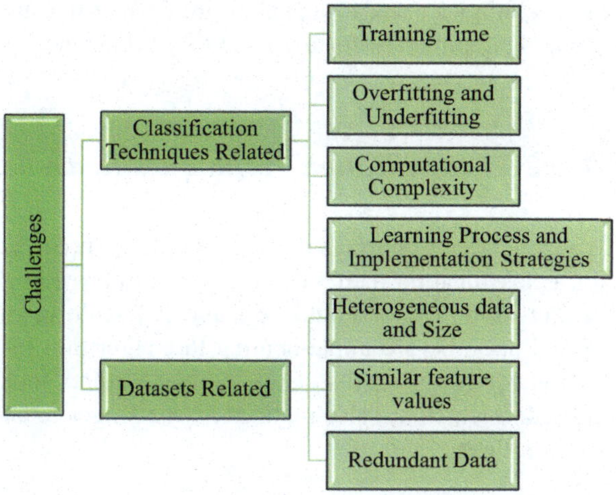

Fig. 10.7 Challenges in classifying mammogram images [2]

processing is to be done with at most care so that important morphological feature that are used for the classification process are preserved. There are many other challenges also exist, that are discussed below.

10.5.1.1 Heterogeneous Data and Dataset Size

Mammogram images are variable in size and heterogeneous in nature. These images may have poor quality, inconsistency and varying resolutions. Hence, there is a need of proper method to pre-process and resize all the images into a uniform scale with out loss of information [46]. Large amount of properly annotated images are needed to train a classifier. It is difficult to train classifier with small size dataset. This may result into lack of generalization in a model.

10.5.1.2 Similarity in Feature Values

Image features are highly similar, thus making them very sensitive in detection [46]. On contrary to the above challenge, sometimes the data is extremely similar making it challenging for the model to determine a class for every data (sample). Many times, due to much similarity, samples are wrongly classified which can lead to consequences.

10.5.1.3 Redundant Data

Multiple occurrences of data make the dataset redundant, less significant, inconsistent, and unnecessarily of large sized [46]. Data redundancy is common in classification techniques. It makes the algorithm exhaustive and inefficient, and the algorithm cannot learn any new feature from this redundancy.

10.5.2 Classification Techniques Related Challenges

We cannot really control data or dataset except preprocessing, so it becomes necessary for a classification method or algorithm to provide good results irrespective of the dataset used. There should be some key features of an algorithm that should be looked upon before implementing that algorithm for analyzing the medical image dataset for better results. Thus the key features looked upon by researchers in an algorithm which are also realized as challenges to choose the best suited algorithm are as follows:

10.5.2.1 Training Time

Training time for any algorithm should be optimized depending on various factors such as layers of architecture, its features and attributes, and computation power needed by the algorithm. It can also depend upon the hardware or software on which the algorithm is implemented. Many algorithms fail to adhere to optimum training time because of the high amount of data, the capacity of hardware or software, redundancy of data, attributes to learn, and other hyper parameters used in architecture.

10.5.2.2 Overfitting and Underfitting

Data Overfitting is a serious issue in any algorithm. It means that an algorithm memorizes the attributes and reacts to every new change in nature attribute. This causes a high variance of an algorithm implemented and on plotting a graph of the data, it shows spikes and abrupt changes (troughs and crests). The algorithm or model that will be implemented should avoid memorizing the data features rather, learn the features. Thus model should avoid overfitting. This issue arises when an algorithm iterates more over the data than the needed number of times. Iterations are not fixed but it should be proportional to the data size and the architectural layers [47]. On contrary to overfitting, underfitting is also an issue where the algorithm does not learn the features (attributes) efficiently and fails to react to any new change in the nature of attribute. This results in a low variance of the algorithm. While plotting a graph for the same, it shows a nearly linear line that rarely covers important data points [47]. The algorithm should also consider this problem and design the architecture in such a way that it does not underfit the dataset. To summarize, the architecture or model to classify an image should not underfit as well as should not overfit. It should just learn adequate features of the dataset that results in an efficient learning

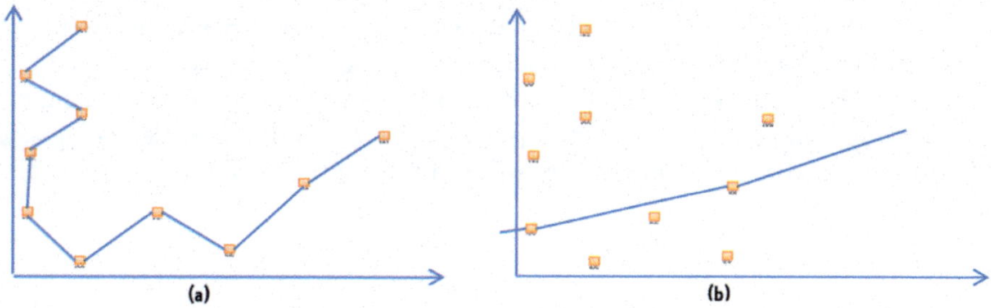

Fig. 10.8 **a** Overfitting, **b** underfitting [47]

and produces generalized output. Sample example of overfitting and underfitting is given in Fig. 10.8.

10.5.2.3 Computational Complexity, Learning Process And Implementation

Due to inconsistency and irregularity in the input, models tend to undergo complex calculations. Complexity increases when the input size is huge and the number of layers to learn the attribute is more. The features should be such that they need not require complex mappings or complex considerations. Hence learning curve should be easily explainable. A model implementation should not be very complex which is normally a problem in a dataset having similar features and inconsistent samples in the dataset. Number of layers, dimensions of the layers, input size, weight initialization methods, all these parameters are equally important for implementing simple yet efficient classifier.

Classification techniques which are discussed above have their on characteristics which are discussed below.

Characteristics of Function-Based Classifiers: These classifiers are generally faster in operation and less time consuming. Non-linear functions have higher accuracy than linear ones. Neural Networks offer the best accuracy among the function-based classifiers in mammographic mass detection or classification. Prior knowledge and internal working of the system are not compulsory to be understood by the user, so its requirement is minimal [48].

Characteristics of Probability-Based Classifiers: If apriori knowledge of the probability distribution of features is required or is present, then these classifiers are the most prominent ones for providing optimal outputs. They offer high accuracy and explainable training process due to apriori knowledge. This method have capacity to resist overfitting situation that makes it more reliable than ANNs [48].

Characteristics of Similarity-Based Classifiers: These are easy to understand as they create a non-parametric classifier with negligible training time. But traditional kNN has a strong dependency on the size of input data and also has high computational complexity. The accuracy of this algorithm is highly depends on number of neighbors and selection of number of neighbors (value of k) is very difficult task.

Characteristics of Rule-Based Classifiers: This method has faster computational capability and implementation as well [46]. Interpretation of the model along with efficiency is an added advantage in rule-based methods. This approach needs less time for training.

10.6 Techniques to Improve Performance of Machine Learning Models

The development of generalized and diagnostically correct and accurate models needs a large amount of properly annotated dataset, which is the biggest challenge in the field of medical imaging due to privacy and legal concerns. Training a model with improper and less amount of data may result in an error called over-fitting. Overfitting may not be fully avoided but there is a solution to combat the effect of it and to make the model more generalizable. This section discusses various approaches used in literature to mitigate the effect of overfitting.

Regularization [49, 50]: Regularization is a technique that is widely used against the problem of overfitting. This technique is also considered as another form of regression. This method helps to regularize the coefficient which approaches to zero during training time. Basically, this technique discourages over learning of very complex as well as flexible data and hence avoid the risk of overfitting. This method depends on the type of ML classifier used, for example for neural networks dropout is a very popular regularization technique. Overfitting by a decision tree can be avoided by the technique called pruning.

Data Augmentation [50, 51]: Data augmentation has been recommended and used by the research community to diminish the effect of overfitting by creating more examples of the training dataset. Data augmentation is performed by applying various geometric transformations on input images like reflection, cropping, translation, scaling, and rotation.

Ensembling: Ensemblers are ML models that are built by combining predictions from multiple other models. Bagging and boosting are two very much popular ensemblers. To reduce the chance of overfitting in complex models bagging is used. Boosting is used to improve the prediction in simple models.

10.7 Towards Deep Learning

Machine learning algorithms have been used for mammogram classification since decades. Though ML methods like support vector machine, random forest, and k-nearest neighbors are extensively used by the research community for breast mam-

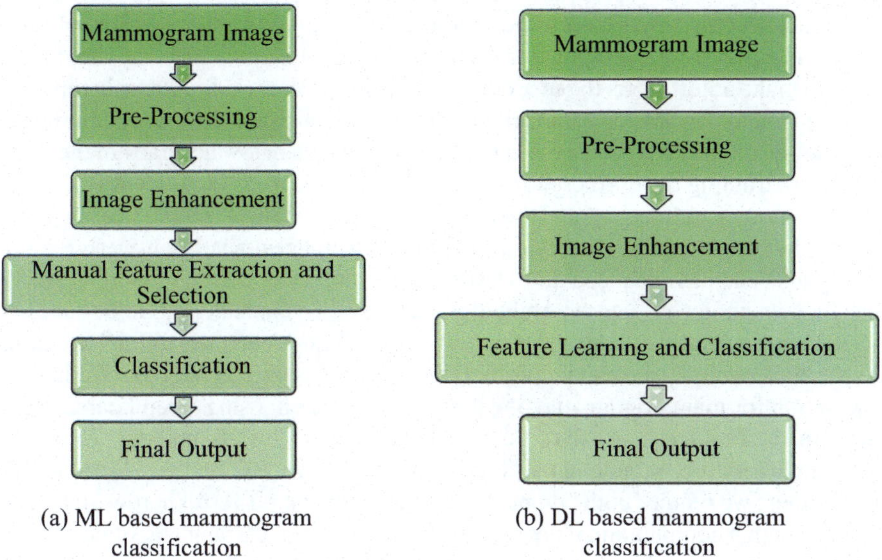

Fig. 10.9 Mammogram classification process using Ml and DL

mogram classification, manual feature extraction is still a problem. ML-based mammogram classification task needs handcrafted features which is a very crucial task as the entire classification process depends on these features. To overcome this limitation, another fascinating field of Artificial Intelligence called Deep Learning (DL) have come up with an idea that, instead of designing feature extractor design feature learner [52]. Deep learning models especially Convolutional Neural Networks (CNNs) are capable of learning low, medium, and high-level features from an image. CNNs have a stack of layers through which it can learn nonlinear features from an input image that is needed to perform a task like classification and detection. DL models require large memory and computational power. With the development of DL approaches there is also an advancement in computation power with the help of GPUs. Various open source toolkits and libraries are also available so that there is no need to build a model from the scratch. Both of these have created a notable revolution for developing DL based models [53]. Figure 10.9 shows the difference between the mammogram classification process using ML and DL method.

10.8 Conclusion

Data Mining and Machine Learning methods have been continuously enhancing and new methods are coming to replace the drawbacks of old ones. Datasets are continuing to become more and more in number and variable size. In this case, it is utmost necessary to correctly identify the methods and the architecture which relates to our application. More computational architectures sometimes may backfire and we need to make adjustments to our architecture based on the requirements. Inconsistent dataset is also a problem. A model trained with Inconsistent and redundant images may not result into correct classification. All this results in low accuracy of the model and more training time. To eradicate these drawbacks, we need to identify the right combination of architecture and the dataset. There are many Machine Learning techniques incorporated for classifying mammogram images, nearly answering all the drawbacks and has enhanced the classification accuracy drastically. Every technique has its own choice of dataset, evaluation parameters and challenges to be faced but these technique thrives to provide the best output. Thus it becomes difficult to choose a specific technique that can fit our requirements by keeping in mind our limitations. An option for mammogram classification can also be choosing Deep Learning over Machine Learning because the procedure to be followed by DL methods are efficient and can learn every minute detail (feature) of the image. But on the selection of DL techniques, we require high computational capacity and powerful hardware that can withstand this computational power needed by the deep learning classifier.

References

1. de Oliveira Silva, L.C., Barros, A.K., Santana, E.E.C.: A telediagnostic system for automatic detection of lesions in digital mammograms. In: 5th ISSNIP-IEEE Biosignals and Biorobotics Conference (2014): Biosignals and Robotics for Better and Safer Living (BRC). IEEE (2014)
2. Mahdikhani, L., Keyvanpour, M.R.: Challenges of data mining classification techniques in mammograms. In: 2019 5th Conference on Knowledge Based Engineering and Innovation (KBEI). IEEE (2019)
3. Jayalakshmi, G.S., Kumar, S.: Performance analysis of convolutional neural network (CNN) based cancerous skin lesion detection system. In: 2019 International Conference on Computational Intelligence in Data Science (2019)
4. Islam, M.S., Kaabouch, N., Hu, W.C.: A survey of medical imaging techniques used for breast cancer detection. In: IEEE International Conference on Electro-Information Technology, EIT 2013. IEEE (2013)
5. Lee, R.S., et al.: A curated mammography data set for use in computer-aided detection and diagnosis research. Sci. Data **4**, 170177 (2017)
6. Suckling, J.P.: The mammographic image analysis society digital mammogram database. Digital Mammo (1994) 375–386
7. Bruno, A., et al.: A novel solution based on scale invariant feature transform descriptors and deep learning for the detection of suspicious regions in mammogram images. J. Med. Signals Sens. **10**(3), 158 (2020)
8. Moreira, Inês C., et al.: Inbreast: toward a full-field digital mammographic data-base. Acad. Radiol. **19**(2), 236–248 (2012)
9. Heath, M., et al.: Current status of the digital database for screening mammography. In: Digital Mammography, pp. 457–460. Springer, Dordrecht (1998)
10. Kashyap, K.L., Bajpai, M.K., Khanna, P.: Breast tissue density classification in mammograms based on supervised machine learning technique. In: Proceedings of the 10th Annual ACM India Compute Conference (2017)

11. Omara, H., Lazaar, M., Tabii, Y.: Classification of breast cancer with improved self-organizing maps. In: Proceedings of the 2nd international Conference on Big Data, Cloud and Applications (2017)
12. Pack, C., et al.: Computer aided breast cancer diagnosis system with fuzzy multiple-parameter support vector machine. In: Proceedings of the 2015 Conference on Research in Adaptive and Convergent Systems (2015)
13. Alkhaleefah, M., Wu., C.-C.: A hybrid CNN and RBF-based SVM approach for breast cancer classification in mammograms. In: 2018 IEEE International Conference on Systems, Man, and Cybernetics (SMC). IEEE (2018)
14. Loizidou, K., et al.: An automated breast micro-calcification detection and classification technique using temporal subtraction of mammograms. IEEE Access **8**, 52785–52795 (2020)
15. Abudawood, T., Al-Qunaieer, F., Alrshoud, S.: An efficient abnormality classification for mammogram images. In: 2018 21st Saudi Computer Society National Computer Conference (NCC). IEEE (2018)
16. Fadil, R., et al.: Classification of microcalcifications in mammograms using 2D discrete wavelet transform and random forest. In: 2020 IEEE International Conference on Electro Information Technology (EIT). IEEE (2020)
17. Laghmati, S., et al.: Classification of patients with breast cancer using neighbourhood component analysis and supervised machine learning techniques. In: 2020 3rd International Conference on Advanced Communication Technologies and Networking (CommNet). IEEE (2020)
18. Bazazeh, D., Shubair, R.: Comparative study of machine learning algorithms for breast cancer detection and diagnosis. In: 5th International Conference on Electronic Devices, p. 2016. Systems and Applications (ICEDSA), IEEE (2016)
19. Ghongade, R.D., Wakde, D.G.: Computer-aided diagnosis system for breast cancer using RF classifier. In: International Conference on Wireless Communications, p. 2017. Signal Processing and Networking (WiSPNET), IEEE (2017)
20. Ghongade, R.D., Wakde, D.G.: Detection and classification of breast cancer from digital mammograms using RF and RF-ELM algorithm. In: 1st International Conference on Electronics, p. 2017. Materials Engineering and Nano-Technology (IEMENTech). IEEE (2017)
21. George, J.: Extreme learning machine based classification for detecting micro-calcification in mammogram using multi scale features. In: 2019 International Conference on Computer Communication and Informatics (ICCCI). IEEE (2019)
22. Loganathan, G.B., Praveen, M., Jamuna Rani, D.: Intelligent classification technique for breast cancer classification using digital image processing approach. In: 2019 International Conference on Smart Structures and Systems (ICSSS). IEEE (2019)
23. El-Sokkary, N., et al.: Machine learning algorithms for breast cancer CADx system in the mammography. In: 2019 15th International Computer Engineering Conference (ICENCO). IEEE (2019)
24. Land, W.H., et al.: New results in breast cancer classification obtained from an evolutionary computation/adaptive boosting hybrid using mammogram and history data. In: SMCia/01. Proceedings of the 2001 IEEE Mountain Workshop on Soft Computing in Industrial Applications (Cat. No. 01EX504). IEEE (2001)
25. Ting, F.F., Sim, K.S.: Self-regulated multilayer perceptron neural network for breast cancer classification. In: International Conference on Robotics, p. 2017. Automation and Sciences (ICORAS). IEEE (2017)
26. El-Naqa, I., et al.: Support vector machine learning for detection of microcalcifications in mammograms. In: Proceedings IEEE International Symposium on Biomedical Imaging. IEEE (2002)
27. de Oliveira, F.S.S., et al.: Classification of breast regions as mass and non-mass based on digital mammograms using taxonomic indexes and SVM. Comput. Biol. Med. **57**, 42–53 (2015)
28. Houssein, E.H., et al.: Deep and machine learning techniques for medical imaging-based breast cancer: a comprehensive review. Expert Syst. Appl. 114161 (2020)
29. Cheng, H., Shi, X., Min, R., Hu, L., Cai, X., Du, H.: Approaches for automated detection and classification of masses in mammograms. Pattern Recogn. **39**(4), 646–668 (2006)

30. Gupta, S., Kumar, D., Sharma, A.: Data mining classification techniques applied for breast cancer diagnosis and prognosis. Indian J. Comput. Sci. Eng. (IJCSE) **2**(2), 188–195 (2011)
31. Lakshmi, I., Padmavathamma, M.: Potential of CAD using image mining techniques for breast cancer screening: a review. Int. J. Innov. Eng. Technol. (IJIET) **7**, 323–329 (2016)
32. Davies, D.H., Dance, D.R.: Automatic computer detection of clustered calcifications in digital mammograms. Phys. Med. Biol. **35**(8), 1111 (1990)
33. Christoyianni, I., Koutras, A., Dermatas, E., Kokkinakis, G.: Computer aided diagnosis of breast cancer in digitized mammograms. Comput. Med. Imaging Graph. **26**, 309–319 (2002)
34. Wang, D., Shi, L., Heng, P.A.: Automatic detection of breast cancers in mammograms using structured support vector machines. Neurocomputing **72**, 3296–3302 (2009)
35. de Oliveira Martins, L., Junior, G.B., Silva, A.C., de Paiva, A.C., Gattass, M.: Detection of masses in digital mammograms using k-means and support vector machine. Electron. Lett. Comput. Vis. Image Anal. **8**(2), 39–50 (2009)
36. Pérez, M., Benalcázar, M.E., Tusa, E., Rivas, W., Conci, A.: Mammogram classification using back-propagation neural networks and texture feature descriptors. IEEE Second Ecuador Tech. Chapters Meeting (ETCM) Salinas **2017**, 1–6 (2017)
37. Pillai, R., Oza, P., Sharma, P.: Review of machine learning techniques in health care. In: Singh, P., Kar, A., Singh, Y., Kolekar, M., Tanwar, S. (eds.) Proceedings of ICRIC: Lecture Notes in Electrical Engineering, vol. 597. Springer, Cham (2019)
38. en:User:Cburnett, CC BY-SA 3.0 http://creativecommons.org/licenses/by-sa/3.0/, via Wikimedia Commons
39. Antony, S.: Linear discriminant analysis algorithm using to detect mammogram image classification with feature selection process. Int. J. Adv. Sci. Tech. Res. **3**, 20–31 (2017)
40. Thawkar, S., Ingolikar, R.: Automatic detection and classification of masses in digital mammograms. Int. J. Intell. Eng. Syst. **10**, 65–74 (2017)
41. Abdalla, A.M.M., Deris, S., Zaki, N., Ghoneim, D.M.: Breast cancer detection based on statistical textural features classification. 2007 Innovations in Information Technologies (IIT), 2007
42. Stetco, A., Dinmohammadi, F., Zhao, X., Robu, V., Flynn, D., Barnes, M., Keane, J., Nenadic, G.: CC0, via Wikimedia Commons, https://commons.wikimedia.org/wiki/File:Support_vector_machines.png
43. Keller, J., Gray, M., Givens, J.: A fuzzy k-nearest neighbor algorithm. IEEE Trans. Syst., Man, Cybern. (SMC) **15**, 580–585 (1985)
44. Hu, H., Li, J.: Using association rules to make rule-based classifiers robust. In: Proceedings of the 16th Australasian Database Conference, vol. 39, pp. 47–54 (2005)
45. Mohanty, A.K., Senapati, M.R., Beberta, S., Lenka, S.K.: Texture-based features for classification of mammograms using decision tree. Neural Comput. Appl. **23**(3–4), 1011–1017 (2012)
46. Olukunle, A., Ehikioya, S.: A fast algorithm for mining association rules in medical image data. In: IEEE CCECE2002. Canadian Conference on Electrical and Computer Engineering. Conference Proceedings, vol. 2, pp. 1181–1187 (2002)
47. Jabbar, H., Khan, R.Z.: Methods to avoid over-fitting and under-fitting in supervised machine learning (comparative study). Comput. Sci., Commun. Instrum. Dev. (2015) 163–172
48. Zheng, B., Chang, Y.-H., Wang, X.-H., Good, W.: Comparison of artificial neural network and Bayesian belief network in a computer-assisted diagnosis scheme for mammography. Proc. Int. Joint Conf. Neural Netw. **6**, 4181–4185 (1999)
49. Abdelhafiz, D., et al.: Deep convolutional neural networks for mammography: advances, challenges and applications. BMC Bioinform. **20**(11), 281 (2019)
50. Yamashita, R., et al.: Convolutional neural networks: an overview and application in radiology. Insights imaging **9**(4), 611–629 (2018)
51. Tariq, M., et al.: Medical image based breast cancer diagnosis: state of the art and future directions. Expert Syst. Appl. 114095 (2020)
52. Burt, J.R., et al.: Deep learning beyond cats and dogs: recent advances in diagnosing breast cancer with deep neural networks. Br. J. Radiol. **91**(1089), 20170545 (2018)
53. Hamidinekoo, A., et al.: Deep learning in mammography and breast histology, an overview and future trends. Med. Image Anal. **47**, 45–67 (2018)

Chapter 11
A Comparative Discussion of Similarity Based Techniques and Feature Based Techniques for Interaction Prediction of Drugs and Targets

Kanica Sachdev and Manoj K. Gupta

Abstract Drug target interaction is the communication between the drug compounds and the proteins present in the human body. The laboratory experiments to verify the drug protein interactions consume a lot of time and are expensive. This creates the requirement of introducing computational approaches for identifying the drug target interactions. The computational approaches that have been introduced for predicting the interactions have broadly been classified into similarity based techniques and feature based techniques. This paper defines the various similarity based as well as the feature based techniques for drug target interaction identification. It highlights the state of the art methods in each category. It also presents a discussion on the comparison of the various categories of similarity and feature based techniques. It also demonstrates their comparison with each other, their merits and demerits.

Keywords Drug compounds · Feature vectors · Drug protein interaction · Target proteins

11.1 Introduction

Determining interactions between the compounds of drugs and protein targets is an arising research topic in the area of pharmacology [1]. The chemical compounds that attach themselves to the proteins in human body to effectuate a change are known as drugs. The proteins present in the human body, which get attached to the drug compounds are known as targets. Thus, drug target interaction can be explained as the procedure of the drug compounds attaching themselves to the target proteins and causing a change in the human body. The prediction of drug protein interaction applies to the procedures that are developed in order to identify if a particular drug will attach itself to a certain protein target or not.

K. Sachdev (✉) · M. K. Gupta
Department of Computer Science and Engineering, Shri Mata Vaishno Devi University, Katra, India

M. Mehta et al. (eds.), *Tracking and Preventing Diseases with Artificial Intelligence*, Intelligent Systems Reference Library 206,
https://doi.org/10.1007/978-3-030-76732-7_11

There are various applications of identifying the drug protein interactions. It aids the procedure of drug discovery. Predicting the drug target communication can help to discover new drug compounds that may treat a particular disease. Also, it aids drug repositioning process. Drug repositioning is the procedure of employing an already existing drug compound to treat another disease/ailment [2]. Identifying novel drug protein interactions also aids the recognition of drug side effects. To identify the drug compound side effects, the negative communication between the compound and the target protein can be examined [3].

Many techniques have been proposed in the recent years that aim to identify the drug target interactions with great accuracy. The existing techniques have been broadly classified into docking based procedures, ligand based procedures and chemogenomic procedures. The ligand based procedures identify the potential interactions by examining the correlation between the target ligands. The main disadvantage of this approach is that the accuracy of these methods substantially decreases if the number of ligands that are known is less [4]. The second category, i.e. docking based procedures, refers to the methods that employ the three dimensional structure of the proteins to predict the possible drug protein interactions [5]. These techniques are however only applicable to the protein targets for which the 3D structure is known. Since presently, there are only a limited number of proteins with known 3D structure; the applicability of these methods is restricted in scope [6]. These limitations were overcome by the introduction of the chemogenomic procedures. These methods utilize the combination of the chemical information of the drug compounds with the genomic information of the protein targets in order to identify interactions [7]. The chemical as well as the genomic information is used simultaneously to identify novel drug protein interactions.

The chemogenomic approaches are further divided into similarity based techniques and feature based techniques. Similarity based techniques use the drug-drug similarity matrix in conjunction with the protein–protein similarity matrix to predict interactions. Contrastively, the feature based techniques compute the drug feature vectors and the target feature vectors. These feature vectors are then used to identify predictions. The similarity based and feature based techniques have further been classified into various categories. The categorization has been depicted in Fig. 11.1.

This paper aims to discuss the eminent works that have been presented under the various categories of similarity based and feature based techniques, and compares the two various categories of methods with each other. It also analyzes and examines the merits and demerits of the similarity based and feature based methods while comparing them with one another. The remaining paper has been arranged as follows. Section 11.2 explains the various categories of similarity based techniques. Section 11.3 describes the various types of feature based techniques. Section 11.4 summarizes the various methods and compares the similarity based techniques with the feature based techniques. Section 11.5 concludes the paper.

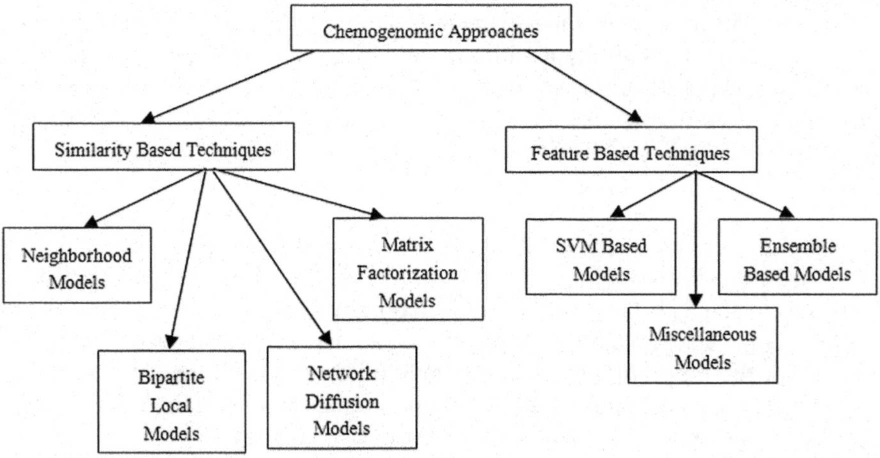

Fig. 11.1 Categorization of chemogenomic methods

11.2 Similarity Based Techniques

The similarity based techniques use the drug similarity matrix along with the protein similarity matrix for predicting the interactions. The prediction is performed using various methods like matrix factorization, network diffusion etc. The similarity based techniques are further divided into four main categories: neighborhood models, bipartite local models, network diffusion models and matrix factorization models. Each of these categories has been explained in the following sub sections.

11.2.1 Neighborhood Models

These methods employ models to predict the interactions of a novel drug or target based on its neighbourhood. A novel drug or target refers to any drug or target whose interactions are not yet known. The neighborhood of a drug or a target refers to the most similar drugs or targets respectively. These methods predict the interactions of a particular drug or a target using the interactions of its neighborhood drugs or targets respectively. Thus the interaction profile of a novel drug is formed by using the interaction profiles of the drugs that are most similar to it. The same procedure is used for the targets.

The simplest approach is to use the interaction profiles of the most similar drug of a novel drug and multiply it with the similarity score. This gives an estimate of the interaction profile of the new drug [8]. The interaction profile of a new drug can be computed as

$$I(d_{new}) = S_d(d_{new}, d_{similar}) \times I(d_{similar}) \tag{11.1}$$

Here, d_{new} refers to the novel drug and $d_{similar}$ refers to the most similar drug to d_{new}. $I(d_{new})$ and $I(d_{similar})$ refer to the interaction profiles of d_{new} and $d_{similar}$ respectively and S_d is the matrix for drug similarity. The interaction of novel targets is also computed accordingly. A variant of this method considers the interaction profiles of the most similar drugs and multiplies them with the similarity score. The weighted average of all these interaction profiles is considered as the predicted profile. The interaction profile of the target is also computed in a similar manner. Another technique computed two indices for each drug-target pair i.e. the tendency of drug to interact with a target and the negative tendency of the drug of not interacting with a particular target. Firstly, the drug neighbourhood is used to compute these two indices. Following this, the target neighbourhood is used to compute the indices. These two indices' ratio are used to compute the final interaction score [9]. The two indices can be combined as follows

$$I(d_m, t_n) = \frac{PI(d_m, t_n)}{NI(d_m, t_n)} \tag{11.2}$$

Here $I(d_m, t_n)$ is the interaction score for the drug d_m and target t_n. PI refers to the positive interaction index and NI is the negative interaction index. A recent method that integrates the neighborhood information with neural networks has also been proposed [10]. This method named NeoDTI can effectively predict novel interactions.

11.2.2 Bipartite Local Models

The methods under this category perform prediction by combining two individual predictions. One prediction is formed from the drug side using the information about the drug similarity. The other prediction is made from the target side. The two predictions are then aggregated to form the final prediction score.

The earliest work in this field employs Support Vector Machine (SVM) to make drug and target predictions respectively [11]. The SVM models were individually trained for the drugs and the targets. These predictions are then averaged to obtain the final result. Another method uses regularized least squares to minimize the objective function for identifying predictions [12]. The objective function to be minimized can be depicted as

$$\min_{\varnothing_d}(||I - S_d\varnothing_d||_F^2 + \partial_d Tr(\varnothing_d^T S_d L_d S_d \varnothing_d)) \tag{11.3}$$

Here $\|...\|_F$ is the Frobenius form, L_d is the normalized Laplacian, $\O_d$ and δ_d are parameters an Tr(..) refers to the trace of the matrix. Kernel Ridge regression has also been used for the prediction process [13]. The Gaussian Interaction Profile (GIP) kernel is used for calculating the interaction profiles. Following this, the RLS-Kron classifier is used. A technique combining bipartite model with nearest neighbor method has also been proposed. It constructs the interaction profiles of novel drugs and targets using the nearest neighbor method. The bipartite model is then used for making predictions [14].

11.2.3 Network Diffusion Models

These methods include the graph based techniques for making predictions. Network diffusion is mostly applied for the identification of interactions from the graphs. It studies the influence propagation in the drug target graphs which are used for predictions.

A drug target bipartite graph is formed in these methods. The drugs and the targets are depicted as the nodes of the graph. Each edge depicts the interactions between the nodes. Network diffusion is applied according to the diffusion rule for interaction prediction [2]. It can be represented as

$$I' = WI \tag{11.4}$$

Here I is the interaction matrix and W is the weight matrix that is formed using the network diffusion rule applied. A technique based on heterogeneous graph has also been proposed [15]. Along with the edges representing interactions, other edges are added to the network that represent drug and target similarity. Network diffusion is then applied to this heterogeneous graph. Another technique based on random walk has been developed [16]. A similar heterogeneous graph is constructed representing the drugs, targets and their interactions. The interaction prediction is performed using random walk on the heterogeneous graph. A method using probabilistic soft logic with heterogeneous graph has also been introduced [17]. In order to predict the interactions; probabilistic rules are applied to the drug target interaction network. A recent method that employs graph based method with machine learning has also been introduced [18]. It uses Random Forest for classification. It improves the performance considerably in terms of decreasing the false negatives in comparison to state of the art methods.

11.2.4 Matrix Factorization Models

The matrix factorization models for drug target prediction have high accuracy. These methods first construct latent matrices using the known drug target interaction matrix. These latent matrices are then multiplied in order to obtain the interaction prediction matrix. The main goal is to identify the value for the missing values in the interaction matrix that are calculated using the matrix factorization.

A technique based on Bayesian probabilistic formulation in conjunction with the matrix factorization method was developed [19]. Another technique performed the weighted matrix factorization [20]. It involved the use of graph regularization terms that were used to learn a manifold for label propagation. A technique based on logistic matrix factorization was also proposed [21]. It modelled the probability of communication between a particular drug and target using a logistic function. An extension of this method uses matrix factorization with network based similarity [22].

11.3 Feature Based Techniques

The feature based techniques identify novel drug protein interactions by determining the discerning and critical features. These methods use feature vectors of the drugs and targets as input. The feature vectors are formed by merging the different properties of the drugs as well as the targets. The feature vectors hence formed are transferred into machine learning frameworks like SVM, Relevance Vector Machines etc. to identify novel communications.

The feature based techniques are further categorized into three types of models i.e. SVM based models, ensemble based models and miscellaneous models [23]. The models using the SVM classifier are compiled under the SVM based models. The methods that employ a combination of more than one classifier are assembled under the ensemble based models. The remaining methods are explained under the miscellaneous models. Each of these models and the prominent works under each of these classifications have been discussed as under.

11.3.1 SVM Based Models

SVMs find their application in various classification and regression problems under the area of data mining. They are used to segregate the available data into different classes on the basis of the training data that is provided and hence facilitate the data analysis procedure. The methods falling under the category of SVM based models calculate the drug compound features and the protein features. These features classify the data based on a kernel function. The drugs and proteins may use the same kernel function or a different kernel function. These methods are simple in execution and perform in less computational time.

One of the initial works in the area of chemogenomic methods that introduced the combination of chemical and genomic properties for interaction prediction used the SVM classifier [24]. It computed the drug feature vector using the drug chemical structure as well as the mass spectrum. Amino acid sequence was used to calculate the protein feature vector. SVM is then used to classify the drug target predictions. It can be depicted by the following function

$$f(x) = sign(\sum_{i=1}^{n}(a_i y_i K(x_i, x) + b)) \quad (11.5)$$

Here x is the object to be classified, n is the number of samples, K is the kernel function and y represents the output class. y may be binding or non binding. Different methods using the molecular signatures and molecular patterns to construct the features were also developed [25]. Another technique proposed the calculation of drug compound similarity and target similarity independently using different kernel functions [4]. For drug similarity computation, Tanimoto kernel was used. Dirac kernel, Multitask kernel and Hierarchy kernel were suggested for computing the protein similarity. To form the drug protein pair vector, the tensor product of the individual vectors is computed. These feature vectors were then used for the classification using SVM. A two layer SVM model was also proposed for the interaction prediction [26]. Recursive feature elimination (RFE) method is used to decide which SVM models will be considered for the second layer.

11.3.2 Ensemble Based Models

Ensemble refers to a combination of various classifiers. These models outperform the traditional SVM based methods as they utilize more than one classifier to train and classify the data. The drug feature vectors and the protein feature vectors are calculated independently, which are then fed to the ensemble classifier. The output that is formed is then merged to obtain the final result by using average, geometric mean etc.

Many ensemble based models have been introduced and implemented for the prediction of novel interactions. Random forest has been used to identify the interactions as it produces satisfactory results on large datasets [27]. In order to calculate the feature importance for classification, Out of Bag (OOB) error is computed. Another random forest based method has been proposed that encodes the proteins using Position-Specific Scoring Matrix (PSSM) descriptor and the drugs using fingerprint feature vector to predict interactions [28]. A similar technique also uses similar features for prediction [29]. It also employs SMOTE to decrease class imbalance in the data and improve the overall performance. Ensemble of decision trees has also been proposed. It has also reduced the bias in the classification process [30]. A variant of this method that uses ensemble of decision trees along with dimensionality reduction has also been suggested to improve the prediction performance [31]. Extremely randomized trees have also been used to enhance the prediction accuracy and decrease bias [32]. The ensemble based models have achieved more efficient performance than SVM classifier.

11.3.3 Miscellaneous Models

Various other feature based techniques exist that do not fall under the category of SVM based or ensemble based models. These methods have been compiled under this subsection.

A predictor model was proposed using the nearest neighbor method [33]. The nearness of the drugs and the targets was calculated based on their feature vectors. It can be computed as

$$N\left(x_i, x_j\right) = 1 - \frac{x_i . x_j}{||x_i|| \, ||x_j||} \tag{11.6}$$

Here x_i and x_j are two samples. $x_i.x_j$ depicts their dot product and $||x_i||$, $||x_j||$ is their modulus. To further improve the efficiency of the technique, Maximum Relevance Minimum Redundancy algorithm was used to select the most relevant features. Another technique based on Sparse Canonical Correspondence Analysis (SCCA) was also developed [34]. It calculates the prediction score of any drug d_i and target t_j as

$$s\left(d_i, t_j\right) = \sum_{k=1}^{n} u_k p_k v_k \tag{11.7}$$

Here n is the total number of canonical components and p_k is the kth singular value. If the prediction score is greater than a predefined threshold, d_i is predicted to interact with t_j. Other techniques included methods based on fuzzy kNN engine and Relevance vector Machines [35].

11.4 Comparison of Similarity Based and Feature Based Techniques

The chemogenomic methods integrate the drug and protein information for the purpose of prediction. Thus, they are able to achieve better accuracy. Both the similarity based techniques and the feature based techniques have certain advantages and certain demerits. Since the similarity based methods and feature based methods require different forms of data for prediction, they cannot be quantitatively compared with each other based on performance metrics. This section discusses and compares the various categories of similarity based and feature based techniques and highlights certain major advantages and disadvantages of employing similarity based or the feature based techniques for prediction (Table 11.1).

The similarity based techniques employ the similarity information of the drugs and the targets to form predictions. The neighborhood based models use the information of the most similar drugs and targets for predicting interactions. The performance of these methods is substantially affected by the number of nearest neighbors considered. However, no optimized value of nearest neighbors has been proposed or established. Thus, these methods are highly dependent on the hit and trial method of choosing the nearest neighbors for different datasets to achieve an optimal performance.

The bipartite local models combine two individual models based on drugs and targets respectively. The major drawback faced by these methods was that they were not able to form predictions for novel drugs and targets. The integration of nearest neighbor model with the bipartite model has been able to achieve good accuracy. The nearest neighbor approach was employed to form the interaction profile of novel drugs and targets. The bipartite local models were used to then make predictions.

The network based models visualize the drug target interaction data in the form of a graph to infer interactions. However, the complexity of these models is generally high due to the formation of complex networks representing the drug, targets and their intercommunication. Also, another drawback is that they employ only local information for prediction.

The similarity based methods are generally computationally faster to execute. Most similarity based techniques like nearest neighbor based prediction, matrix factorization etc. are computationally simple and hence take lesser time for implementation. In general, the matrix factorization methods perform better than the other methods [7]. Also, the method based on kernel ridge regression yields accurate results [12] and is not as sensitive to parameters as the matrix factorization based methods. It is also relatively faster to execute. However, it does not provide as good accuracy as the matrix factorization based methods.

The feature based methods construct drug and target feature sets to classify the interactions. The SVM based models infer the interactions using SVM classifier. Different properties have been used to form features and classify the interactions. However, the SVM based models usually take high computational time to execute. The complexity and time also increases substantially as the feature dimensions

Table 11.1 The merits and demerits of various chemogenomic methods for drug target interaction prediction

Technique	Model	Merits	Demerits	Merits	Demerits
Similarity based techniques	Neighborhood models	These are easy to execute	The performance is dependent on number of nearest neighbors that have no optimised value	These are faster in execution	The similarity of drugs and targets needs to be computed first based on some criteria to form prediction They do not preserve the properties of the data to be used for further understanding and exploration
	Bipartite local models	Performance is better than neighborhood models	They cannot be used to predict interactions for novel drugs or targets directly		
	Network diffusion models		The complexity of the methods is high due to the formation of complex networks They employ only local information for prediction		
	Matrix factorization models	The performance of these methods is better than other methods	They cannot be used to predict interactions for novel drugs or targets directly		

(continued)

Table 11.1 (continued)

Technique	Model	Merits	Demerits	Merits	Demerits
Feature based techniques	SVM based models	These are easy to execute	The performance is not good in comparison to other methods They take time to execute	They provide a direct method of using the drug target properties for constructing feature vectors The features preserve the properties of the data that can be used for further understanding and exploration	These methods take more time to execute Selecting optimal features for classifiers is a complex task
	Ensemble based models	The performance is good in comparison to other methods	The use of various classifiers increases complexity		
	Miscellaneous models		The various scoring methods need to tune parameters The performance of these methods is not that good		

increase. Thus, they have been replaced by more efficient classifiers that are faster to execute and yield highly accurate results.

The ensemble based methods combine two or more ensembles for improving the classifier performance. These methods perform better than the other two categories of methods. The better performance can be attributed to the fact that a combination of classifiers is usually better than any single classifier. The various distinct individual classifiers in the ensemble lead to more stability than single classifier [36]. Also, they reduce any kind of bias or variance in the results [37, 38]. Certain miscellaneous models using SCCA or fuzzy engine have also been proposed, but they have not been able to achieve any significant improvement in performance.

The feature based techniques provide a direct method of using the drug target properties for constructing feature vectors that are used for interaction prediction. They preserve the properties of the data that can be used for further understanding and exploration [21]. The continuing study in the area of classifiers and ensemble learning directly help in developing more efficient feature based techniques for prediction. In general, as mentioned earlier, the ensemble based models have better accuracy than the SVM based or miscellaneous models as they employ more than a single classifier in the prediction process. This helps to improve the overall accuracy.

The feature based techniques, however, take more time to execute than the similarity based techniques [21]. The training and testing process on a large number of features take time. Also, training complex classifiers and ensembles is also a complex and time consuming task. Moreover, efficient classifiers require the selection of an optimal feature set. The process of identifying relevant features and optimal feature set is also a complex task.

11.5 Conclusion

This paper explains the chemogenomic approaches for the prediction of interaction between drug compounds and proteins. The approaches have been further classified into similarity based techniques and feature based techniques. These methods and their further categorization have been explained in the paper. Although the performance of the methods cannot be directly compared to each other due to the varying data requirements, a discussion has been presented to analyze the difference in the two approaches. These methods have been compared and their advantages and demerits have been outlined.

References

1. Chen, X., et al.: Drug–target interaction prediction: databases, web servers and computational models. Brief. Bioinform. **17**(4), 696–712 (2016)
2. Cheng, F., et al.: Prediction of drug-target interactions and drug repositioning via network-based inference. PLoS Comput. Biol. **8**(5), e1002503 (2012)
3. Atias, N., Sharan, R.: An algorithmic framework for predicting side-effects of drugs. In: Annual International Conference on Research in Computational Molecular Biology. Springer, Berlin (2010)
4. Jacob, L., Vert, J.-P.: Protein-ligand interaction prediction: an improved chemogenomics approach. Bioinformatics **24**(19), 2149–2156 (2008)
5. Li, H., et al.: TarFisDock: a web server for identifying drug targets with docking approach. Nucleic Acids Res. **34**(suppl_2), W219–W224 (2006)
6. Mousavian, Z., Masoudi-Nejad, A.: Drug–target interaction prediction via chemogenomic space: learning-based methods. Expert Opin. Drug Metab. Toxicol. **10**(9), 1273–1287 (2014)
7. Ezzat, A., et al.: Computational prediction of drug–target interactions using chemogenomic approaches: an empirical survey. Briefings Bioinform. bby002–bby002 (2018)
8. Yamanishi, Y., et al.: Prediction of drug–target interaction networks from the integration of chemical and genomic spaces. Bioinformatics **24**(13), i232–i240 (2008)
9. Shi, J.-Y., Yiu, S.-M.: SRP: a concise non-parametric similarity-rank-based model for predicting drug-target interactions. In: 2015 IEEE International Conference on Bioinformatics and Biomedicine (BIBM). IEEE (2015)

10. Wan, F., et al.: NeoDTI: neural integration of neighbor information from a heterogeneous network for discovering new drug–target interactions. Bioinformatics **35**(1), 104–111 (2019)
11. Bleakley, K., Yamanishi, Y.: Supervised prediction of drug–target interactions using bipartite local models. Bioinformatics **25**(18), 2397–2403 (2009)
12. Xia, Z., et al.: Semi-supervised drug-protein interaction prediction from heterogeneous biological spaces. In: BMC Systems Biology. BioMed Central (2010)
13. van Laarhoven, T., Nabuurs, S.B., Marchiori, E.: Gaussian interaction profile kernels for predicting drug–target interaction. Bioinformatics **27**(21), 3036–3043 (2011)
14. Mei, J.-P., et al.: Drug–target interaction prediction by learning from local information and neighbors. Bioinformatics **29**(2), 238–245 (2012)
15. Wang, W., Yang, S., Li, J.: Drug target predictions based on heterogeneous graph inference. In: Biocomputing, pp. 53–64. World Scientific (2013)
16. Chen, X., Liu, M.-X., Yan, G.-Y.: Drug–target interaction prediction by random walk on the heterogeneous network. Mol. BioSyst. **8**(7), 1970–1978 (2012)
17. Fakhraei, S., et al.: Network-based drug-target interaction prediction with probabilistic soft logic. IEEE/ACM Trans. Comput. Biol. Bioinform. (TCBB) **11**(5), 775–787 (2014)
18. Olayan, R.S., Ashoor, H., Bajic, V.B.: DDR: efficient computational method to predict drug–target interactions using graph mining and machine learning approaches. Bioinformatics **34**(7), 1164–1173 (2018)
19. Gönen, M.: Predicting drug–target interactions from chemical and genomic kernels using Bayesian matrix factorization. Bioinformatics **28**(18), 2304–2310 (2012)
20. Cobanoglu, M.C., et al.: Predicting drug–target interactions using probabilistic matrix factorization. J. Chem. Inf. Model. **53**(12), 3399–3409 (2013)
21. Zheng, X., et al.: Collaborative matrix factorization with multiple similarities for predicting drug-target interactions. In: Proceedings of the 19th ACM SIGKDD International Conference on Knowledge Discovery and Data Mining. ACM (2013)
22. Ezzat, A., et al.: Drug-target interaction prediction with graph regularized matrix factorization. IEEE/ACM Trans. Comput. Biol. Bioinform. (TCBB) **14**(3), 646–656 (2017)
23. Sachdev, K., Gupta, M.K.: A comprehensive review of feature based methods for drug target interaction prediction. J. Biomed. Inform. 103159 (2019)
24. Nagamine, N., Sakakibara, Y.: Statistical prediction of protein–chemical interactions based on chemical structure and mass spectrometry data. Bioinformatics **23**(15), 2004–2012 (2007)
25. Faulon, J.-L., et al.: Genome scale enzyme–metabolite and drug–target interaction predictions using the signature molecular descriptor. Bioinformatics **24**(2), 225–233 (2007)
26. Nagamine, N., et al.: Integrating statistical predictions and experimental verifications for enhancing protein-chemical interaction predictions in virtual screening. PLoS Comput. Biol. **5**(6), e1000397 (2009)
27. Yu, H., et al.: A systematic prediction of multiple drug-target interactions from chemical, genomic, and pharmacological data. PLoS ONE **7**(5), e37608 (2012)
28. Wang, L., et al.: Rfdt: a rotation forest-based predictor for predicting drug-target interactions using drug structure and protein sequence information. Curr. Protein Pept. Sci. **19**(5), 445–454 (2018)
29. Shi, H., et al.: Predicting drug-target interactions using Lasso with random forest based on evolutionary information and chemical structure. Genomics **111**(6), 1839–1852 (2019)
30. Ezzat, A., et al.: Drug-target interaction prediction via class imbalance-aware ensemble learning. BMC Bioinform. **17**(19), 509 (2016)
31. Ezzat, A., et al.: Drug-target interaction prediction using ensemble learning and dimensionality reduction. Methods **129**, 81–88 (2017)
32. Huang, Y.-A., You, Z.-H., Chen, X.: A systematic prediction of drug-target interactions using molecular fingerprints and protein sequences. Curr. Protein Pept. Sci. **19**(5), 468–478 (2018)
33. He, Z., et al.: Predicting drug-target interaction networks based on functional groups and biological features. PLoS ONE **5**(3), e9603 (2010)
34. Yamanishi, Y., et al.: Extracting sets of chemical substructures and protein domains governing drug-target interactions. J. Chem. Inf. Model. **51**(5), 1183–1194 (2011)

35. Xiao, X., et al.: iGPCR-Drug: a web server for predicting interaction between GPCRs and drugs in cellular networking. PLoS ONE **8**(8), e72234 (2013)
36. Lessmann, S., et al.: Benchmarking state-of-the-art classification algorithms for credit scoring: an update of research. Eur. J. Oper. Res. **247**(1), 124–136 (2015)
37. Kim, M.-J., Min, S.-H., Han, I.: An evolutionary approach to the combination of multiple classifiers to predict a stock price index. Expert Syst. Appl. **31**(2), 241–247 (2006)
38. Tsai, C.-F., Hsiao, Y.-C.: Combining multiple feature selection methods for stock prediction: union, intersection, and multi-intersection approaches. Decis. Support Syst. **50**(1), 258–269 (2010)

Printed by Printforce, the Netherlands